J. Polster, H. Lachmann

Spectrometric Titrations

© VCH Verlagsgesellschaft mbH, D-6940 Weinheim (Federal Republic of Germany), 1989

Distribution
VCH Verlagsgesellschaft, P.O. Box 101161, D-6940 Weinheim (Federal Republic of Germany)
Switzerland: VCH Verlags-AG, P.O. Box, CH-4020 Basel (Switzerland)
Great Britain and Ireland: VCH Publishers (UK) Ltd., 8 Wellington Court, Wellington Street, Cambridge CB1 1HW (Great Britain)
USA and Canada: VCH Publishers, Suite 909, 220 East 23rd Street, New York NY 10010-4606 (USA)

ISBN 3-527-26436-1 (VCH Verlagsgesellschaft)   ISBN 0-89573-570-9 (VCH Publishers)

Jürgen Polster
Heinrich Lachmann

Spectrometric Titrations

Analysis of
Chemical Equilibria

Professor Dr. Jürgen Polster
Lehrstuhl für Allgemeine Chemie
und Biochemie
Technische Universität München
D-8050 Freising-Weihenstephan
Federal Republic of Germany

Dr. Heinrich Lachmann †
Institut für Physikalische und
Theoretische Chemie
Universität Tübingen

This book was carefully produced. Nevertheless, authors and publisher do not warrant the information contained therein to be free of errors. Readers are advised to keep in mind that statements, data, illustrations, procedural details or other items may inadvertently be inaccurate.

Published jointly by
VCH Verlagsgesellschaft, Weinheim (Federal Republic of Germany)
VCH Publishers, New York, NY (USA)

Editorial Director: Dr. Michael G. Weller
Production Manager: Claudia Grössl
Composition: Bliefert Computersatz und Software, D-4437 Schöppingen
Printing: betz-druck gmbh, D-6100 Darmstadt
Bookbinding: J. Schäffer, D-6718 Grünstadt

Library of Congress Card No. applied for

British Library Cataloguing in Publication Data
Polster, Jürgen
Spectrometric titrations.
1. Chemical analysis, Spectroscopy
I. Title II. Lachmann, Heinrich
543'0858
ISBN 3-527-26436-1

CIP-Titelaufnahme der Deutschen Bibliothek:
Polster, Jürgen:
Spectrometric Titrations: Analysis of Chemical Equilibria/Jürgen Polster; Heinrich Lachmann. – Weinheim; Basel (Switzerland); Cambridge; New York, NY:
VCH, 1989
   ISBN 3-527-26436-1
    (Weinheim...)
   ISBN 0-89573-570-9
    (Cambridge...)
NE: Lachmann, Heinrich

© VCH Verlagsgesellschaft, D-6940 Weinheim (Federal Republic of Germany), 1989
All rights reserved (including those of translation into other languages). No part of this book may be reproduced in any form – by photoprint, microfilm, or any other means – nor transmitted or translated into a machine language without written permission from the publishers. Registered names, trademarks, etc. used in this book, even when not specifically marked as such, are not to be considered unprotected by law.
Printed in the Federal Republic of Germany

Dedicated to
Professor HEINZ MAUSER,
our distinguished teacher

Preface

Upon hearing the word *titration* most analytical chemists think first of the technique introduced by Gay-Lussac for the quantitative determination of a dissolved material, in which a *titration endpoint* is established with the aid of a colored indicator. It has more recently been recognized, however, that there are other sorts of problems that lend themselves to investigation by titration. Indeed, it is now common to utilize titration techniques in the search for fundamental information regarding a wide variety of equilibrium systems. Thus, if stepwise changes are made in the position of an equilibrium through the addition of successive portions of an appropriate titrant, physical characteristics of the altered system at each "step" may be defined in terms of a recording of appropriate potentiometric or spectroscopic parameters. In many cases the resulting record permits accurate computation of an equilibrium constant, the quantity that is usually of greatest interest.

Equilibria play a fundamental role in a wide array of problems in both chemistry and biochemistry, so it is little wonder that equilibria also constitute a fertile field of investigation for chemists, biochemists, biologists, pharmacologists, toxicologists, and others. These researchers represent the target audience for our monograph.

The purpose of the book is to summarize recently developed methods for the spectroscopic analysis of chemical equilibria. The methods themselves are applicable to virtually any spectroscopic technique that produces an output signal linearly dependent upon the concentrations of certain solution components, a criterion met by UV/VIS, IR, fluorescence, CD, ORD, NMR, and ESR. Methods of evaluation applicable to each of these types of spectroscopy are dealt with explicitly in the text. The term *spectrometry* implies the recording of spectra of any type, so we elected to underscore the generality of this experimental approach by assigning our book the broad title *Spectrometric Titration*.

The book itself is divided into three parts, each of which may be read independently. Nevertheless, since Part I provides the relevant theoretical and methodological background, as well as definitions for key technical terms, we recommend that it be read first.

Part II is a systematic examination of an assorment of titration systems, including equilibria of the acid-base, metal complex, association (or bonding), and redox types. The treatment is limited to homogeneous phases; problems posed by precipitation or other phase separations are deliberately ignored. The central theme of each chapter is a search for answers to two questions: (1) what types of information can one derive solely on the basis of spectrometric measurements, and (2) how can this information be correlated with electrometric (e.g. pH) data in order to establish equilibrium constants?

Part III is devoted to experimental considerations related to UV/VIS, fluorescence, CD/ORD, IR, Raman, and NMR, as well as the application of these tools to spectrometric titration. At least one concrete example is provided with respect to each of the corresponding spectroscopic methods. The literature is covered fully until the end of 1986. The book concludes with an appendix listing two computer programs, EDIA and TIFIT, which we have found useful in interpreting our data.

The starting point for the analytical methods described is the general theory of isosbestic points, originally developed by Heinz Mauser for the study of closed thermal and photochemical reactions. Mauser also introduced a series of graphic techniques [absorbance (A-) diagrams, absorbance difference (AD-) diagrams, absorbance difference quotient (ADQ-) diagrams] that provide rapid, simple, yet effective means for establishing the number of linearly independent steps constituting a reaction system. The first application of this methodology to the study of acid-base equilibria, generalized to include various types of spectroscopy, was by Heinrich Lachmann, who recognized the value of this route to establishing the number of independent steps characterizing a titration system. In a subsequent collaboration with Rüdiger Blume the authors were then able to introduce a number of additional techniques for spectrometric determination of the pK values that characterize acid-base equilibria.

The authors are deeply indebted to their distinguished mentor, Prof. Heinz Mauser, to whom this book is dedicated, for many critical and stimulating discussions and for his continued support of their research efforts over a great many years. Special thanks are also due to Dr. Gabriele Lachmann, who shares the responsibility for editing those chapters (chapters 1–4, 6, 10–12, and 16–19) prepared originally by her husband, a task necessitated by his sadly premature and untimely death.

We also wish to thank Prof. Rüdiger Blume for his many valuable contributions and for placing at our disposal much useful material; Prof. Günter Jung, who provided us with experimental data and spectra; and Drs. Friedrich Göbber and Klaus Willamowski for numerous suggestions and contributions, including the basis for the two computer programs EDIA and TIFIT. Thanks are due as well to Christina Nagy-Huch-Hallwachs, Jutta Konstanczak and Brigitte Moratz for their technical assistance.

We are particularly grateful to Prof. Hanns-Ludwig Schmidt for his continued and active interest in and support of our work. We have also appreciated the cooperation of VCH Verlagsgesellschaft—particularly that of Dr. Weller—in providing a critical

Preface

review of the translated manuscript and then ensuring its swift publication. Finally, we wish to thank Prof. Claus Bliefert, who carried the burden of desktop publishing this book to an excellent end, and Dr. William Russey (Juniata College, Huntingdon, PA) for his rapid and skillful translation of the German manuscript, and for his useful suggestions and observations.

Jürgen Polster

Contents

Part I The Theoretical and Methodological Basis of Spectrometric Titration

1	**Basic Principles and the Classification of Titration Methods**	3
1.1	Definitions	3
1.2	The Classification of Spectrometric Titration Methods	5
	Literature	7
2	**Graphic Treatment of the Data from a Spectrometric Titration**	9
	Literature	12
3	**Basic Principles of Multiple Wavelength Spectrometry as Applied to Titration Systems**	13
3.1	Generalized Lambert-Beer-Bouguer Law	13
3.2	Matrix-Rank Analysis of a Titration System	15
3.2.1	Numeric Matrix-Rank Analysis	15
3.2.2	Graphic Methods of Matrix-Rank Analysis	16
	Literature	20
4	**Thermodynamic and Electrochemical Principles of Spectrometric Titration Systems**	23
	Literature	25

Part II The Formal Treatment and Evaluation of Titration Systems – Analysis of Chemical Equilibria

5	**The Goal of a Spectrometric Titration**	29

6	**Single-Step Acid-Base Equilibria**	33
6.1	Basic Equations	33
6.2	Graphic Matrix-Rank Analysis	34
6.3	Determination of Absolute pK Values	39
6.3.1	Inflection Point Analysis	39
6.3.2	Determining the Midpoint of Overall Change in Absorbance (the "Halving Method")	42
6.3.3	Linearization of Spectrometric Titration Curves	42
6.3.4	Numeric Evaluation of pK Through the Use of Linear Equations	44
6.3.5	Non-linear Curve-Fitting Methods	46
6.3.6	Photometric Determination of pK in the Absence of pH Measurements	47
6.4	The Treatment of Non-overlapping, Multi-Step Titration Systems as Combinations of Single-Step Subsystems	48
6.5	Summary	60
6.5.1	Single-Step ($s = 1$) Titrations	60
6.5.2	Non-overlapping, Multi-Step Titration Systems	64
	Literature	66
7	**Two-Step Acid-Base Equilibria**	67
7.1	Basic Equations	67
7.2	Graphic Matrix-Rank Analysis (ADQ-Diagrams)	69
7.3	The Determination of Absolute pK Values	75
7.3.1	Inflection Point Analysis (First Approximate Method)	75
7.3.2	Uniform Subregions in an Absorbance Diagram (Second Approximate Method)	80
7.3.3	Dissociation Diagrams	84
7.3.4	Bisection of the Sides of an Absorbance Triangle	88
7.3.5	Pencils of Rays in an Absorbance Triangle	90
7.3.6	Linear and Non-linear Regression Methods	92
7.4	Determining the Quotient K_1/K_2 ($\Delta pK = pK_2 - pK_1$)	95
7.4.1	Distance Relationships Within an Absorbance Triangle	96
7.4.2	Non-linear Regression Analysis	98
7.5	The Characteristic Concentration Diagram	105
7.5.1	Second-Order Curves (Conic Sections)	106
7.5.2	Poles and Polars of Conic Sections	108
7.5.3	Pencils of Rays, and Lines Bisecting the Sides of the Triangle	109
7.5.4	Distance Relationships	112
7.6	Geometric Relationships Between Concentration Diagrams and Absorbance Diagrams	115
7.7	Concentration Difference Quotient (CDQ) Diagrams	117
7.8	Geometric Relationships Between ADQ-, CDQ-, and A-Diagrams	121

7.9	Summary and Tabular Overview	126
	Literature	129
8	**Three-Step Acid-Base Equilibria**	**133**
8.1	Basic Equations	133
8.2	Graphic Matrix-Rank Analysis (ADQ3-Diagrams)	135
8.3	Absolute pK Determination	136
8.3.1	Analysis of Inflection Points	136
8.3.2	Uniform Subregions Within Absorbance Diagrams	136
8.3.3	Dissociation Diagrams	141
8.3.4	Pencils of Rays in an Absorbance Tetrahedron	144
8.3.5	Pencils of Planes in the Absorbance Tetrahedron	145
8.3.6	Correlation of Two-dimensional A- and ADQ-Diagrams	148
8.3.6.1	Determining the Molar Absorptivities of BH_2^- and BH^{2-}	148
8.3.6.2	Reducing the System by One Stage of Equilibrium	152
8.3.6.3	Determining pK Values from the Three Separate Equilibria	153
8.3.7	Linear and Non-linear Regression Analysis	155
8.4	The Determination of K_1/K_2 and K_2/K_3 with the Aid of the Absorbance Tetrahedron	160
8.5	The Characteristic Concentration Diagram	162
8.5.1	Tangents and Osculating Planes	162
8.5.2	Pencils of Rays	166
8.5.3	Pencils of Planes	168
8.5.4	Determining K_1/K_2 and K_2/K_3 from the Concentration Tetrahedron	171
8.6	Geometrical Relationships Between A- and ADQ-Diagrams	173
8.7	Summary and Tabular Overview	180
	Literature	180
9	**s-Step Overlapping Acid-Base Equilibria**	**183**
9.1	Basic Equations	183
9.2	Characteristic Concentration Diagrams and Absorbance Diagrams	185
9.3	The Generalized Form of an ADQ-Diagram	189
9.4	The Establishment of Absorbance Polyhedra	191
9.5	The Determination of Absolute pK-Values	194
9.6	Summary	197
	Literature	198
10	**Non-Linear Curve-Fitting of Spectrometric Data from s-Step Acid-Base Equilibrium Systems**	**201**
10.1	Basic Equations	201
10.2	Iterative Curve-Fitting Procedures	202

10.3	Statistical Analysis	207
10.4	Summary	208
	Literature	210

11 Simultaneous Titrations … 211

11.1	Basic Equations	211
11.2	Graphic Matrix-Rank Analysis	214
11.3	The Determination of Relative pK Values	214
11.3.1	Creation of a Dissociation-Degree Diagram	215
11.3.2	Points of Bisection in a Dissociation-Degree Diagram	219
11.3.3	Terminal Slope in a Dissociation-Degree Diagram	219
11.3.4	Complete Evaluation of a Dissociation-Degree Diagram	220
11.3.5	The Diagonal of a Dissociation-Degree Diagram	220
11.3.6	A Comparison of the Various Methods	221
	Literature	222

12 Branched Acid-Base Equilibrium Systems … 223

12.1	Macroscopic and Microscopic pK Values: The Problem of Assignment	223
12.2	pK Assignments Based on Extrapolations from Unbranched Systems	226
12.3	Assignment of Spectral Features to Specific Dissociable Groups	227
12.3.1	UV–VIS Spectroscopy	227
12.3.2	IR and Raman Spectroscopy	229
12.3.3	NMR Spectroscopy	230
12.3.4	The Dependence of Spectra on Solvent and Temperature; Band Shape Analysis	232
12.4	Summary	237
	Literature	238

13 Metal-Complex Equilibria … 241

13.1	The System BH $\rightleftarrows$ B (H), M + B $\rightleftarrows$ MB	241
13.1.1	Concentration Diagrams	241
13.1.2	Absorbance Diagrams	245
13.2	Determination of Overall Stability Constants for Metal Complexes Related to Multi-Step Acid-Base Equilibria	252
13.3	The Method of Continuous Variation ("Job's Method")	256
13.4	The System M + B $\rightleftarrows$ MB, MB + B $\rightleftarrows$ MB	259
13.4.1	Concentration Diagrams	259
13.4.2	Excess-Absorbance Diagrams	261
13.4.3	The System Daphnetin/FeCl$_3$	265
13.5	Ion-Selective Electrodes in the Determination of Stability Constants	270

13.6	Summary	272
	Literature	273

14 Association Equilibria . . . 277

14.1	Establishing Association Constants for Dimerism Equilibria Related to Protolysis Systems	277
14.2	Determining Association Constants for the System $2B \leftrightarrows C$ by Successively Increasing the Dye Concentration	280
14.2.1	Absorbance Diagrams	280
14.2.2	A Practical Example Based on the Eosine System	284
14.2.3	Excess Absorbance and Reduced Absorbance Diagrams	287
14.3	Reduced Absorbance Difference Quotient Diagrams and Excess Absorbance Quotient Diagrams ($2B \leftrightarrows C$, $C + B \leftrightarrows E$)	290
14.4	Determining Association Constants of Charge-Transfer Complexes	292
14.5	Summary	296
	Literature	297

15 Redox Equilibria . . . 299

15.1	Determining Standard Electrode Potentials	299
15.2	The Redox System $Red_1 + Ox_2 \leftrightarrows Ox_1 + Red_2$	303
15.2.1	Evaluating the Equilibrium Constant by Successively Increasing the Concentration of One Component (First Method)	303
15.2.2	Determining the Equilibrium Constant by the Method of Continuous Variation (Second Method)	306
15.2.3	Determining the Equilibrium Constant for an NADH-Dependent Dehydrogenase Reaction	310
15.3	Summary	313
	Literature	313

Part III Instrumental Methods

16 Suitable Apparatus for Spectrometric Titration . . . 317

16.1	m Experiments, Each Performed in a Different Buffer Medium	317
16.2	Discontinuous Spectrometric Titration in the Strictest Sense	319
16.2.1	Titration Outside the Spectrometer	319
16.2.2	Titration Inside the Spectrometer	324
16.3	Optimal Sample Size for Spectrometric Titration	325
16.4	Introducing the Titrant: Burets and Other Alternatives	326
	Literature	327

17	**Electrochemical Methods**	329
17.1	Problems Associated with the Conventional pH Scale	329
17.2	Calibrating a pH Meter Using Standard Buffers	331
17.3	pH Measurements	334
	Literature	336
18	**Spectrometric Methods**	337
18.1	UV–VIS Absorption	337
18.1.1	Spectrometric Data Collection	337
18.1.2	The Lambert-Beer-Bouguer Law and its Associated Error Function	340
18.2	Fluorescence	344
18.2.1	Spectrometric Data Collection	344
18.2.2	Quantitative Fluorescence Spectrometry and Predictable Sources of Error	346
18.2.3	Protolysis Equilibria: The Electronic Ground State vs. Excited States	350
18.3	Circular Dichroism (CD) and Optical Rotatory Dispersion (ORD)	356
18.3.1	Spectrometric Data Collection	356
18.3.2	Quantitative CD and ORD Spectrometry	359
18.4	IR Absorption	365
18.4.1	Spectrometric Data Collection	365
18.4.2	Quantitative Spectrometry in Aqueous Solution	366
18.5	Raman Scattering	368
18.5.1	Spectrometric Data Collection	368
18.5.2	Quantitative Spectrometry in Aqueous Solution	370
18.6	Nuclear Magnetic Resonance (NMR)	372
18.6.1	Basic Principles and Spectrometric Data Collection	372
18.6.2	Quantitative NMR Analysis of Titration Equilibria	379
	Literature	386
19	**Automation and Data Processing**	391
	Literature	394
Appendix		395
A	Source Listing of EDIA	395
B	Source Listing of TIFIT	405
Index		425

Part I

The Theoretical and Methodological Basis of Spectrometric Titration

1 Basic Principles and the Classification of Titration Methods

1.1 Definitions

Titration is a common technique in analytical chemistry. It is most often used for the purpose of quantitative (usually mass) determination, and involves stoichiometric, stepwise addition of a particular substance B to a solution containing an unknown amount of a substance A, the material under investigation (known as the *analyte*).[1–4] The *endpoint* of a titration (often taken to be the equivalence point) is sometimes determined visually, taking advantage of the change in appearance of an indicator, but any of several physicochemical methods may also be employed.[1,3] For example, if the course of a titration is followed by means of a photometer, the endpoint may be apparent from a graph of absorbance versus added volume of B (A_λ vs. mL(B); see below). In this case one speaks of a *photometric titration*.[3,5–7]

In contrast to pure stoichiometric mass titrations of the type described above, many protein and other biochemical determinations have long been conducted by the method known as *spectrophotometric titration*.[8–10] Here the goal is quite different. For example, an investigation of the pH-dependence of an absorption spectrum, or the determination of a series of pH/absorbance values at a single wavelength (A_λ vs. pH; see below), may be used to obtain information about the number and type of dissociable groups present in a substance, or to establish the corresponding dissociation constants. Such spectrophotometric titrations are applicable to both low- and high-mass molecular structures, and they are not limited to acid-base equilibrium systems. Indeed, just as with mass analysis, a wide range of reaction types is amenable to investigation (e.g., redox reactions, complex formation, etc.). Moreover, the approach extends well beyond absorption spectrophotometry to include such techniques as spectrofluorometry, CD, ORD, etc. (see below and also Chapter 18).

Since photometry and spectrophotometry differ only with respect to the dispersing element placed in front of the source of radiation (i.e., a filter or a monochromator, respectively), and thus—perhaps—in the bandwidth of the incident light,[11,12] it is clear that from a physicochemical standpoint the terms "photometric titration" and "spectrophotometric titration" are unfortunate choices. In fact, the two are often confused. Furthermore, the IUPAC definition of the word "titration" (with respect to mass analysis) is inconsistent with the way it is used in the term "spectrophotometric titration".

In the context of this book we shall employ a generalized definition of the concept "titration", one that includes both the above meanings. Whenever reference is here made to "titration in the general sense" it is to be understood as synonymous with "controlled displacement of a chemical equilibrium by means of a titration reagent (R), the *titrant*".[5,13] We shall also assume that the rate at which equilibrium is established in a titration is rapid with respect to the measurement process. Titration systems so defined constitute a special case within the realm of classical thermodynamics. Since equilibrium displacement (state change) is here occasioned by the introduction of material (the titrant), titrations must be regarded as examples of open chemical systems. Nevertheless, there are certain conditions under which a titration may be treated as if it constituted a closed system (see below).

The *titration variable* may be either the volume of titrant V_R or some characteristic system variable dependent upon it, such as pH. Considering the process of titration from the perspective of regulation and control engineering, V_R is the *manipulated variable*, while pH, for example, would be a *controlled variable*.

The term "spectrometric titration" will be used here as a generalization of the concept introduced above as "spectrophotometric titration": *spectrometric titration* consists of recording and evaluating spectra derived from a titration system as a function of some titration variable. The term "spectrometric"[14,15] thus includes all quantifiable spectroscopic methods, covering such non-photometric techniques as NMR spectroscopy (cf. Sec. 18.6). The complete set of spectra acquired in the course of a spectrometric titration will be referred to as the corresponding *titration spectra* (cf. Fig. 2–2).

In the event that the titrant is "optically transparent" (i.e., nonabsorbent throughout the spectral range under investigation), and provided that dilution accompanying the addition of titrant is either negligible or somehow compensated (usually mathematically), the titration system behaves spectrometrically as if it were a closed system; e.g.,

$$BH \xrightarrow{OH^-} B^- (+ H_2O)$$

Such apparently closed spectrometric titration systems lend themselves heuristically and mathematically to unusually straightforward analysis. Table 1–1 suggests a number of formal analogies that may be drawn between various titration and reaction systems.

1.2 Classification of Spectrometric Titration Methods

Table 1–1. Formal analogies between titration systems and reaction systems.

Type of System	Independent Variable	Reaction Scheme
Spectrometric Titration	Amount of titrant R (e.g., OH^-)	
(a) (apparently) closed	Titrant spectroscopically transparent	$BH \xrightarrow{(R)} B^-$
(b) open	Titrant spectroscopically observable	$j_R \downarrow$ $R + BH \xrightarrow{(R)} B^- + \ldots$
Chemical Reaction		
(a) Closed thermal reaction[16]	Reaction time t (after mixing)	$A \xrightarrow{(t)} B$
(b) Open thermal reaction[17]	Mean reactor residence time t	$j_A \downarrow$ $A \xrightarrow{(t)} B$ $\downarrow j_B$
(c) Photochemical reaction[18]	Irradiation time (or number of light quanta)	$A \xrightarrow{(h\nu)} B$

Note: The symbols j_i imply constant or controlled addition or removal of material.

1.2 The Classification of Spectrometric Titration Methods

Spectrometric titrations, like electrochemical titrations,[3,19] can be characterized according to a number of criteria, and these in turn may serve as the basis for classification schemes, as shown in Table 1–2. The schemes in Table 1–2 are related to the organizational pattern of this book, and relevant chapter numbers are indicated.

Table 1–2. Various ways of categorizing spectrometric titrations.

1. **By relationships among the titration equilibria**

    (a) Sequential, unbranched equilibria   Chapters 6–9
    (b) Sequential, branched equilibria   Chapter 12
    (c) Simultaneous (parallel) equilibria   Chapter 11

2. **By chemical reaction type**

    (a) Protolysis equilibria (acid-base titrations)   Chapters 6–12
    (b) Metal complex equilibria   Chapter 13
    (c) Association equilibria   Chapter 14
    (d) Redox equilibria   Chapter 15

3. **By type of spectrometry employed**

    (a) UV-VIS absorption   Section 18.1
    (b) UV-VIS fluorescence   Section 18.2
    (c) Spectropolarimetry (optical rotatory dispersion, ORD)   Section 18.3
    (d) Circular dichroism (CD)   Section 18.3
    (e) IR and Raman spectroscopy   Sections 18.4, 18.5
    (f) Nuclear magnetic resonance spectrometry (NMR)   Section 18.6

4. **By the nature of the electrochemical titration variable**

    (a) Hydronium ion activity or pH   Chapters 6–12
    (b) Ionic activity or p_{Ion} Chapter 13
    (c) Redox potential   Chapter 15

5. **By titration control mode**

    (a) Continuous   Chapters 16, 19
    (b) Discontinuous   Chapters 16, 19

6. **By data acquisition mode**

    (a) Analog (continuous data)   Chapters 16, 19
    (b) Digital (discrete data)   Chapters 16, 19

The major part of the book is devoted to applications of absorption spectrometry to protolytic equilibria, so it is appropriate that most of the illustrations in the theoretical section also involve acid-base titrations monitored by absorption spectrometric means. However, in later chapters, and with appropriate modifications, the same basic principles are applied to other types of spectrometry and to other modes of reaction. The fact that the scope of the book has been limited to titration in homogeneous solution dictates omission of precipitation reactions.

Electrochemical titrations are normally treated as *continuous, analog titrations* (cf. entries 5a and 6a in Table 1–2). That is, titrant is added continuously, and the titration curve (e.g. pH vs. mL) is recorded with an analog device. By contrast, spectrometric titrations are nearly always *discontinuous* (cf. entries 5b and 6b in Table 1–2). One either records the spectra of discrete solutions buffered to various pH values,[8] or else titrant is added stepwise with samples removed for spectral investigation after each step (cf. Chapter 16). Since both approaches provide discrete values for the variable pH, both are examples of *discontinuous, digital titration*, even though the spectra themselves are recorded on an analog device.[20]

Literature

1. *Römpps Chemielexikon*, 7. Aufl. Franckh'sche Verlagsbuchhandlung, Stuttgart 1977, Vol. 6, p. 3613.
2. *Römpps Chemielexikon*, 7. Aufl. Franckh'sche Verlagsbuchhandlung, Stuttgart 1974, Vol. 4, p. 2078.
3. E. B. Sandell, T. S. West, *(IUPAC) Pure Appl. Chem.* 18 (1969) 429.
4. R. Winkler-Oswatitsch, M. Eigen, *Angew. Chem.* 91 (1979) 20.
5. R. Blume, H. Lachmann, M. Mauser, F. Schneider, *Z. Naturforsch.* 29 b (1974) 500.
6. R. F. Goddu, D. N. Hume, *Anal. Chem.* 26 (1954) 1679, 1740.
7. J. B. Headridge, *Photometric Titrations*. Pergamon Press, New York 1961.
8. R. E. Benesch, R. Benesch, *J. Amer. Chem. Soc.* 77 (1955) 5877.
9. J. L. Crammer, A. Neuberger, *Biochem. J.* 1943, 302.
10. C. Tanford, *Adv. Prot. Chem.* 17 (1962) 69.
11. G. Kortüm, *Kolorimetrie, Photometrie und Spektrometrie*. Springer Verlag, Berlin 1962, Chap. I.
12. *Römpps Chemielexikon*, 7. Aufl. Franckh'sche Verlagsbuchhandlung, Stuttgart 1975, Vol. 5, p. 3274.
13. H. Lachmann, *Dissertation*. Tübingen 1973.
14. H. Lachmann, *Z. Anal. Chem.* 290 (1978) 117.
15. H. Lachmann, *Biophys. Struct. Mech.* 7 (1981) 268.
16. H. Mauser, *Formale Kinetik*. Bertelmann Universitätsverlag, Düsseldorf 1974, Chap. I.
17. K. G. Denbigh, J. C. R. Turner, *Einführung in die Chemische Reaktionskinetik*. Verlag Chemie, Weinheim 1973.

18. H. Mauser, *Formale Kinetik*. Bertelmann Universitätsverlag, Düsseldorf 1974, Chap. II.
19. IUPAC – Analytical Chemistry Division, *Pure Appl. Chem.* 45 (1976) 81.
20. S. Ebel, W. Parzefall, *Experimentelle Einführung in die Potentiometrie*. Verlag Chemie, Weinheim 1975.

2 Graphic Treatment of the Data from a Spectrometric Titration

The data associated with a spectrometric titration may take many forms. A typical example in the case of a protolysis equilibrium would be a set of titration spectra acquired as a function of pH. Moreover, the acquired data lend themselves to various forms of graphic presentation.

Stereospectrograms[1] represent a particularly effective display mode (cf. Fig. 2–1), based on a pseudo three-dimensional depiction of the relationships linking absorbance A, wavelength λ, and pH. The positions of absorption bands, as well as the ways in

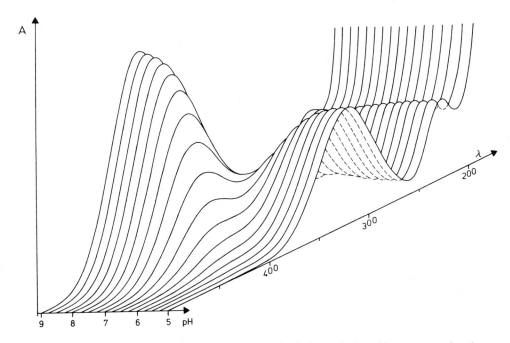

Fig. 2–1. Example of a *stereospectrogram* depicting relationships among the three variables absorbance (A), pH, and wavelength (λ), and based on measurements taken at discrete pH intervals.

which these change in the course of dissociation, are here very apparent, although the position and relative sharpness of an isosbestic point (cf. Chapter 3) may be difficult to ascertain. The most desirable arrangement and scaling of the three coordinates is often hard to foresee, and it is advisable that a stereospectrogram be optimized empirically by varying its appearance in the course of several computer runs. Clarity of presentation requires the omission of so-called "hidden lines" (i.e. those portions of curves that are covered by others in the foreground). Consequently, preparation of a stereospectrogram inevitably results in the elimination of some data.

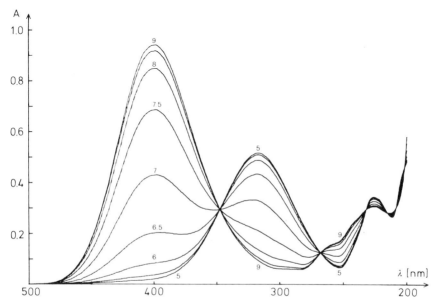

Fig. 2–2. Example of a set of *titration spectra*. Spectra were obtained at several values of pH during a spectrometric titration and are plotted in a single (A,λ) plane, thus providing clear indication of all isosbestic points.

For reasons such as these it is usually preferable to choose a different mode of presentation, one in which all spectra are projected in the same $A(\lambda)$ plane—producing a set of *titration spectra* in the strict sense of the term (cf. Fig. 2–2). The titration variable (e.g., pH) accompanies each spectrum in the form of a label. This particular representation has the advantage that it permits ready investigation of shifts in positions of absorption bands, and it also emphasizes the presence and relative definition of isosbestic points. However, such graphs may become very cumbersome in the case of a titration with more than two steps, or if spectra are taken over very small pH intervals. One potential remedy to the multi-step problem is use of more than one color (cf. Fig. 6–15); alternatively, the whole may be divided into a series of partial representations, each reflecting only one or two titration steps (e.g., Figures 6–19 and

6–20). In the case of more than three titration steps, effective graphic presentation of all titration spectra in a *single* diagram requires preparation of a stereospectrogram.

Another useful way to present spectrometric titration data is through a set of *spectrometric titration curves*. These are lines that result when absorbance A is plotted as a function of pH at several wavelengths (cf. Fig. 2–3). Each curve must then be labeled with the corresponding wavelength. In the case of a real (discontinuous) titration the available data will not suffice to produce a true "curve" (like that seen in a spectrum), resulting instead in sets of discrete points from which the corresponding curve must be inferred (see, for example, Fig. 6–8).

There is one other convenient, partial representation of the three-dimensional space defined by the variables (A, λ, pH) in a spectrometric titration: the so-called *A-diagram* A_{λ_1} vs. A_{λ_2}, which is described more fully in Sec. 3.2.2.

In the case of *mass analytical titration*, a photometric indicator is used to establish the functional relationship between the absorbance A, an electrochemical titration variable (e.g., pH), and the stoichiometric titration variable V_R (the *manipulated variable*; e.g., volume of titrant) (cf. Chapter 1). The function A_λ(pH,mL) and such partial representations as A vs. mL (*photometric titration*) and pH vs. mL (*potentiometric titration*) will not be subjects of further discussion in this book since existing literature coverage is adequate.[2–4] Nevertheless, it is worth noting that the complete data set from a *spectrometric titration*—for example, $A(\lambda, \text{pH}, \text{mL})$—contains all the information normally associated with *both* potentiometric and photometric titrations. Only at the stage of data reduction is the stoichiometric variable V_R (mL) discarded or restricted to the role of entering into calculations that compensate for the effect of dilution.

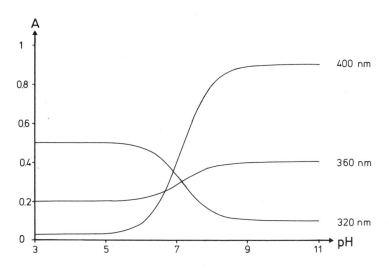

Fig. 2–3. Example of a set of *spectrometric titration curves*, in which the absorbance (A) measured at each of several wavelengths is plotted as a function of pH.

Literature

1. G. W. Ewing, A. Maschka, *Physikalische Analysen- und Untersuchungsmethoden in der Chemie*. R. Bohmann Verlag, Wien 1961.
2. J. B. Headridge, *Photometric Titrations*. Pergamon Press, New York 1961.
3. S. Ebel, W. Parzefall, *Experimentelle Einführung in die Potentiometrie*. Verlag Chemie, Weinheim 1975.
4. S. Ebel, E. Glaser, R. Kantelberg, B. Reyer, *Z. Anal. Chem.* 312 (1982) 604.

3 Basic Principles of Multiple Wavelength Spectrometry as Applied to Titration Systems

3.1 Generalized Lambert-Beer-Bouguer Law

In most applications of UV-VIS absorption spectrometry, it is assumed that a linear relationship exists between the concentration c of an absorbing species present in dilute solution and the absorbance A at a particular wavelength λ. This relationship is expressed by the *Lambert-Beer-Bouguer law*:

$$A_\lambda = l \cdot \varepsilon_\lambda \cdot c \tag{3-1}$$

where l = pathlength (cm)
  ε_λ = molar absorptivity of the absorbing species

The principle of additivity can usually be applied to cases in which multiple species undergo absorption simultaneously, leading through linear combination to the *generalized Lambert-Beer-Bouguer law*:

$$A_\lambda = l \cdot \sum_{i=1}^{j} \varepsilon_{\lambda_i} \cdot c \tag{3-2}$$

There are several *prerequisites* that must be fulfilled if the Lambert-Beer-Bouguer law is to be valid in a particular situation:[1-3]

(1) no (spectroscopically observable) concentration-dependent interactions are permitted among the absorbing species, or between these and the solvent or other non-absorbing species;

(2) adequate monochromaticity must be assured with respect to the incident radiation;

(3) there must be no "saturation" or "depletion" of the energy levels associated with the absorption phenomenon (as often occurs, for example, in laser- and NMR-spectroscopy; cf. Chapter 18).

Failure adequately to satisfy criterion (1) leads to so-called chemical or "true" deviations from the Lambert-Beer-Bouguer law. Problems related to criteria (2) and (3) are referred to as physical or "apparent" deviations.[1,4-7] Chemical deviations can be recognized and investigated with the aid of dilution series A_λ vs. c_i, as well as on the basis of A-diagrams (cf. Sec. 3.2.2) In practice, aqueous solutions usually adhere to the Lambert-Beer-Bouguer law to a degree sufficient for most purposes at concentrations $\leq 10^{-3}$ M, and often up to 10^{-2} M.

In the simplest case of a single-step acid-base titration, such as

$$BH \xrightarrow{OH^-} B^- (+ H_2O) \tag{3-3}$$

provided that the titrating reagent (OH⁻) and the solvent (H₂O) are transparent, and assuming that dilution by the titrant can be either ignored or arithmetically compensated, the Lambert-Beer-Bouguer law may be rewritten as follows:

$$A_\lambda = l \cdot \left(\varepsilon_{\lambda BH} \cdot bh + \varepsilon_{\lambda B} \cdot b \right) \tag{3-4}$$

where $bh = c_{BH}$ and $b = c_{B^-}$

To prevent confusion between the absorbance A_λ and the activity a_i, proton donors will be designated in this book by BH..., CH... instead of the usual AH. The corresponding concentrations (in mol/L) will be abbreviated by lower case letters and without explicit reference to charges. However, for typographical reasons the abbreviation "h" is used to refer to the hydronium ion, and specifically to its *activity*, a (H₃O⁺), not its concentration, c (H₃O⁺), since activity is the quantity most directly related to the measured parameter pH (cf. equation 4–5b). The pathlength will normally be taken as 1 cm and thus omitted.

The *dilution effect* caused by addition of a titrant will be dealt with by means of the following correction function:

$$\frac{V_0 + V_R}{V_0} = 1 + \frac{V_R}{V_0} \tag{3-5}$$

where V_0 = initial volume of the system to be titrated
V_R = total volume of titrant added up to the titration step in question

Provided that the Lambert-Beer-Bouguer or some analogous law is applicable, recorded values from spectrometric measurements (e.g., A_λ) will be extrapolated to zero dilution through multiplication by the correction function expressed in equation 3–5; i.e., they will be adjusted to correspond to a concentration equivalent to the initial concentration b_0. The dilution effect cannot be readily eliminated mathematically from titration spectra, since these are recorded in analog form, and this is an important reason for keeping the effect negligibly small (cf. also Chapters 16 and 18).

Stoichiometric boundary conditions for protolyte BH introduced at an initial concentration b_0 into the titration system described by equation 3–3 require that

$$b_0 = bh + b \tag{3-6}$$

from which it follows that

$$A_\lambda = \varepsilon_{\lambda BH} \cdot (b_0 - b) + \varepsilon_{\lambda B} \cdot b =$$
$$= \varepsilon_{\lambda BH} \cdot b_0 + (\varepsilon_{\lambda B} - \varepsilon_{\lambda BH}) \cdot b \qquad (3\text{--}7a)$$

or

$$A_\lambda = \varepsilon_{\lambda BH} \cdot bh + [\varepsilon_{\lambda B} \cdot (b_0 - bh)] =$$
$$= \varepsilon_{\lambda B} \cdot b_0 + (\varepsilon_{\lambda BH} - \varepsilon_{\lambda B}) \cdot bh \qquad (3\text{--}7b)$$

Thus, unambiguous description of a single-step acid-base titration system requires specification of only *one* concentration variable; equation 3–6 shows that bh and b are linearly dependent upon each other.

3.2 Matrix-Rank Analysis of a Titration System

The number of linearly independent concentration variables required for unambiguous specification of a titration system is referred to as the *rank* of the system, s. Thus, for a single-step titration $s = 1$. For a titration consisting of an unbranched *sequence* of steps the rank is always equal to the number of such steps (i.e., to the total number of components, less one).

In the event that the number and order of steps in a given titration is unknown, the rank s can be determined from experimental titration data by carrying out a so-called "matrix-rank analysis".

3.2.1 Numeric Matrix-Rank Analysis

Using the example of an acid-base titration, if the absorbance A is measured at $m \geq s$ wavelengths λ_j for $n \geq s$ pH values pH_k, the data may be summarized in matrix form, as shown below (where $j = 1, 2, 3, ..., m$ and $k = 1, 2, 3, ..., n$):

$$A = \begin{pmatrix} A_{11} & A_{12} & \cdots & A_{1n} \\ A_{21} & A_{22} & \cdots & \cdot \\ \cdot & \cdot & & \cdot \\ \cdot & \cdot & & \cdot \\ A_{m1} & A_{m2} & \cdots & A_{mn} \end{pmatrix} \qquad (3\text{--}8)$$

Since the number of wavelengths employed need not be identical to the number of pH measurements, the general form of A is rectangular rather than square.

Various techniques are available for establishing the rank of a matrix.[8–18] In the simplest case of a square matrix (i.e., $m = n \geq s$), the rank is equal to the number of rows or columns (the *order*) of the smallest non-vanishing subdeterminant.

With very large matrices of this type—provided they contain *precise* data—an exact determination of rank can be carried out with the aid of a computer, but a matrix that contains *real* data subject to experimental error presents problems, because none of the many subdeterminants will exactly equal zero. For this reason, numeric matrix-rank analysis is only applicable in practice if a suitable statistical criterion can be provided for determining when the above condition has been *adequately* fulfilled (i.e., to within the limits of a specific degree of measurement error). Thus, the calculated rank s for a matrix of real data must be stated in statistical terms. The literature contains a wealth of recommendations regarding appropriate statistical criteria, but their relative advantages and disadvantages remain a subject of debate.[17,19,20]

Recently, the method of *factor analysis* has been proposed as an alternative to matrix-rank analysis,[18,19] but again, suitable statistical tests must be available.

Yet another possible approach to the determination of the rank s is a systematic *curve-fitting analysis* of all conceivable titration mechanisms (cf. Chapter 10).

All the above-described methods require extensive calculation and access to appropriate computer facilities.

3.2.2 Graphic Methods of Matrix-Rank Analysis

The literature of the past 20 years describes numerous attempts to obtain information about the number of components in a multi-component system either from the existence of isosbestic points[21–25] or from observed systematic displacements of absorption maxima.[26] In 1968, H. Mauser[27] proposed a general theory of isosbestic points for closed, thermally controlled reactions and photoreactions. He also developed various graphic methods (A-, AD-, and ADQ-diagrams, see below) that permit rapid, straightforward, and precise determination from spectral data of the number s of linearly independent reaction steps present. The same diagrams were derived independently two years later by a group of American authors,[15] who recommended their use as an alternative to, and extension of, the numeric method of matrix-rank analysis.* In 1971 yet a third independent derivation of A-diagrams appeared.[29] To distinguish these methods from numeric matrix-rank analyses, they will be referred to in this book as examples of *graphic matrix-rank analysis*.[30] The following graphic methods are applicable to the investigation of spectrometric data sets:

---

*A numeric method requiring constancy of quotients $A_{\lambda_i}/A_{\lambda_j}$, which is equivalent to the A-diagram approach, had already been utilized in the early 1960s;[10,11] cf. also ref. 28.

3.2 Matrix-Rank Analysis of a Titration System

- analysis of *titration spectra* with respect to isosbestic points and other types of spectral intersection;

- analysis of *spectrometric titration curves* A_λ vs. pH with respect to inflection points, maxima, and minima;

- establishment of linearity with respect to *absorbance diagrams (A-diagrams)* A_{λ_1} vs. A_{λ_i} or *absorbance difference diagrams (AD-diagrams)* ΔA_{λ_1} vs. ΔA_{λ_i}; and

- demonstration of linear *absorbance difference quotient diagrams (ADQ-diagrams)*.

The above methods are treated only in an introductory way in this chapter since their application to titration systems of rank $s = 1$ to $s = 4$ is discussed thoroughly in Parts II and III.

In the case of a single-step titration system examined at two different wavelengths λ_1 and λ_2, two relationships corresponding to equation 3–7a can be formulated. Collecting the constants $\varepsilon_{\lambda_B} \cdot b_0$ on the left leads to:

$$A_{\lambda_1} - \varepsilon_{\lambda_1 BH} \cdot b_0 = \Delta A_{\lambda_1} = (\varepsilon_{\lambda_1 B} - \varepsilon_{\lambda_1 BH}) \cdot b$$
$$A_{\lambda_2} - \varepsilon_{\lambda_2 BH} \cdot b_0 = \Delta A_{\lambda_2} = (\varepsilon_{\lambda_2 B} - \varepsilon_{\lambda_2 BH}) \cdot b \tag{3–9}$$

The term ΔA_λ expresses for any pH the change observed in absorbance relative to the absorbance at the "acid side" (i.e., at the acidic limit) of the system, where $bh = b_0$ and $b = 0$. If the first of the equations 3–9 is divided by the second, the pH-dependent concentration function $b(\text{pH})$ is eliminated, as shown by equation 3–10.

$$\Delta A_{\lambda_1} = \frac{\varepsilon_{\lambda_1 B} - \varepsilon_{\lambda_1 BH}}{\varepsilon_{\lambda_2 B} - \varepsilon_{\lambda_2 BH}} \cdot \Delta A_{\lambda_2} = \frac{Q_{\lambda_1}}{Q_{\lambda_2}} \cdot \Delta A_{\lambda_2} \tag{3–10}$$

In the case where $s = 1$, a plot of ΔA_{λ_1} vs. ΔA_{λ_2} at equivalent pH values (known as an *absorbance difference diagram*, abbreviated *AD-diagram*) leads to a straight line passing through the origin (cf. Fig. 3–1).

If one now adds the function $\varepsilon_{\lambda_2 BH} \cdot b_0$ to both sides of equation 3–10 and makes appropriate substitutions from equations 3–7a and 3–9, the result can be expressed as

$$A_{\lambda_1} = \frac{\varepsilon_{\lambda_1 B} - \varepsilon_{\lambda_1 BH}}{\varepsilon_{\lambda_2 B} - \varepsilon_{\lambda_2 BH}} \cdot \Delta A_{\lambda_2} + (\varepsilon_{\lambda_1 BH} - \varepsilon_{\lambda_2 BH}) \cdot b_0 \tag{3–11}$$

Thus, an *absorbance diagram (A-diagram)* of A_{λ_1} vs. A_{λ_2} is also linear in the case of a single-step titration, although the resulting line normally does not pass through the origin (Fig. 3–2). Linearity of A- and AD-diagrams, regardless what wavelength combination is chosen, is therefore a necessary condition for any system of rank $s = 1$.

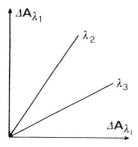

Fig. 3–1. Example of an *absorbance difference diagram* (*AD-diagram*) constructed for two wavelength combinations. Both lines pass through the origin.

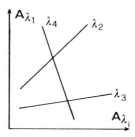

Fig. 3–2. Example of an *absorbance diagram* (*A-diagram*) presenting data for three different wavelength pairs.

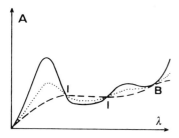

Fig. 3–3. Isosbestic points (I) formed by intersecting spectra, together with a point of contact (B) formed where the spectra approach each other tangentially.

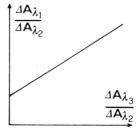

Fig. 3–4. Example of an *absorbance difference quotient diagram* (*ADQ-diagram*) constructed from data acquired at three wavelengths.

The conditions required for the appearance of *isosbestic points* in a single-step titration system can be read directly from equation 3–10: at any wavelength for which $\varepsilon_{\lambda BH} = \varepsilon_{\lambda B}$ the change in absorbance during titration must be zero. Thus, for titration spectra that intersect, sharp isosbestic points must arise at the wavelengths of intersection (cf. Fig. 2–2). This is another necessary condition for a system of rank $s = 1$. If the corresponding spectra do not actually cross, only approaching each other tangentially, then $\varepsilon_{\lambda BH} = \varepsilon_{\lambda B}$ still applies, but what results are so-called *points of contact* (cf. Fig. 3–3). If no contact whatsoever occurs it is not possible to draw conclusions about rank from titration spectra themselves, although this information is still derivable from A-, AD-, and ADQ-diagrams.

To prepare *absorbance difference quotient diagrams*, abbreviated *ADQ-diagrams*, two quotients must be developed from changes in absorbance measured at three wavelengths, where one of the three changes serves as the denominator quantity.

3.2 Matrix-Rank Analysis of a Titration System

These quotients are then plotted against one another; i.e., a graph is constructed of $\Delta A_{\lambda_1}/\Delta A_{\lambda_2}$ vs. $\Delta A_{\lambda_3}/\Delta A_{\lambda_2}$ (Fig. 3–4). Linear ADQ-diagrams are a necessary condition for the rank $s = 2$, as will be shown in Sec. 7.2. It is also possible in principle to derive appropriate ("higher order") ADQs-diagrams applicable to titration systems of rank $s > 2$. Indeed, linear ADQ3-diagrams for the rank $s = 3$ have been described in the literature.[31] Nevertheless, development of the required multiple quotients demands critical analysis of error functions and of the significance to be ascribed to apparent conclusions, the same problem encountered with numeric methods of matrix-rank analysis.

Table 3–1 provides an overview of the criteria applicable to graphic matrix-rank analysis of various systems with rank $s < 3$. It will be noted that matrix-rank analysis permits determination *only* of the rank s of a titration system; it does not not allow one

Table 3–1. Characteristics associated with graphic matrix-rank analyses for spectrometric titrations of rank $s < 3$.

System	Rank	Titration spectra	Spectrometric titration curves	A-Diagrams	ADQ-Diagrams	Chapter
$BH \rightleftarrows B^-$	$s = 1$	isosbestic points (if spectra intersect)	1 inflection point (at pK value)	linear	cluster of points	6.1–6.3
$BH_2 \rightleftarrows BH^- \rightleftarrows B^{2-}$	$s = 2$	no isosbestic points	2 inflection points 1 maximum or minimum	curved	linear	7
$BH \rightleftarrows B^-$ $CH \rightleftarrows C^-$ (simultaneous)	$s = 2$	no isosbestic points	2 inflection points 1 maximum or minimum	curved	linear	11
$B{\sim}C{-}H$ / $HB{\sim}CH$ / $B{\sim}C^-$ / $H{-}B{\sim}C^-$ (branched)	$s = 2$	no isosbestic points	2 inflection points 1 maximum or minimum	curved	linear	12

Note: This table does not take into account special cases arising from unusual limiting conditions with respect to ΔpK and $\varepsilon_{\lambda i}$ (cf. the corresponding cited chapter).

to choose between different titration mechanisms corresponding to the same rank. For example, in the branched titration system shown, only two of the five so-called "microdissociation equilibria" are linearly independent. This system behaves with respect to spectrometric titration precisely as would an unbranched sequential system of rank $s = 2$ (for more complete discussion see Chapter 12).

The methods described above by no means presuppose the use of UV-VIS spectrometry; they are equally applicable to other quantifiable spectrometric methods (Chapter 18).

In contrast to numeric matrix-rank analysis, graphic methods entail little mathematical effort. All quotients needed for ADQ-diagrams, as well as any desired regression lines, can be generated quite simply using a pocket calculator.

Literature

1. G. Kortüm, *Kolorimetrie, Photometrie und Spektrometrie*. Springer-Verlag, Berlin 1962, Chap. I.
2. G. W. Ewing, A. Maschka, *Physikalische Analysen- und Untersuchungsmethoden der Chemie*. R. Bohmenn Verlag, Wien 1961, Chap. 7.
3. M. Pestemer, D. Brück, *Houben-Weyl*, Vol. 3/2. Georg Thieme Verlag, Stuttgart 1955, p. 567.
4. W. Luck, *Z. Elektrochem.* 64 (1960) 676.
5. J. Agterdenbos, J. Vink, *Talanta* 18 (1971) 676.
6. J. Agterdenbos, J. Vlogtmen, L. v. Broekhoven, *Talanta* 21 (1974) 225, and J. Agterdenbos, J. Vlogtmen, *Talanta* 21 (1974) 231.
7. R. B. Cook, R. Jankow, *J. Chem. Educ.* 49 (1972) 405.
8. L. Newman, D. N. Hume, *J. Amer. Chem. Soc.* 79 (1957) 4571, 4576, 4581.
9. R. M. Wallace, *J. Amer. Chem. Soc.* 82 (1960) 899.
10. W. Liptai, *Z. Elektrochem.* 65 (1961) 375.
11. G. Briegleb, *Elektronen-Donator-Acceptor-Komplexe*. Springer Verlag, Berlin 1961, Chap. XII.
12. S. Ainworth, *J. Phys. Chem.* 65 (1961) 1968; 67 (1963) 1613.
13. D. Katakis, *Anal. Chem.* 37 (1965) 867.
14. J. J. Kankare, *Anal. Chem.* 42 (1970) 1322.
15. J. S. Coleman, L. P. Varga, S. H. Mastin, *Inorg. Chem.* 9 (1970) 1015.
16. Z. Z. Hugus, A. A. El-Awady, *J. Phys. Chem.* 75 (1971) 2954.
17. M. E. Magar, *Data Analysis in Biochemistry and Biophysics*. Academic Press, New York 1972, Chap. 9.
18. M. E. Magar, P. W. Chun, *Biophys. Chem.* 1 (1973) 18.
19. J. T. Bulmer, H. F. Shurvell, *J. Phys. Chem.* 77 (1973) 256.
20. D. J. Legett, *Anal. Chem.* 49 (1977) 276.
21. H. L. Schläfer, O. Kling, *Angew. Chem.* 68 (1956) 667; *Z. El. Chem.* 65 (1961) 142.
22. M. D. Cohen, E. Fischer, *J. Chem. Soc.* 1962, 3044.
23. C. F. Timberlake, P. Brodle, *Spectrochim. Acta* 23 A (1967) 313.
24. J. Brynestad, G. P. Smith, *J. Phys. Chem.* 72 (1968) 296.

25. T. Nowicka-Jankowska, *J. Inorg. Nucl. Chem.* 33 (1971) 2043.
26. R. E. Benesch, R. Benesch, *J. Amer. Chem. Soc.* 77 (1955) 5877.
27. H. Mauser, *Z. Naturforsch.* 23 b (1968) 1021, 1025.
28. G. L. Ellmann, *Arch. Biochem. Biophys.* 82 (1959) 70.
29. C. Chylewski, *Angew. Chem.* 83 (1971) 214.
30. H. Lachmann, *Z. Anal. Chem.* 290 (1978) 117.
31. R. Blume, *Dissertation.* Tübingen 1975.

4 Thermodynamic and Electrochemical Principles of Spectrometric Titration Systems

Titration data are unaffected by the kinetics associated with establishment of equilibrium, since the reactions involved are exceedingly rapid. Consequently, the titration process is conveniently treated as a series of changes of state, in the course of which the system is always at thermodynamic and chemical equilibrium. The corresponding thermodynamic description proceeds from the *law of mass action* (Guldberg and Waage, 1867) or the more exact treatment due to Gibbs.[1,2] Thus, for an equilibrium reaction in general:

$$\nu_B \cdot B + \nu_C \cdot C + \ldots \leftrightarrows \nu_M \cdot M + \nu_N \cdot N + \ldots$$

$$K = \frac{a_M^{|\nu_M|} \cdot a_N^{|\nu_N|} \ldots}{a_B^{|\nu_B|} \cdot a_C^{|\nu_C|} \ldots} \tag{4-1}$$

where K = thermodynamic equilibrium constant (at constant temperature and pressure)
a_i = activity of the constituent i
$|\nu_i|$ = absolute value of the stoichiometric coefficient of the constituent i

In the case of a single-step protolysis equilibrium in highly dilute aqueous solution, i.e.,

$$BH + H_2O \leftrightarrows B^- + H_3O^+$$

the *thermodynamic dissociation constant* may thus be expressed as:

$$K = \frac{h \cdot a_{B^-}}{a_{BH}} = \frac{c_{H_3O^+} \cdot b \cdot f_{H_3O^+} \cdot f_{B^-}}{bh \cdot f_{BH}} \tag{4-2}$$

where f_i = activity coefficient of the ith constituent
$|\nu_i|$ = 1 for all constituents
h = hydronium ion activity

The activity of the pure solvent H_2O is here taken to be unity (or included in the equilibrium constant).[3,4]

Assuming the dilution effect of the titrant is negligible or can be compensated mathematically, the total concentration of substrate $b_0 = b + bh$ remains unchanged throughout the titration. Given this assumption, the titrant serves only to shift the protolysis equilibrium, and the more generalized treatment required by an open system can be avoided (cf. Chapter 1).

Spectrometric methods do not provide measurements of activities, but rather of the concentrations bh and b. For this reason, it is common to utilize so-called *classical dissociation constants*:

$$K_c = \frac{c_{H_3O^+} \cdot b}{bh} = \frac{\beta^2}{1-\beta} \cdot b_0 \tag{4-3}$$

In the event that the *degree of dissociation* β is directly available from spectrophotometric measurement, or if the pH plays no role in the analysis, the absolute or relative value of the classical dissociation constant K_c can be determined solely from a combination of spectrophotometric data and the initial weight of sample, expressed as the concentration b_0 (Sec. 6.3.6).

In contrast to the thermodynamic constant of equation 4–2, the classical dissociation constant is only approximately independent of concentration, even in highly dilute solution, as shown by combining equations 4–2 and 4–3:

$$K_c = \frac{f_{BH}}{c_{H_3O^+} \cdot f_{B^-}} \cdot K = g(b_0) \tag{4-4}$$

For this reason, reports of classical dissociation constants should always be accompanied by the precise conditions of measurement, paying particular attention to b_0 and perhaps to any large excess of neutral salts present. A much more exact approach, albeit a more difficult one, involves experimental determination of the function $K_c = g(b_0)$. Extrapolation to $b_0 \to 0$ then makes it possible to obtain the thermodynamic dissociation constant K. The various empirical and theoretical extrapolation formulas used in the determination of K will not be discussed in this book since existing literature on the subject is quite adequate.[3,5-7]

One often finds evaluations of spectrometric titration data that utilize—explicitly or implicitly—yet a third definition of the dissociation constant. This has been referred to in the literature as a *"mixed"* (sometimes *"mixed-mode"* or *"apparently mixed"*) *dissociation constant*.[1,8-10] In the above example of a one-step protolysis equilibrium such a constant would be expressed:[1]

$$K = h \cdot \frac{b}{bh} = h \cdot \frac{\beta}{1-\beta} \tag{4-5a}$$

or, after taking logarithms (to give what is known as the Henderson-Hasselbalch equation[11]):

$$pK = pH - \log \frac{b}{bh} = pH - \log \frac{\beta}{1-\beta} \tag{4-5b}$$

Equations 4–5a and 4–5b contain only quantities that can be obtained directly from spectrometric titration data (pH or h, as well as b/bh). The relationship between this function K and the thermodynamic dissociation constant $\mathbf{K}$ becomes evident upon combining equations 4–2 and 4–5a:

$$K = \frac{f_{BH}}{f_{B^-}} \cdot \mathbf{K} = g(b_0) \tag{4-6}$$

Since the activity coefficient f_{BH} of the undissociated acid in highly dilute solution can be approximated with a high degree of accuracy by unity,[1] the function $K_c = g(b_0)$ is largely determined by f_{B^-}.

The "mixed" dissociation constant K is most frequently used in spectrometric titrations involving large initial concentrations b_0 and high concentrations of neutral salts (e.g., in NMR titrations, Sec. 18.6). As with classical dissociation constants, it is again important that measurement conditions be fully specified, especially in cases involving comparisons with literature data or results obtained by other analytical methods (Chapter 18).

Various theoretical and practical problems involved in carrying out electrochemical determinations are discussed in Chapter 17, especially problems associated with the conventional pH scale.

Literature

1. G. Kortüm, H. Lachmann, *Einführng in die Chemie Thermodynamik*. 7. Aufl. Verlag Chemie, Weinheim 1981, Chap. V.
2. H. Rau, *Kurze Einführung in die Physikalische Chemie*. Vieweg Verlag 1977. Chap. II/6.
3. S. Ebel, W. Parzefall, *Experimentelle Einführung in die Potentiometrie*. Verlag Chemie, Weinheim 1975.
4. G. Kortüm, *Lehrbuch der Elektrochemie*. Verlag Chemie, Weinheim 1972, Chap. II
5. G. Kortüm, *Lehrbuch der Elektrochemie*. Verlag Chemie, Weinheim 1972, Chap. IV.
6. G. Kortüm, W. Vogel, K. Andrussow, *Pure Appl. Chem.* 1 (1960) 190.
7. H. Faldenhagen, *Theorie der Elektrolyte*. S. Hirzel Verlag, Leipzig 1971.
8. D. H. Rosenblatt, *J. Phys. Chem.* 58 (1954) 40.
9. E. J. King, *Acid-Base-Equilibria*. Pergamon Press, London 1965.
10. D. L. Rabenstein, M. S. Greenberg, C. A. Evans, *Biochemistry* 16 (1977) 977.
11. L. J. Henderson, M. Tennenbaum, *Blut, seine Pathologie und Physiologie*. Steinkopf-Verlag, Dresden 1932.

Part II

The Formal Treatment and Evaluation of Titration Systems – Analysis of Chemical Equilibria

5 The Goal of a Spectrometric Titration

Hypotheses and experiments are of equal importance in the spectrometric analysis of titration equilibria. Experiment shows how the concentrations of the components or other selected variables change during the course of a titration, while hypothesis provides an intuitive picture of possible equilibria comprising the system. A spectrometric analysis involves comparing hypothesis with experiment. If the two conflict, the current hypothesis must be replaced with a new one. If they are consistent, one may wish to devise additional experiments to further substantiate the hypothesis.

A titration system is characterized by the number and types of equilibria present, as well as by the relative or absolute values of the corresponding equilibrium constants. Thus, the primary goal is establishing these quantities. However, it is also important to determine certain characteristic spectrometric parameters of the system, such as molar absorptivities and the spectra of various components.

This, then, is the challenge with which one is normally confronted in undertaking a typical spectrometric titration analysis, whether of an acid-base, metal complex, association, or redox equilibrium system. Apart from heterogeneous equilibria, these are the most important types of equilibria encountered in chemistry and biochemistry, and a few examples may help clarify why they warrant investigation.

In the case of *acid-base equilibria*, one useful piece of information that can be derived from a knowledge of pK values is the extent of charge on the protolytes as a function of pH. This plays a key role, for example, with respect to intensity of coloration, as with the colors of blossoms or synthetic dyes. Extent of dissociation is also important in pharmacology and toxicology, especially since the binding of drugs and metabolites to membrane receptors is often strongly dependent on surface charge. The phenomena of resorption and excretion are dominated by similar considerations, as are the forces of interaction between enzymes and such low molecular weight materials as substrates, coenzymes, inhibitors, or effectors (ca. 2/3 of all enzymes play host to anionic substrates or cofactors). With "branched" equilibrium systems, the spatial distribution of charges within certain molecules is strongly affected by the positions of the constituent equilibria. Such systems may be fully characterized by so-called "microscopic dissociation constants" that are in turn related to the directly measurable "macroscopic dissociation constants" (pK values).

A knowledge of pK values is also important with respect to *metal complex equilibria*. Both protonated and deprotonated ligands can undergo binding with metals, and metal-complex formation of this type can often facilitate the identification and quantitative determination of both metals and ligands. Analytical chemists can put such equilibria to good use, but only if reliable information is available about the constitution and number of metal complexes present in a given system, together with accurate measures of stability (e.g., formation constants). Metal complexes are widely distributed in nature. For example, many of the colors displayed by flowers result from the formation of aluminum or iron complexes. In addition, the biological activity of many proteins depends on the presence of metal ions. Respiration, for example, depends on a series of electron transfers to molecular oxygen, and these reactions are mediated by iron- and copper-containing proteins.

Equilibria related to electron donor-acceptor and charge-transfer complexes are often described as *association equilibria*. Examples include the self-association equilibria characteristic of certain dyes (e.g., eosin, acridine orange, etc.), many of which are of interest in histology. The techniques of tissue staining rely heavily on the formation of dye-molecule aggregates within the cellular structure of tissue, which alters the absorption properties of the cells. Stains of this type provide one important way of distinguishing, for example, between normal and cancerous cells. The binding of a low molecular weight compound to a protein can also be regarded as a form of association equilibrium, as can the typical antigen-antibody reaction seen in the immune system.

Redox reactions play an especially important role in biological systems, including in photosynthesis, the most fundamental of the routes by which energy is transferred into the life cycle of terrestrial organisms. Electron-transfer reactions always involve two redox pairs whose electron affinities differ. Each pair can be characterized electrometrically in terms of its standard electrode potential, a parameter that has applications with respect to any oxidation-reduction reaction. For example, the hydrogen-transfer step in the alcohol dehydrogenase reaction

$$\text{ethanol} + NAD^+ \leftrightarrows \text{acetaldehyde} + NADH + H^+$$

involves the two redox pairs (couples) ethanol/acetaldehyde and NAD^+/NADH. The position of equilibrium of the NADH-dependent dehydrogenase reaction can be calculated directly provided standard electrode potentials are available for the two half-reactions. Alternatively, the standard electrode potential of one half of a redox system can be determined from a known position of overall equilibrium provided the standard electrode potential of the second half has been established. The availability of the latter approach can be of great use in a case in which one redox pair fails to provide a stable redox potential at the electrode of a galvanic cell even though it reacts in a fully reversible fashion when joined with other redox pairs. Spectrometric determination of the equilibrium constant of the overall reaction offers a straightforward means of ascertaining the value of the unknown standard potential.

The chapters that follow illustrate how spectrometric data acquired at various wavelengths (*multiple wavelength analysis*) can be used to examine a wide variety of equilibria important in chemistry and biochemistry. Absorbance diagrams and related graphical treatments will be seen to occupy a central place in the development of the arguments.

6 Single-Step Acid-Base Equilibria

6.1 Basic Equations

Since most of the fundamental equations required for studying protolysis equilibria of rank $s = 1$ have already been encountered in the examples of Part I they will be summarized here only briefly.

At every point on the spectrometric titration curve of a single-step acid-base titration, proton-transfer equilibrium is established very rapidly:

$$BH + H_2O \leftrightarrows B^- + H_3O^+ \tag{6-1}$$

A relationship thus exists between the electrochemical parameter pH ($= -\log h$) and the concentrations of undissociated and dissociated acid (cf. Chapter 4):

$$K_1 = h \cdot \frac{b}{bh} = h \cdot \frac{\beta}{1-\beta} \tag{6-2}$$

where K_1 = mixed dissociation constant, single-step titration system
h = hydronium ion activity
β = degree of dissociation

Solving for the concentrations b and bh:

$$b = \frac{K_1 \cdot bh}{h} \tag{6-3}$$

$$bh = \frac{h \cdot b}{K_1} \tag{6-4}$$

Assuming that only BH and B⁻ absorb, and that the Lambert-Beer-Bouguer law is fully applicable, then in any absorption spectrometric titration it must also be true that

$$A_\lambda = \varepsilon_{\lambda BH} \cdot b_0 + (\varepsilon_{\lambda B} - \varepsilon_{\lambda BH}) \cdot b$$
$$= \varepsilon_{\lambda B} \cdot b_0 + (\varepsilon_{\lambda BH} - \varepsilon_{\lambda B}) \cdot bh \tag{6-5}$$

Equations 6–2 and 6–5 can be combined to provide a relationship between the parameters actually measured during a spectrometric titration (A_λ and either pH or h):[1]

$$K_1 = h \cdot \frac{b}{bh} = h \cdot \frac{\beta}{1-\beta} = h \cdot \frac{A_\lambda - A_{\lambda BH}}{A_{\lambda B} - A_\lambda} \tag{6-6}$$

where $A_\lambda = f(\text{pH})$ and is the absorbance corresponding to a given titration point
$A_{\lambda BH}$ = absorbance at the "acidic end" of the titration ($bh = b_0$)
$A_{\lambda B}$ = absorbance at the "alkaline end" of the titration ($b = b_0$)

Taking logarithms, one obtains directly the Henderson-Hasselbalch equation:

$$pK = \text{pH} - \log \frac{b}{bh} = \text{pH} - \log \frac{\beta}{1-\beta} = \text{pH} - \log \frac{A_\lambda - A_{\lambda BH}}{A_{\lambda B} - A_\lambda} \tag{6-7}$$

Virtually all methods for the spectrometric determination of (absolute) pK values for single-step protolysis equilibria are based on some reformulation of equations 6–6 and 6–7; that is, they entail a combination of spectrophotometric *and* electrochemical data (cf. Sec. 6.3)

6.2 Graphic Matrix-Rank Analysis

The rank s of an unknown titration system presumed to consist of a single step is best ascertained through the use of graphic methods of matrix-rank analysis (cf. Sec. 3.2.2). Even if it is known that the protolyte under investigation contains only a *single* dissociable group it is still useful to carry out a graphic matrix-rank analysis, for the following reasons:

(1) to test whether impurities or foreign substances are affecting the spectrometric titration at specific wavelengths or pH values (e.g., isomers; cf. Chapter 11);

(2) to test for potential chemical or photochemical instability of the sample over the pH and wavelength ranges of interest (i.e., to rule out the formation of impurities during the titration);

(3) to test whether, and in what pH and wavelength ranges, the titrant absorbs, a common cause of apparent deviations from the rank $s = 1$ (see below);

(4) to see if interfering substances may be entering the system from the vapor phase during the course of titration (e.g., CO_2, or, in the case of oxidizable samples, O_2);

(5) to investigate whether the high concentrations of acid or base required at the extreme ends of the pH scale (2 > pH > 12) cause disruptive changes in the solvent (so-called "medium effects");[2] and

(6) to determine optimum wavelengths for use in further analysis of the system.

6.2 Graphic Matrix-Rank Analysis

A numeric approach to matrix-rank analysis is far too complicated to warrant its use if one only wishes to establish that a system is indeed of rank $s = 1$. The numeric and statistical issues discussed in Sec. 3.2.1 also render numeric analysis unsuitable for performing tests (1)–(6) above.

The following criteria, drawn from graphic matrix-rank analysis, are *necessary conditions* if a spectrometric titration system is to be assigned the rank $s = 1$ (cf. Sec. 3.2.2 and Table 3–1):

- Wherever *titration spectra* intersect, sharp *isosbestic points* must result (cf. Fig. 6–1). This is a direct consequence of the fact that intersection implies $\varepsilon_{\lambda BH} = \varepsilon_{\lambda B}$. A special case of this condition covers spectra that only meet tangentially rather than intersecting (cf. Figures 6–12 and 6–15). If the spectra fail to meet at any point, no immediate conclusions with respect to rank can be drawn.

- *Absorbance diagrams* (*A-diagrams*) constructed for all wavelength combinations must be *linear* (cf. Fig. 6–2). An A-diagram for a single-step acid-base titration is always described by a linear equation of the following form (cf. Sec. 3.2.2):

$$A_1 = \frac{\varepsilon_{1B} - \varepsilon_{1BH}}{\varepsilon_{2B} - \varepsilon_{2BH}} \cdot A_2 + (\varepsilon_{1BH} - \varepsilon_{2BH}) \cdot b_0 \tag{6-8}$$

- *Absorbance difference diagrams* (*AD-diagrams*) for all wavelength combinations must consist of straight lines passing through the origin. Thus, choice of the acidic end of the titration as a reference point for calculating differences (i.e., $\Delta A_\lambda = A_\lambda - A_{\lambda BH}$) results in the following relationship (Sec. 3.3.2):

$$\Delta A_1 = \frac{\varepsilon_{1B} - \varepsilon_{1BH}}{\varepsilon_{2B} - \varepsilon_{2BH}} \cdot \Delta A_2 \tag{6-9}$$

A-diagrams and AD-diagrams actually contain the same information. If one wishes to display a large number of wavelength combinations as clearly as possible in a *single* diagram (cf. Fig. 6–3), then an AD-diagram is preferable, since here the various lines do not intersect.

Spectrometric titration under nitrogen of *p-nitrophenol* with *0.1 N sodium hydroxide* provides an illustration of the above criteria for the rank $s = 1$.[1,3]

$$O_2N\text{-}C_6H_4\text{-}OH \underset{}{\overset{OH^-}{\rightleftarrows}} O_2N\text{-}C_6H_4\text{-}O^- \tag{6-10}$$

$$\quad\quad BH \quad\quad\quad\quad\quad\quad\quad\quad B^-$$

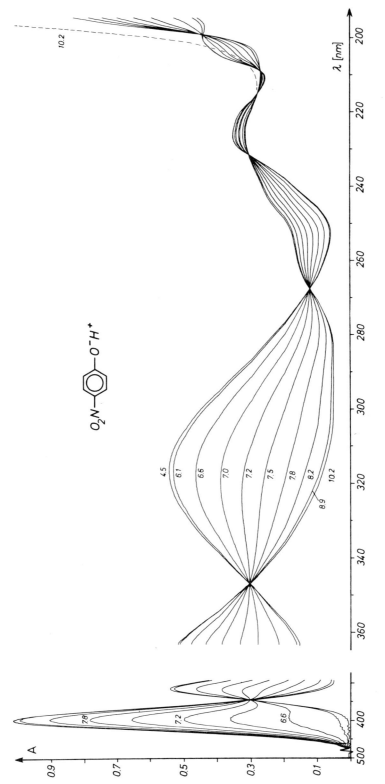

Fig. 6–1. Titration spectra resulting from spectrometric titration under nitrogen at 25 °C of p-nitrophenol ($b_0 = 5 \cdot 10^{-5}$ mol/L) with NaOH (0.1 mol/L)

6.2 Graphic Matrix-Rank Analysis

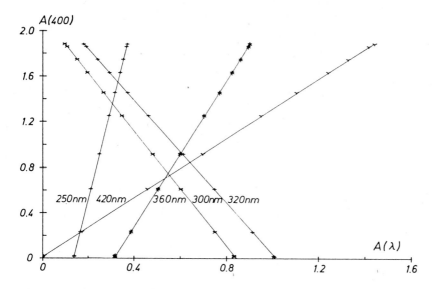

Fig. 6–2. A-diagrams from titration of *p*-nitrophenol with NaOH (0.1 mol/L). For titration conditions see Fig. 6–1.

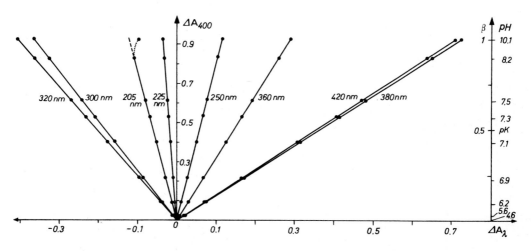

Fig. 6–3. AD-diagrams from titration of *p*-nitrophenol with NaOH (0.1 mol/L). For titration conditions see Fig. 6–1.

The corresponding *titration spectra* (Fig. 6–1) show six sharp isosbestic points at pH ≤ 9. Due to the design of the spectrometer, the region from about 365 nm to 320 nm is depicted twice at two different expansions, so the longest wavelength isosbestic point also appears twice. In more strongly alkaline solution, self-absorption[1,4] by excess titrant (OH$^-$) becomes a problem, since the basic assumption of an apparently closed spectrometric titration system (cf. Sec. 1.1) ceases to be applicable. This interference by titrant is strictly reproducible, and it disappears upon back-titration with HCl. Unless the titration is carried out under nitrogen, other *non-reproducible* interferences arise in the alkaline region as a result of dissolved CO_2 from the air (absorption by $NaHCO_3$ below ca. 240 nm).

A-diagrams constructed for the region between 420 and 250 nm (Fig. 6–2) are strictly linear through the entire range of pH. In the interest of clarity, diagrams are shown here for only five wavelength combinations. The non-linear pH scale has been reproduced along the right-hand edge of the diagram to simplify identification of the corresponding titration curves.

AD-diagrams (cf. Fig. 6–3) are also linear for eight wavelength combinations at pH ≤ 9. Once again, the data at 205 nm for pH > 9 show the influence of OH$^-$ absorption, so the corresponding line deviates from linearity.

Straightforward analysis of the results under the assumption of an apparently closed system is only possible with respect to data obtained at pH ≤ 9 or at wavelengths ≥ 220 nm. Both the wavelength region showing the greatest change in absorbance and the most favorable region for photometric measurement can be determined with ease by examination of the titration spectra and the A- and AD-diagrams.

It should be readily apparent that the graphic method of matrix-rank analysis can be quite useful even in an investigation of a single protolyte so long as it is certain that the molecule in question contains only one dissociable group.

The most important technical prerequisite to the successful application of this approach—as in any form of quantitative multi-wavelength spectrophotometry—is very good mechanical and photometric reproducibility with respect to the spectrometer (cf. Sec. 18.1). Indeed, spectrometric titration of *p*-nitrophenol serves as a valuable test of the performance of a UV/VIS-spectrometer.

It is important to recognize that criteria for the rank $s = 1$ such as those described above and based on graphic—or numeric—matrix-rank analysis represent only *necessary conditions*.[5] Many multi-step titration systems ($s > 1$) contain certain wavelength regions in which only one titration step is spectroscopically observable (for examples see Sec. 6.4). To prevent errors in interpretation it is therefore advisable that the widest possible spectral region be subjected to matrix-rank analysis, and that the use of multiple spectrometric methods at least be considered (cf. Chapter 18).

The methods so far described for matrix-rank analysis utilize only the spectrophotometric data from a spectrometric titration, not the electrochemical data.

6.3 Determination of Absolute pK Values

Therefore, consistency of a proposed titration mechanism with the electrochemical data must be tested separately, as discussed in the following section.

6.3 Determination of Absolute pK Values

Sec. 6.3 is limited to a consideration of methods for determining *absolute pK values*. Chapter 11 examines the assignment of *relative* pK values to single-step protolytes by simultaneous titration with a second acid-base system of known pK.

Almost all spectrometric methods for absolute pK determination utilize both spectrophotometric *and* electrometric titration data, although the method of Sec. 6.3.6 represents one exception to this generalization. Assuming that the rank $s = 1$ is applicable over the entire spectral range under investigation, methods of pK determination have as their point of departure an equation such as 6–6 or 6–7.

6.3.1 Inflection Point Analysis

Inflection points are most often regarded as indicating stoichiometric equivalence in potentiometric titrations conducted for mass analytical purposes.[6] Nevertheless, they can also provide pK values in the case of certain potentiometric or spectrometric acid-base titrations (cf. Figures 6–4 and 6–5).[1,6,7]

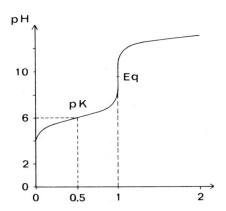

Fig. 6–4. Typical potentiometric titration curve from the titration of a weak acid with a strong base (schematic representation).

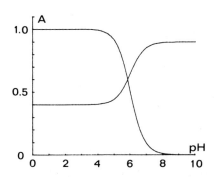

Fig. 6–5. Spectrometric titration curves generated by equation 6–11 using the following parameters: $pK = 6$; $A_{\lambda_1 BH} = 1$; $A_{\lambda_1 B} = 0$; $A_{\lambda_2 BH} = 0.4$; $A_{\lambda_2 B} = 0.9$.

Prerequisite to a pK determination based on inflection point analysis is a titration curve that is symmetrical with respect to the corresponding pK value (or equivalence point), another way of saying that the inflection point and the relevant pK (or equivalence point) must coincide.[6] Provided the titrant is non-absorbing (i.e., $\varepsilon_{\lambda R} = 0$), and if error due to dilution is eliminated mathematically, this symmetry criterion is much more likely to be met by a spectrometric titration than by a potentiometric titration. Solution of equation 6–6 for A_λ shows that

$$A_\lambda = \frac{A_{\lambda BH} \cdot h + A_{\lambda B} \cdot K_1}{h + K_1} = \frac{A_{\lambda BH} \cdot 10^{-pH} + A_{\lambda B} \cdot 10^{-pK_1}}{10^{-pH} + 10^{-pK_1}} \quad (6\text{–}11)$$

Figure 6–5 illustrates two spectrometric titration curves generated by the use of this equation.

Standard procedures for locating inflection points on titration curves include the following:

(a) Tangent method[6]

In this approach, two parallel tangents to the titration curve are constructed, as shown in Fig. 6–6. A third parallel line centered between these tangents intersects the titration curve at the point of inflection.

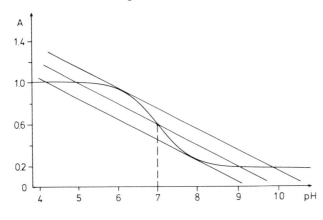

Fig. 6–6. Use of the tangent method to locate the inflection point in a plot of A_λ vs. pH.

(b) Tubbs ring method[8]

Circles of the smallest possible size are inscribed in the two sharply curved regions flanking the inflection point, as shown in Fig. 6–7. This is best accomplished with the aid of a template consisting of a set of concentric circles. The line joining the centers of the inscribed circles intersects the titration curve at the point of inflection. In the event that the titration curve is asymmetric, the two circles will have different radii. Variants on this procedure are described in ref. 6, which also identifies sources of suitable templates.

6.3 Determination of Absolute pK Values

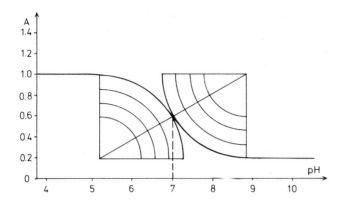

Fig. 6–7. Use of the Tubbs ring method to locate the inflection point in a plot of A_λ vs. pH.

(c) Inflection points may also be dectected by seeking a maximum or minimum in the first derivative of the titration curve (so-called *differential methods*).[2,6,9]

All the above methods for locating inflection points rely heavily on data taken in the vicinity of the desired pK value. Methods (a) and (b) also utilize the regions of maximum curvature. A knowledge of the limiting values $A_{\lambda B}$ and $A_{\lambda BH}$ is not required, however.

An inflection point analysis of the titration curves for *p*-nitrophenol at six wavelengths is shown in Fig. 6–8, leading to the result pK_1 = 7.2 ± 0.1.

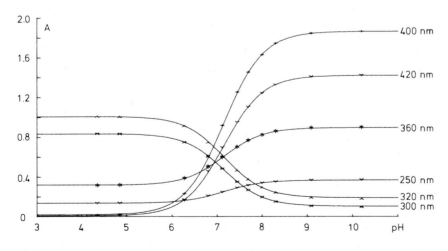

Fig. 6–8. Titration curves for the titration of *p*-nitrophenol with NaOH (0.1 mol/L; for other conditions see Fig. 6–1). The solid lines represent control simulations (see Sec. 6.3.5).

6.3.2 Determining the Midpoint of Overall Change in Absorbance (the "Halving Method")

The expression $(A_\lambda - A_{\lambda BH})/(A_{\lambda B} - A_\lambda)$ in equations 6–6 and 6–7 has the value unity in the event that

$$A_\lambda = 0.5 \cdot (A_{\lambda BH} + A_{\lambda B}) \quad \text{or} \quad \beta = 1 - \beta = 0.5 \qquad (6\text{–}12)$$

Therefore, if one takes the mean of the initial and final absorbances in a titration, the corresponding point on the titration curve must reflect a pH equivalent to the pK value.[1] Clearly, use of this procedure requires knowledge of *both* $A_{\lambda BH}$ and $A_{\lambda B}$.

In order adequately to establish the necessary titration boundary values one is forced to employ a large excess of an appropriate titrant. For pK values < 3–4 or > 9–10 this requires that at least one of the points be determined in strongly acidic or alkaline medium, which in turn increases the likelihood of medium-effect deviations from the Lambert-Beer-Bouguert law [cf. rationale (5) in Sec. 6.2]. Moreover, other interferences described in Sec. 6.2 may present problems under extreme pH conditions (e.g., absorption by the titrant, oxidation, or time-dependent side reactions). It is thus important that the linearity of the A- or AD-diagrams in the limiting regions be carefully verified. In the case of both A- and AD-diagrams, the pK value is associated with the half-way point relative to the overall titration.

Assuming the titration end values $A_{\lambda BH}$ *and* $A_{\lambda B}$ are known with certainty, and provided they are not subject to interference, this "halving method" represents a very rapid and simple approach to pK determination. Just as in the inflection point method, data obtained near the actual pK play an important role (as do, of course, the end values).

Application of the halving method to the titration curves shown in Fig. 6–8, with arithmetic means taken at six wavelengths, leads to the result pK_1 = 7.16 ± 0.07.

6.3.3 Linearization of Spectrometric Titration Curves

(a) Equation 6–7 can be rewritten in the form

$$\log \frac{A_\lambda - A_{\lambda BH}}{A_{\lambda B} - A_\lambda} = \text{pH} - \text{p}K_1 \qquad (6\text{–}13)$$

A graph of $\log [(A_\lambda - A_{\lambda BH})/(A_{\lambda B} - A_\lambda)]$ vs. pH (Fig. 6–9) produces a straight line with a slope of one and an intercept along the abscissa at pH = pK_1.[1,10,11] As with the halving method of Sec. 6.3.2, this approach requires knowledge of the limiting values $A_{\lambda BH}$ *and* $A_{\lambda B}$.

6.3 Determination of Absolute pK Values

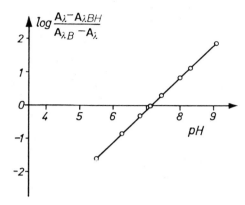

Fig. 6–9. Linearization of spectrometric titration data obtained from *p*-nitrophenol at $\lambda = 400$ nm by application of equation 6–13.

(b) An alternative type of linearization[1,12–14] is preferable if one of the titration limits falls at an extreme along the pH scale, or is otherwise difficult to measure or inaccessible. For this purpose equation 6–6 is rewritten in the form

$$(A_\lambda - A_{\lambda BH}) \cdot 10^{-pH} = - K_1 \cdot A_\lambda + K_1 \cdot A_{\lambda B} \qquad (6\text{–}14a)$$

or

$$(A_\lambda - A_{\lambda B}) \cdot 10^{+pH} = - \frac{1}{K_1} \cdot A_\lambda + \frac{1}{K_1} \cdot A_{\lambda BH} \qquad (6\text{–}14b)$$

Evaluation of such a function requires knowledge of only *one* of the titration end values; the other can be obtained from the corresponding abscissa intercept (cf. Fig. 6–10).

Either of the above linearization techniques may be applied to the entire set of $A_\lambda(\mathrm{pH})$ data constituting a spectrometric titration curve. Nevertheless, the regions near the ends of the titration are characterized by small absorbance differences and a correspondingly high degree of scatter about the most probable line. In method (a) it is pH that serves as the scale for the abscissa, and the corresponding data can be plotted

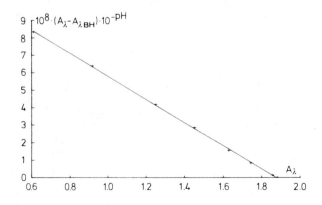

Fig. 6–10. Linearization of spectrometric titration data obtained from *p*-nitrophenol at $\lambda = 400$ nm by application of equation 6–14.

directly, while A_λ data must be incorporated into a complex expression. The reverse is true of method (b). Thus, the two methods differ in terms of their error functions; i.e., in the relative weighting of photometric and electrometric data.

Evaluation of the spectrometric titration data from Fig. 6–8 for *p*-nitrophenol and sodium hydroxide at 400 nm gives pK_1 values of 7.16 ± 0.01 and 7.14 ± 0.02 by methods (a) and (b), respectively (cf. Figures 6–9 and 6–10).

Evaluation of the *entire* titration curve in the course of linearization by methods (a) and (b) provides a significant test of the validity of equations 6–13 and 6–14. Assuming one has previously verified the applicability of the rank $s = 1$ by matrix-rank analysis based on photometric data, subsequent use of the linearization method can assure the mutual consistency of the photometric *and* the electrometric data with the proposed mechanism.

Method (b) is particularly well suited not only to single-step titrations, but also to multi-step titration systems in which A-diagrams contain regions that are linear (cf. Sec. 6.4).

6.3.4 Numeric Evaluation of p*K* Through the Use of Linear Equations

(a) The methods in Sec. 6.3.2 and method (a) in Sec. 6.3.3 require a knowledge of the two titration end values $A_{\lambda BH}$ and $A_{\lambda B}$. If this information is available to a sufficient degree of accuracy, equations 6–6 or 6–7 can be used to obtain a numerical value of K_1 or pK_1 at each titration point $A_\lambda(pH)$—for example, using the relationship

$$Y = \frac{1}{K_1} \cdot h \quad \text{where} \quad Y = \frac{A_{\lambda B} - A_\lambda}{A_\lambda - A_{\lambda BH}} \tag{6–15}$$

Here again, the entire data set from a spectrometric titration curve may be utilized. As with graphic linearization—methods (a) and (b) in Sec. 6.3.3—very precise results can be expected from the center of the titration curve, while the small absorbance differences $A_{\lambda B} - A_\lambda$ and $A_\lambda - A_{\lambda BH}$ characteristic of the extremes lead to more scatter.

A calculation of this type based on the A_λ vs. pH titration curve for *p*-nitrophenol (Fig. 6–8) at 400 nm between pH 5.6 and 7.5 indicates $pK_1 = 7.16 \pm 0.03$. Increasing deviations result from inclusion of data below pH 5 and above pH 8.

According to equation 6–6, the quotient $(A_\lambda - A_{\lambda BH})/(A_{\lambda B} - A_\lambda)$ is equal to b/bh, which in turn equals $\beta/(1 - \beta)$. Thus, calculations like those above can also be used to obtain the degree of dissociation corresponding to any particular titration point. This permits transformation of the spectrometric titration curve $A_\lambda(pH)$ by a simple change of scale into a so-called *dissociation diagram*[1,10] of β vs. pH (Fig. 6–11). The p*K* can then be read directly from $\beta = 0.5$. Such a diagram provides an especially clear presentation of relative dissociation as a function of pH, particularly in the case of sequential (Sec. 7.3) and simultaneous (Chapter 11) multi-step titrations.

6.3 Determination of Absolute pK Values

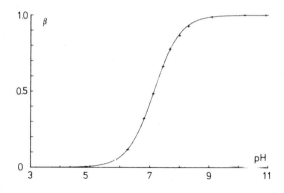

Fig. 6–11. Extent of dissociation of p-nitrophenol as a function of pH; calculated on the basis of A_λ and pH data ($\lambda = 400$ nm).

(b) If the limiting values $A_{\lambda BH}$ and $A_{\lambda B}$ are not experimentally accessible, method (a) can be modified such that the missing quantities are introduced as additional unknowns.[15] For example, if equation 6–6 is reformulated as

$$A_\lambda \cdot h \cdot \boxed{\frac{1}{K_1}} - A_\lambda \cdot \boxed{A_{\lambda BH} \cdot \frac{1}{K_1}} - \boxed{A_{\lambda B}} = -A_\lambda \qquad (6\text{–}16)$$

the three unknowns K_1, $A_{\lambda BH}$, and $A_{\lambda B}$ are confined to the three terms surrounded by boxes. Construction of three simultaneous equations like 6–16 for three different $A_\lambda(\text{pH})$ data points produces a system that can be solved by the standard techniques of linear algebra (e.g., Cramer's rule) to give numeric values for the three unknowns.

However, a more sensible approach is to utilize all the data points $A_\lambda(\text{pH})$ from a given spectrometric titration and then evaluate by linear regression the entire set of equations corresponding to 6–16. This provides as additional information an estimate of the standard deviations applicable to the calculated values of K_1, $A_{\lambda BH}$, and $A_{\lambda B}$.

In the event that *one* of the titration end values is known, only two unknowns remain in the corresponding system of linear equations, solution of which is equivalent to graphic procedure (b) of Sec. 6.3.3. Analogous approaches can also be devised for dealing with multi-step titration systems (cf. ref. 16–18 and Sec. 7.3.6).

As already noted in the context of numeric matrix-rank analysis (Sec. 3.3.1), large matrices composed of real, inexact data may introduce arithmetic difficulties. In the case of linear equations one speaks of "conditioning problems".[19,20] The literature discusses various ways to circumvent such problems when they arise in the application of method (b). The alternatives differ primarily in terms of the way one structures the system of linear equations (i.e., in how one formulates the measured quantities A_λ and pH), or in the way data sets are chosen. Techniques have also been described for avoiding the need for conditioning.

6.3.5 Non-linear Curve-Fitting Methods

As seen in equation 6–10, the functional relationship between that which is sought, a pK value, and the data available (A_λ and pH) is a non-linear one. It is therefore tempting to refrain altogether from linearization techniques, which involve complicated reformulations of the data, and turn instead to non-linear regression methods, working directly with the function $A_\lambda(\mathrm{pH})$.

Chapter 10 describes a universal iterative curve-fitting program applicable to any sequential spectrometric titration system.[21] It is of course also applicable to the simplest case of a single-step titration. The program simultaneously evaluates all available $A_\lambda(\mathrm{pH})$ curves; i.e., all the data acquired during a titration are treated in combination. The process completely eliminates conditioning problems such as those caused by non-absorbing species ($A_{\lambda i} = 0$). Moreover, no data need be supplied for the titration limits ($A_{\lambda \mathrm{BH}}$ and $A_{\lambda \mathrm{B}}$); indeed, these values normally emerge from the calculation. Entire segments of the titration curve may even be excluded provided the missing information is supplemented at other wavelengths, and the procedure is applicable to cases in which the protolyte decomposes above or below some particular pH.

To initiate the iteration process it is necessary that one supply a rough estimate (within ca. ± 1 pH unit) of the unknown pK_1 value together with $A_\lambda(\mathrm{pH})$ data acquired during a titration. The program then calculates pK_1, $A_{\lambda \mathrm{BH}}$, and $A_{\lambda \mathrm{B}}$ at every wavelength. Finally, all $A_\lambda(\mathrm{pH})$ curves are recalculated, and these are displayed as a control simulation along with the experimental data, all in a single diagram.

Fig. 6–8 presents data and such a control simulation for the titration of *p*-nitrophenol with 0.1 N sodium hydroxide, analysis having been conducted at six wavelengths. The calculated value for pK_1 is 7.15 ± 0.01.

The precision of the data and the resulting solution can be estimated not only by means of a standard error analysis (least squares summation), but also with a Monte Carlo approach (cf. Sec. 7.4.2 and Chapter 10) that introduces random error into the control simulation. In the event that graphic matrix-rank analysis has *already* confirmed the rank $s = 1$, a subsequent non-linear regression analysis provides a powerful test of the postulated titration mechanism by subjecting *all* $A_\lambda(\mathrm{pH})$ curves to *simultaneous* evaluation.

Most iterative non-linear regression procedures require the use of a rather large computing facility, and the effort is seldom warranted if one simply wishes to evaluate a set of single-step titration curves. However, if a computer and the software necessary for studying multi-step titration mechanisms happen to be available it is useful to perform a coupled final analysis of a series of $A_\lambda(\mathrm{pH})$ data even from a simple titration system.

6.3.6 Photometric Determination of pK in the Absence of pH Measurements

Under certain conditions it is possible using *only* optical data to make precise determinations of classical and thermodynamic dissociation constants for single-step protolytes.[22-25]

If one assumes that solution of the pure substance BH in highly purified water results exclusively in establishment of the dissociation equilibrium

$$BH + H_2O \leftrightharpoons B^- + H_3O^+ \tag{6-1}$$

then, from the corresponding condition of electrical neutrality

$$c_{H_3O^+} = b = \beta \cdot b_0 \tag{6-17}$$

it follows (cf. Chapter 4) that the classical dissociation constant K_c is described by

$$K_c = \frac{c_{H_3O^+} \cdot b}{bh} = \frac{b^2}{b_0 - b} = \frac{\beta^2}{1 - \beta} \cdot b_0 \tag{6-18}$$

Evaluation is simplified if one further assumes that there exists a wavelength region in which the pure acid BH is non-absorbant ($\varepsilon_{\lambda BH} = 0$, or $A_{\lambda BH} = 0$)*. If, in addition, one can specify the hypothetical starting concentration b_0 (from the weight of substance introduced), the pathlength l, and the alkaline limiting absorbance value $A_{\lambda B}$ (and thus $\varepsilon_{\lambda B}$), then a single absorbance measurement on a pure aqueous solution of protolyte provides a value for the equilibrium concentration b, and equation 6-18 permits calculation of β and K_c. If, in a series of experiments, one next varies the initial concentration b_0, extrapolation to ionic strength $J = 0$ (e.g., using the limiting Debye-Hückel equation) leads to the thermodynamic equilibrium constant K.[2,23]

Strictly speaking, the procedure just described is not a titration method, since only two states of a titration have been examined photometrically:

(1) the alkaline limit, where $bh = 0$, and
(2) the pure aqueous solution, where $b = c_{H_3O^+}$

Thus, precise measurements of the initial sample mass and of the molar absorptivity $\varepsilon_{\lambda B}$ have taken the place of pH measurements, and the difficulties associated with the conventional pH-scale (cf. Chapter 17) have been circumvented.
However, this approach is restricted to single-step protolytes, as well as to samples and solvents that are highly purified and stable. Any impurity in the form of foreign protolytes (e.g., CO_2 from the air, water contaminants arising from an ion exchange system, traces of isomeric substances) diminishes the rigor with which equation 6-16 describes the overall system. For this reason it is advisable to subject the fundamental premises to experimental test, paying special attention to the possibility of apparent

---

* This assumption is not required; it can be avoided by experimental determination of $\varepsilon_{\lambda BH}$, or by use of suitable approximation methods.[25]

changes in the molar absorptivities $\varepsilon_{\lambda B}$ and $\varepsilon_{\lambda BH}$ under extreme pH conditions (due to medium effects,[22] contact ion pairs, double anions,[25] etc.). This is best done by conducting a separate spectrometric titration, complete with graphic matrix-rank analysis.

Apart from the investigation of highly dilute aqueous solutions, the method described above is also suited to the determination of dissociation constants in non-aqueous solutions, including solvent mixtures,[25] a subject beyond the scope of this book.

Literature values[23] for the thermodynamic dissociation constant at 25 °C of *p*-nitrophenol, the compound repeatedly cited in the foregoing examples, range from $7.08 \cdot 10^{-8}$ to $7.24 \cdot 10^{-8}$, corresponding to a pK_1 value between 7.15 and 7.14 (see ref. 23, p. 450, and the sources cited therein).

6.4 The Treatment of Non-overlapping, Multi-Step Titration Systems as Combinations of Single-Step Subsystems

In many multi-step titration systems, the extent of overlap of the various dissociation equilibria is sufficiently small to permit a reasonable analysis in terms of a set of single-step subsystems. Normally one can assume that a *pK* difference *ΔpK > 3.5–4* will suffice to reduce overlap until it is no longer detectable with currently available techniques.

Beyond assuring that this particular criterion is fulfilled it is also necessary that one examine the molar absorptivities $\varepsilon_{\lambda i}$ relevant to all the associated equilibria, since these determine overall changes in absorbance A_λ that will accompany each step in the titration, and thus the extent of relative measurement errors that must be anticipated. The corresponding error functions are discussed in detail in Sec. 7.3.1 and ref. 26.

The *A-diagrams* corresponding to non-overlapping multi-step titrations are composed of a series of linear segments; i.e., all data points necessarily fall on mutually intersecting straight lines. In practice this is an exceptionally simple and precise criterion by which to establish the absence of measurable overlap. The points of intersection of the linear segments correspond to solutions of the various pure ampholytes (cf. Sec. 7.3.3).

If, for a titration system of *s* steps, *none* of the titration equilibria overlap to a measurable degree, the overall titration can be divided into *s* single-step subsystems. The methods described in Sec. 6.3 (with the exception of that in Sec. 6.3.6) can then be applied to the determination of *all individual pK values.*

It is often the case that *only certain* sub-equilibria in a multi-step titration defy simple analysis, while others can be evaluated readily. For example, a four-step titration equilibrium such as

6.4 Non-overlapping, Multi-Step Titration Systems

$$BH_4 \rightleftarrows BH_3^- \rightleftarrows BH_2^{2-} \rightleftarrows BH^{3-} \rightleftarrows B^{4-}$$
$$pK_1 = 1 \quad pK_2 = 3 \quad pK_3 = 7.5 \quad pK_4 = 12$$
$$\Delta pK = 2 \quad \Delta pK = 4.5 \quad \Delta pK = 4.5 \tag{6-19}$$

can be divided into the following parts:

$BH_4 \rightleftarrows BH_3^- \rightleftarrows BH_2^{2-}$	$BH_2^{2-} \rightleftarrows BH^{3-}$	$BH^{3-} \rightleftarrows B^{4-}$
$pK_1 = 1 \quad pK_2 = 3$	$pK_3 = 7.5$	$pK_4 = 12$
Evaluated as an $s = 2$ system	Evaluated as an $s = 1$ system	Evaluated as an $s = 1$ system

This results in a considerable simplification of the evaluation process.

The "dissection" technique is illustrated in the paragraphs that follow with three titration systems from the *vitamin B_6 series*.

(1) 4-Deoxypyridoxine

The *titration spectra* (Fig. 6–12) obtained starting with this compound are divisible into two subsystems (pH ≈ 2–8 and pH ≈ 8–12). Both subsystems demonstrate sharp isosbestic points. The first titration step illustrates a special case of an isosbestic point: in the region 230–235 nm the two spectra touch rather than intersecting, producing a so-called "point of contact" (cf. Sec. 3.2.2).

4-Deoxypyridoxine

Pyridoxal

Cyclization product of pyridoxal with histamin

The corresponding *A-diagrams* (Fig. 6–13) show the presence of two strictly linear segments for various wavelength combinations in the region 240–320 nm. To within the limits of error, all data points fall on one of the two linear segments. Thus, the overall titration system has the rank $s = 2$ and it is non-overlapping.

A-pH curves for the system (Fig. 6–14) display flat maxima, minima, and plateaus near pH 7.5, and the titration curves can clearly be divided into two single-step subsystems.

50 6 Single-Step Acid-Base Equilibria

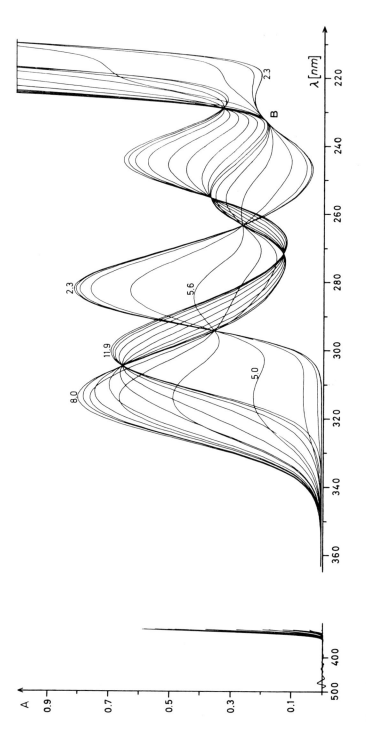

Fig. 6-12. Titration spectra for 4-deoxypyridoxine; titrated under nitrogen at 25 °C, $b_0 = 10^{-4}$ mol/L. The point B is a point of contact.

6.4 Non-overlapping, Multi-Step Titration Systems

The methods described in Sections 6.3.1 to 6.3.4 can all be used to establish pK values provided the appropriate acidic and basic limiting titration values are known. Each titration step must be evaluated independently. Only the non-linear curve-fitting method of Sec. 6.3.5 permits simultaneous analysis of all subsystems at all wavelengths. Evaluation according to this method leads to the following results (25 °C):

$pK_1 = 5.26 \pm 0.02$

$\Delta pK = 4.56$

$pK_2 = 9.82 \pm 0.02$

Corresponding values reported in the literature fall within the ranges $pK_1 = 5.35$–5.4 and $pK_2 = 9.73$–9.98.[27,28]

The two pK values are normally taken to reflect dissociation of the pyridinium ion $=NH^+$ and the phenolic $-OH$ group, the assumption being that in aqueous solution at pH < 12 the CH_2OH group would not dissociate to a significant extent and/or such dissociation would not be detectable spectroscopically. This is a convenient place to point out that it is not *a priori* possible to assign pK values like these to particular dissociable groups. The case under consideration is one example of a *branched titration system* (cf. Chapter 12). The fact that two subsystems overlap only to a small extent may *not* be taken as evidence that the order of dissociation of the two groups is unambiguously established (though such conclusions often appear in the literature).

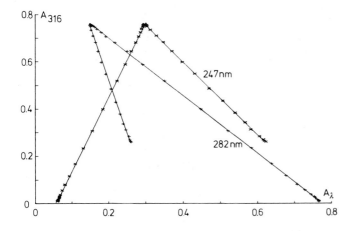

Fig. 6–13. A-diagrams for 4-deoxypyridoxine. For titration conditions see Fig. 6–12.

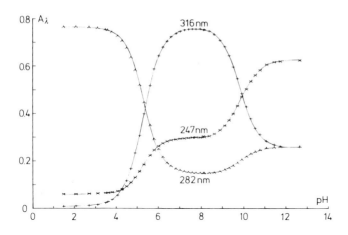

Fig. 6–14. A_λ vs. pH curves for 4-deoxypyridoxine. For titration conditions see Fig. 6–12.

(2) Pyridoxal

The complete set of *titration spectra* in this particular case has been divided into two parts for purposes of clarification. Fig. 6–15 contains two subsystems from the region pH = 2.4–10.6, each depicted in a different color. Both reveal sharp isosbestic points. The spectra approach each other at about 230–235 nm, but they produce no point of contact comparable to that seen with *4-deoxypyridoxine* (Fig. 6–12). The change in absorbance accompanying the third titration step (in the region pH = 11–13, cf. Fig. 6–16) is so slight that the sharpness of isosbestic points is difficult to judge.

A-diagrams (Fig. 6–17) for all wavelength combinations from the pH range 1–13 consist of three linear segments; i.e., the system involves three non-overlapping titration steps ($s = 3$).

The *A-pH-diagrams* (Fig. 6–18) demonstrate flat maxima, minima, or plateaus in the vicinity of pH 6 and 11, and the titration curves can be divided into three single-step subsystems. The third stage of titration remains incomplete, since medium effects due to the high NaOH concentration gain the upper hand above pH 13–14. For this reason the third step should be analyzed with one of the linearization methods of Sec. 6.3.3 or the non-linear curve-fitting approach of Sec. 6.3.5. Use of the latter with data collected at seven wavelengths between 235 and 380 nm leads to the following pK values:[21,29]

$pK_1 = 4.07 \pm 0.01$

$\Delta pK = 4.41$

$pK_2 = 8.48 \pm 0.01$

$\Delta pK = 4.83$

$pK_3 = 13.31 \pm 0.02$

6.4 Non-overlapping, Multi-Step Titration Systems

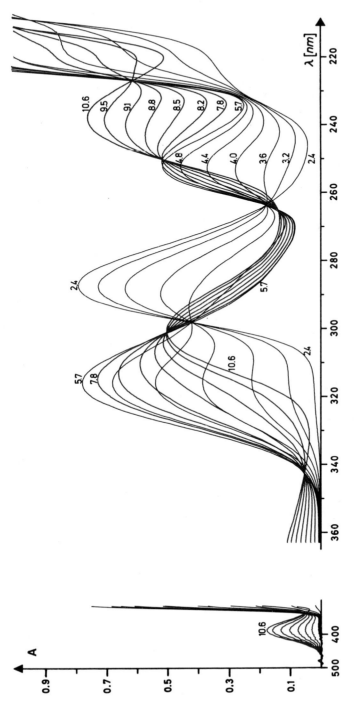

Fig. 6–15. Titration spectra for pyridoxal over the pH range 2.4–10.6; titrated under nitrogen at 25 °C, $b_0 = 10^{-4}$ mol/L.

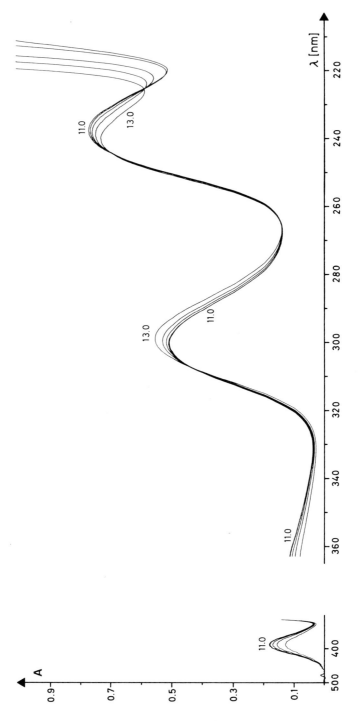

Fig. 6-16. Titration spectra for pyridoxal over the pH range 11–13. For titration conditions see Fig. 6–15.

6.4 Non-overlapping, Multi-Step Titration Systems

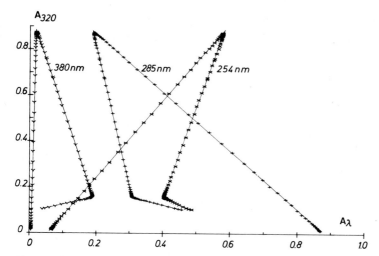

Fig. 6–17. A-diagrams for pyridoxal. For titration conditions see Fig. 6–15.

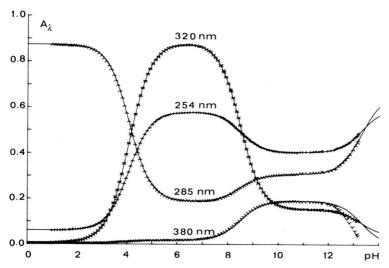

Fig. 6–18. A_λ vs. pH curves for pyridoxal. For titration conditions see Fig. 6–15.

Absorbances A_λ and molar absorptivities ε_{λ_i} are also generated for the pure limiting species and all other ampholytes present (cf. Chapter 10). Reconstructed spectrometric titration curves based on the calculated results are in very good agreement with experimental data. Only at 380 nm and above pH 12 is there significant deviation of experiment from theory, presumably as a result of spectroscopic detection of a modification of the titration mechanism: at 380 nm the spectra are affected by tautomeric equilibrium between aldehyde and cyclic hemiacetal (cf. Sec. 12.3.4).

The pK values listed above correspond within experimental error to results obtained by linearization (Sec. 6.3.3). This method also shows significant deviations at 380 nm above pH 12. The linearization approach entails considerably more effort than does the non-linear regression method; use of data at seven wavelengths in conjunction with three pK values means that there are ≥ 21 single-step subsystem calculations to be performed. Values reported in the literature[30–33] fall within the following ranges:

   pK_1 3.93 to 4.23
   pK_2 8.37 to 8.70
   pK_3 13.0 to 13.1

The three titration steps of the pyridoxal system are due to the pyridinium ion, the phenolic –OH group, and the –CH$_2$OH group. In contrast to 4-deoxypridoxine, pyridoxal's side-chain aliphatic –OH group dissociates to a spectroscopically measurable extent because the state of dissociation is coupled with the position of tautomeric equilibrium between aldehyde and cyclic hemiacetal. In fact, therefore, pyridoxal provides an example of a rather complicated branched titration system, further discussion of which is deferred to Chapter 12.

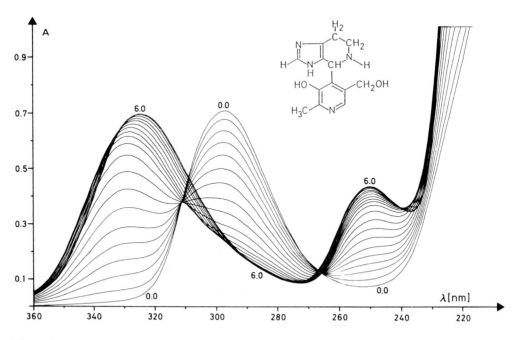

Fig. 6–19. Titration spectra over the pH range 0–6 for the tetrahydropyridine derivative of equation 6–20; titrated under nitrogen at 25 °C, $b_0 = 10^{-4}$ mol/L.

6.4 Non-overlapping, Multi-Step Titration Systems

(3) Cyclization Product from Reaction between Pyridoxal and Histamine

To conclude, we examine a weakly overlapping four-step titration system. Pyridoxal reacts with histamine to form an intermediate Schiff base, which is in turn converted to the final product, a tetrahydropyridine derivative. The last step is essentially irreversible over the pH range < 1 to 12, so the final solution can be titrated spectrometrically as if it consisted only of a single pure substance incapable of undergoing the reverse reaction.[34]

The *titration spectra* are again depicted in two sets. In the pH region ≈ 0–6 (Fig. 6–19), gradual divergence of the initially observed isosbestic points indicates that the overall titration system must be composed of overlapping subsystems. The second part of the titration (Fig. 6–20), from pH 6 to ca. 12, apparently consists of two non-overlapping titration steps. Several spectra taken between pH 8.1 and 9.8 have been omitted in the interest of clarity.

Clarity has also dictated that *A-diagrams* (Fig. 6–21) be constructed on the basis of only two wavelength combinations. It will be noted that the spectroscopic data reveal a total of four titration steps. None of the points of intersection of the linear segments,

Aldehyde + Histamine ⇆ Schiff Base

Tetrahydropyridine Derivative

(6–20)

58 6 Single-Step Acid-Base Equilibria

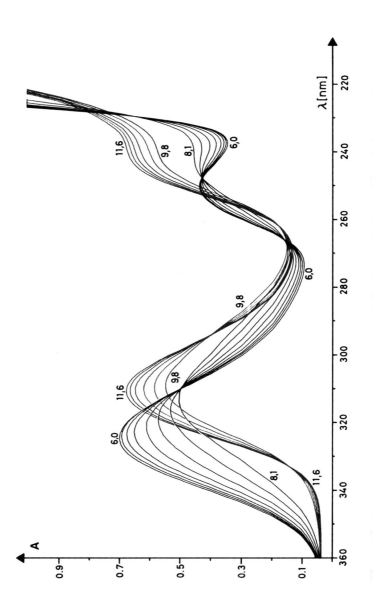

Fig. 6-20. Titration spectra over the pH range 6–12 for the tetrahydropyridine derivative of equation 6-20. For titration conditions see Fig. 6-19.

6.4 Non-overlapping, Multi-Step Titration Systems

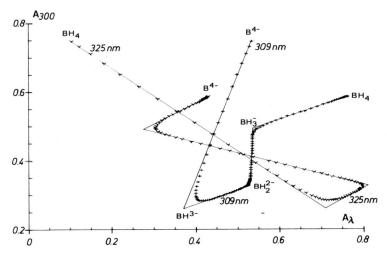

Fig. 6–21. A-Diagrams for the tetrahydropyridine derivative of equation 6–20 (cf. Figs. 6–19 and 6–20).

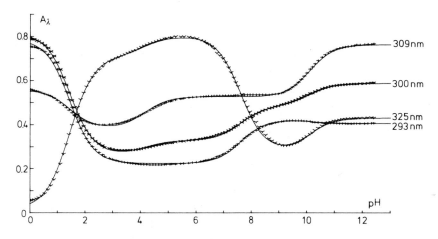

Fig. 6–22. A_λ vs. pH curves for the tetrahydropyridine derivative of equation 6–20 (cf. Figs. 6–19 and 6–20).

corresponding to the pure ampholytes, coincides exactly with experimental data; thus, A-diagrams indicate that all four titration steps suffer some degree of overlap.

It is difficult to discern four separate titration steps in the corresponding A-pH *curves* (Fig. 6–22), prepared with data recorded at four wavelengths; several of the steps are characterized by small changes in absorbance. Considerable caution is required in applying evaluation methods based on rank $s = 1$ to the individual sub-equilibria in a titration of this type. Only the linearization method of Sec. 6.3.3 is appropriate, and even here a certain amount of experience and extensive effort are required if all the linear segments are to be examined. By contrast, the non-linear regression method (Sec. 6.3.5) is unaffected by overlap among individual steps, so a coupled evaluation of all the titration data by this means is the preferred procedure. A calculation based on five wavelengths over the pH range ≈ 0–12.5 leads to the following results:

$pK_1 = 1.46 \pm 0.02$

$\Delta pK = 2.69$

$pK_2 = 4.15 \pm 0.02$

$\Delta pK = 3.46$

$pK_3 = 7.61 \pm 0.02$

$\Delta pK = 2.49$

$pK_4 = 10.10 \pm 0.02$

The simulated results correspond rather well to experiment (Fig. 6–22) considering the complicated nature of the system. Deviations in the region pH < 0.5 are presumably due to medium effects.

No published information is available in this case concerning correlations between pK values and dissociable groups, nor is it known whether the titration system is branched.

6.5 Summary

The methods described in Sections 6.2–6.4 for graphic matrix-rank analysis and pK determination are here briefly compared, and they are summarized in a pair of tables.

6.5.1 Single-Step ($s = 1$) Titrations (cf. Table 6–1)

Graphic matrix-rank analysis makes it possible to verify that the mechanism of a titration is indeed consistent with the rank $s = 1$. The method also reveals the most appropriate wavelengths for use in further investigations. In addition, graphic approaches provide extremely sensitive tests of whether impurities or other interferences are responsible for deviations from rank $s = 1$ in a particular spectral region.

Analysis of *inflection points* (Sec. 6.3.1) permits the estimation of *pK values*. Rough estimates can be obtained very rapidly, especially if appropriate templates are

6.5 Summary

available [cf. methods (a) and (b) in Sec. 6.3.1] or if the spectrometer (or a computer to which it is interfaced) is capable of providing first derivatives of the acquired $A_\lambda(\text{pH})$ curves [Sec. 6.3.1, method (c)]. Otherwise, the *halving method* (Sec. 6.3.2) represents the simplest and most rapid approach to obtaining pK estimates, although this method requires a knowledge of absorbances at the two titration limits.

Two *graphic linearization methods* [Sec. 6.3.3, methods (a) and (b)] permit very precise determination of pK. The second method is usually preferred, since only one limiting absorbance is required (the other is calculated). Computations for both methods can easily be performed with the aid only of a pocket calculator. Linearity of the final diagrams in each case provides a significant test of the validity of the assumption that the mechanism in question is of rank $s = 1$. Graphic matrix-rank analysis also tests this assumption, but only in terms of photometric data, while the methods of Sec. 6.3.3 incorporate both photometric and electrochemical measurements.

The various *numeric methods of pK determination* described in the literature [including methods (a) and (b) of Sec. 6.3.4] are all based on solving appropriate sets of linear equations. Compared to graphic methods (Sec. 6.3.3), numeric methods present more disadvantages than advantages. The solution of a large set of linear equations requires a programmable calculator or computer, and problems related to "conditioning" must always be anticipated.

If one occasionally has the need to analyze complex multi-step titrations in addition to single-step systems, and if appropriate computing facilities are available, the *non-linear curve-fitting method* (Sec. 6.3.5) is advantageous. This provides a coupled analysis of a complete set of $A_\lambda(\text{pH})$ curves, and it permits estimation of pK values in a way that is particularly accurate and certain. The associated statistical analysis (cf. Chapter 10) also offers a sensitive test of the proposed titration mechanism. Absorbance data at the titration limits are not required, but are instead calculated. It should also be noted that this method is applicable to segments of titration curves.

Whichever method is chosen, it is strongly recommended that the results be used in conjunction with equation 6–11 to construct simulated $A_\lambda(\text{pH})$ curves, and that these be compared with original data.

The method of Sec. 6.3.6 is unique in that it may permit precise *photometric evaluation of pK in the absence of pH data*. An actual titration need not be carried out, since in the simplest case ($\varepsilon_{\lambda\text{BH}} = 0$) data are examined from only solutions (a solution of the protolyte in pure water and one corresponding to the alkaline limit). The method has limitations, however; it is only applicable to extemely pure, single-step protolytes, and it demands a photometer of very high quality.

Table 6-1. Summary of the methods for evaluating single-step ($s = 1$) acid-base equilibria.

Graphic matrix-rank analysis	Methods for (absolute) pK determination	Required data		Assessment of the method	Relationship to the titration mechanism
		Titration limits	Source(s) of primary data		
(1) *Titration spectra* → Isosbestic points (for intersecting spectra)	(1) *Inflection points in $A\lambda$-pH-curves* (a) Tangent method (b) Tubbs ring method (c) Differential method	*Neither* $A_{\lambda BH}$ *nor* $A_{\lambda B}$ required	Middle portions of A_λ-pH-curves	Rapid, crude determination of pK; template, etc., required	—
(2) *A- and AD-diagrams* → Linear for all wavelength combinations	(2) *Bisection of overall changes in $A\lambda$-pH-curves* (≜ bisecting points in A-diagrams)	$A_{\lambda BH}$ *and* $A_{\lambda B}$ needed	Data near the pK value	Simplest means of rough estimation	—
(3) *Aλ-pH-Curves* → Each contains one inflection point	(3) *Linearization of A_λ-pH-curves* (a) $\log \dfrac{A_\lambda - A_{\lambda BH}}{A_{\lambda B} - A_\lambda} = \text{pH} - \text{p}K$ (6-13) $\underbrace{}_{y} \underbrace{}_{\ddot{x}}$ and/or (b) $(A_\lambda - A_{\lambda BH}) \cdot 10^{-\text{pH}} = -K \cdot A_\lambda + K \cdot A_{\lambda B}$ (6-14a) $\underbrace{}_{y} \underbrace{}_{\ddot{x}}$ $(A_\lambda - A_{\lambda B}) \cdot 10^{+\text{pH}} = -\dfrac{1}{K} \cdot A_\lambda + \dfrac{1}{K} \cdot A_{\lambda BH}$ (6-14b) $\underbrace{}_{y} \underbrace{}_{\ddot{x}}$	$A_{\lambda BH}$ *and* $A_{\lambda B}$ needed $A_{\lambda BH}$ *or* $A_{\lambda B}$ needed (the other is calculated)	Entire A_λ-pH-curve Entire A_λ-pH-curve	Simple and precise pK determination; pocket calculator suffices	Provides a significant test of mechanism, complements graphic matrix-rank analysis

6.5 Summary

Table 6-1 (continued).

Methods for (absolute) pK determination	Required data		Assessment of the method	Relationship to the titration mechanism
	Titration limits	Source(s) of primary data		
(4) Linear equations				
(a) $\dfrac{A_{\lambda B} - A_\lambda}{A_\lambda - A_{\lambda BH}} = -\dfrac{1}{K} \cdot a_{H_3O^+}$ (6–15) $\underbrace{}_{y} \qquad \underbrace{}_{x}$		$A_{\lambda BH}$ and $A_{\lambda B}$ needed	Entire A_λ-pH-curve	Like (3a), but numerical and with different error function	Little significance
(b) System of linear equations		$A_{\lambda BH}$ or $A_{\lambda B}$ may be needed, or neither	Entire A_λ-pH-curve	Risk of conditioning problems	Significant if coupled with a statistical analysis
(5) Non-linear curve fitting of all experimental A_λ-pH data		Neither $A_{\lambda BH}$ nor $A_{\lambda B}$ required	Coupled analysis of all A_λ-pH-curves	Very precise pK determination; large computer required	Highly significant due to: (a) statistical analysis (b) control simulation
(6) Photometric pK determination without pH measurements $K_c = \dfrac{c_{H_3O^+} \cdot c_B}{c_{BH}} = \dfrac{\beta^2}{1 - \beta} \cdot b_0$ (6–18) with $c_{H_3O^+} = c_B$		*Prerequisites:* (1) (usually): $\varepsilon_{\lambda BH} = 0$ → *only* $A_{\lambda B}$ needed (2) b_0 required (3) Highly pure substance ($s = 1$) (4) Purest water		Precision method for determining classical and thermodynamic pK using *only* photometric data and weights	*No significance;* → spectrometric titration and graphic matrix-rank analysis should be added

6.5.2 Non-overlapping, Multi-Step Titration Systems (cf. Table 6–2)

In principle, systems of this type can be evaluated in essentially the same way as single-step titrations. It is first necessary to verify by *graphic matrix-rank analysis*, preferably *A-diagrams*, that there is indeed no overlap among the various subequilibria. Only if this criterion is met is it possible to dissect the system into a set of subsystems of rank $s = 1$. Success at this stage is normally assured if $\Delta pK > 3.5$–4. Depending on the relative magnitudes of the molar absorptivities, pK differences of 3–3.5 may suffice.

Experience with a large number of spectrometric titration systems has shown the *determination of pK values* to be best conducted by a combination of the methods in Sections 6.3.2, 6.3.3 [method (b)], and 6.3.5. The *halving method* of Sec. 6.3.2 is particularly recommended for rapid estimations.

Linearization method (b) described in Sec. 6.3.3 is especially well suited to accurate graphic analysis of non-overlapping (or slightly overlapping) multi-step titration systems. In theory, knowledge of a single limiting absorbance, or a single point of interception for two linear segments, should permit analysis of an entire s-step titration system using the pair of equations 6–14. In practice, however, it is advisable to carry out several concurrent calculations based on the equations 6–14, starting with various well-established end values or inflection points as determined from A-diagrams. Such a procedure is only feasible for a titration system composed of three or more steps if one has access to a microcomputer with graphics capability or a large mainframe computer.

In cases such as these, and more generally for the final evaluation of entire sets of titration data obtained from multi-step systems, the preferred analytical approach is *non-linear curve fitting* (Sec. 6.3.5), a method applicable to both overlapping and non-overlapping titration systems (cf. Chapter 10).

Finally, it is important to point out once again that even non-overlapping titration systems may, in principle, be complicated by branching (cf. Chapter 12). Moreover, it is not safe to assume (as the authors of many published reports do!) that a pK difference > 4–5 ensures the correct assignment of calculated pK values to specific dissociable groups.

6.5 Summary

Table 6–2. Analysis of an s-step non-overlapping titration system as a series of s single-step subsystems.

Graphic matrix-rank analysis	Recommended methods for (absolute) pK determination (cf. Table 6–1)	Required data — Titration limits	Required data — Source(s) of primary data	Assessment of the method	Relationship to the titration mechanism
(1) *Titration spectra* → s subsystems with isosbestic points (for intersecting spectra)	(2) *Bisection of overall change in absorbance* for a given titration step (i.e., the mid-point of each straight-line segment in an A-diagram)	*Both* limiting values *and all* ampholyte absorbances required	Data near the pK value	Rapid, crude, simple means of estimating pK values	—
(2) *A- and AD-diagrams* → s intersecting straight line segments whose intersection points correspond to the pure ampholytes	(3b) *Linearization of each individual titration step*	Requires *in principle* only one limiting value *or* one point of intersection of two straight-line segments	Entire $A\lambda$-pH-curve	Precise method for graphic determination of pK; partially overlapping systems are evaluable; multi-step cases very labor-intensive	Provides a significant test of the mechanism, complements graphic matrix-rank analysis
(3) $A\lambda$-pH-Curves → can be dissected into s single-step subsystems	(5) *Non-linear curve-fitting of the entire s-step titration*	*Neither* the limiting values *nor* the ampholyte absorbances required	Coupled analysis of *all* $A\lambda$-pH-curves	Provides very precise value for pK; applicable to overlapping systems; requires large computer	Highly significant due to: (a) statistical analysis (b) control simulation
	[Method (6) essentially inapplicable, cf. p. 63]				

Literature

1. R. Blume, H. Lachmann, H. Mauser, F. Schneider, *Z. Naturforsch.* 29 b (1974) 500.
2. G. Kortüm, *Lehrbuch der Elektrochemie.* Verlag Chemie, Weinheim 1972, Chap. XI.
3. H. Lachmann, *Dissertation.* Tübingen 1973.
4. H. Ley, B. Arends, *Z. Phys. Chem.* B 6 (1929) 240.
5. H. Mauser, *Z. Naturforsch.* 23 b (1968) 1021, 1025.
6. S. Ebel, W. Parzefall, *Experimentelle Einführung in die Potentiometrie.* Verlag Chemie, Weinheim (1975).
7. C. Bliefert, *pH-Wert-Berechnungen.* Verlag Chemie, Weinheim, New York 1978.
8. C. F. Tubbs, *Anal. Chem.* 26 (1954) 1670.
9. R. Winkler-Oswatitsch, M. Eigen, *Angew. Chem.* 91 (1979) 20.
10. H. H. Perkampus, T. Rössel, *Z. Elektrochem.* 60 (1956) 1102; 62 (1958) 94.
11. W. Preetz, G. Schätzel, *Z. Anorg. Chem.* 423 (1976) 117.
12. J. H. Baxendale, H. R. Hardy, *Trans. Farad. Soc.* 49 (1953) 1140.
13. V. Böhmer, R. Wamsser, H. Kämmerer, *Monatsh. Chem.* 104 (1973) 1315.
14. R. Blume, J. Polster, *Z. Naturforsch.* 29 b (1974) 734.
15. D. H. Rosenblatt, *J. Phys. Chem.* 58 (1954) 40.
16. B. J. Thamer, *J. Phys. Chem.* 59 (1955) 450.
17. B. Roth, J. F. Bunnett, *J. Amer. Chem. Soc.* 87 (1965) 334.
18. G. Heys, H. Kinn, D. Perrin, *Analyst* 97 (1972) 52.
19. M. E. Magar, *Data Analysis in Biochemistry and Biophysics.* Academic Press, New York 1972, Chap. 9.
20. G. E. Forsythe, C. B. Moler, *Computer-Verfahren für lineare algebraische Systeme.* R. Oldenbourg Verlag, München 1971, Chap. 8.
21. F. Göbber, H. Lachmann, *Hoppe-Seylers Z. Physiol. Chem.* 359 (1978) 269.
22. G. Kortüm, *Kolorimetrie, Photometrie und Spektrometrie.* Springer-Verlag, Berlin 1962, Chap. I.
23. G. Kortüm, W. Vogel, K. Andrussow, *Pure Appl. Chem.* 1 (1960) 190.
24. G. Kortüm, H. v. Halban, *Z. Phys. Chem.* A 170 (1934) 212, 351.
25. G. Kortüm, H. C. Shih, *Ber. Bunsenges. Phys. Chem.* 81 (1977) 44.
26. R. Blume, H. Lachmann, J. Polster, *Z. Naturforsch.* 30 b (1975) 263.
27. A. K. Lunn, R. A. Morton, *Analyst* 77 (1952) 718.
28. D. E. Metzler, C. M. Harris, R. J. Johnson, D. B. Siano, J. S. Thomson, *Biochemistry* 12 (1973) 5377.
29. A. Binder, S. Ebel, *Z. Anal. Chem.* 272 (1974) 16.
30. D. E. Metzler, E. E. Snell, *J. Amer. Chem. Soc.* 77 (1955) 2431.
31. K. Nagano, D. E. Metzler, *J. Amer. Chem. Soc.* 89 (1967) 2891.
32. Y. V. Morozov, N. P. Bazhulina, L. P. Cherkashina, M. Y. Karpeiskii, *Biophysics* 12 (1968) 454.
33. C. M. Harris, R. J. Johnson, D. E. Metzler, *Biochem. Biophys. Acta* 421 (1976) 181.
34. H. Lachmann, *Z. Anal. Chem.* 290 (1978) 117.

7 Two-Step Acid-Base Equilibria

7.1 Basic Equations

Overlap plays an important role in the investigation of two-step acid-base equilibrium systems. That is, such systems are likely to be characterized by certain pH regions within which it is necessary to take into account the concentrations of *all* the species, and it is in these regions that the equilibria are said to "overlap".

The techniques for analyzing one- and two-step acid-base equilibria are of considerable practical importance because even protolysis systems comprised of more than two equilibrium steps often behave much like sets of independent, successive one- and/or two-step (overlapping) equilibria.[1-13] Examination of potentiometric data remains the dominant approach to analyzing these equilibria, and modern evaluation is greatly facilitated by the application of powerful computer methods.[14-38]

Unbranched two-step protolysis systems (cf. also the *branched* systems discussed in Chapter 12) consist of equilibria of the following form:

$$BH_2 \stackrel{K_1}{\leftrightarrows} BH^- \; (H^+) \qquad BH^- \stackrel{K_2}{\leftrightarrows} B^{2-} \; (H^+) \qquad (7-1)$$

The quantities for which spectrometric titration is expected to supply values include the following:

- the number s of linearly independent equilibria (in the present case, $s = 2$);
- the dissociation constants K_1 and K_2;
- the quotient $K_1/K_2 \; (= \kappa)$;
- the molar absorptivities of BH_2, BH^-, and B^{2-}; and
- the concentrations of the individual components.

A wide variety of methods can be invoked for obtaining solutions to the problems posed, and these are summarized for purposes of comparison in Table 7–6 of Sec. 7.9.

Absorbance diagrams (A-diagrams) occupy a central role in a contemporary approach to two-step equilibrium systems (cf. Chapters 2 and 6), and their use is strongly recommended. Individual techniques are examined in detail in subsequent sections of this chapter with the aid of the example *o*-phthalic acid. This substance (and its isomers) is frequently cited in the literature as well-suited to the testing of a variety of spectrometric titration procedures.[3,39–46]

The principles underlying the recommended methods are discussed in Sections 7.5 and 7.6, but we begin by noting that the relevant relationships are most easily visualized when data are presented in the form known as a "characteristic concentration diagram",[47,48,43–45] in which the concentrations of two protolytes are plotted against each other. This is in fact analogous to examining a plot of absorbances determined at two wavelengths, as in an A-diagram, since an "affine transformation" can be used to convert concentrations into absorbances, leaving many basic relationships unchanged (see below). Thus we may avoid the chore of supplying independent derivations for conclusions based on A-diagrams.

All the relationships necessary for our argument arise out of the following basic equations for the mixed dissociation constants K_1 and K_2:

$$K_1 = a_{H_3O^+} \cdot \frac{bh}{bh_2} = h \cdot \frac{bh}{bh_2} = 10^{-pH} \cdot \frac{bh}{bh_2}$$

$$K_2 = a_{H_3O^+} \cdot \frac{b}{bh} = h \cdot \frac{b}{bh} = 10^{-pH} \cdot \frac{b}{bh}$$

(7–2)

where pH $= -\log a_{H_3O^+} = -\log h$

$bh_2 = $ (molar) concentration of the component BH_2 at equilibrium (the same convention is employed for the symbols bh and b)

The following definitions are also required:

$$pK_1 = -\log K_1$$

$$pK_2 = -\log K_2$$

$$\Delta pK = pK_1 - pK_2 = \log \frac{K_1}{K_2} = \kappa$$

(7–3)

where $\kappa = \dfrac{K_1}{K_2}$

If the initial (molar) concentration of protolyte is represented by b_0, then mass balance considerations require that

$$b_0 = bh_2 + bh + b \tag{7–4}$$

Combination of equations 7–2 through 7–4 permits the expression of individual concentrations as functions of h:

$$bh_2 = \frac{b_0 h^2}{H_2}, \quad bh = \frac{b_0 K_1 h}{H_2}, \quad b = \frac{b_0 K_1 K_2}{H_2} \tag{7–5}$$

where $H_2 = h^2 + K_1 h + K_1 K_2$

Successive additions of acid and base (e.g., HCl, NaOH) are normally used to shift the positions of the two equilibria 7–1. After each such addition it is necessary to measure the pH of the system and to record the corresponding spectrum. Within the range of its validity, the Lambert-Beer-Bouguer law ensures that

$$A_\lambda = l \cdot (\varepsilon_{\lambda BH_2} \cdot bh_2 + \varepsilon_{\lambda BH} \cdot bh + \varepsilon_{\lambda B} \cdot b) \tag{7-6}$$

where l = pathlength through the cuvette
  $\varepsilon_{\lambda BH_2}$ = molar absorptivity of BH_2 at the wavelength of observation (and analogously for $\varepsilon_{\lambda BH}$ and $\varepsilon_{\lambda B}$)

As before, we assume that the pH titration in question has been so designed that neither titrant nor solvent is subject to absorption, and that dilution effects are negligible. Equations 7–6 and 7–5 then lead to the following analytical expression for the titration curve A vs. pH:

$$A_\lambda = l\,b_0 \cdot \frac{\varepsilon_{\lambda BH_2} \cdot h^2 + \varepsilon_{\lambda BH} \cdot h + \varepsilon_{\lambda B} \cdot K_1 K_2}{h^2 + K_1 h + K_1 K_2} \tag{7-7}$$

In many cases, $\varepsilon_{\lambda BH_2}$ and $\varepsilon_{\lambda BH}$ can be determined directly. For example, if the titration solution is made so acidic that BH_2 is essentially the only protolyte present (i.e., the acidic titration limit is established), then $b_0 = bh_2$ and $A_\lambda = l \cdot \varepsilon_{\lambda BH_2} \cdot b_0$. Alternatively, if the solution is treated with sufficient base so that only the component B^{2-} is present (equivalent to the basic titration limit), then $b_0 = b$ and $A_\lambda = l \cdot \varepsilon_{\lambda B} \cdot b_0$.

In general, direct determination of the molar absorptivity of the intermediate product BH^- (an *ampholyte*) is not possible, since if the pK differences for the system are small there will be no pH region in which the ampholyte BH^- is the only species present. Nevertheless, A-diagrams provide a remarkably easy solution to the problem of determining $\varepsilon_{\lambda BH}$ (cf. Sec. 7.3.3).

7.2 Graphic Matrix-Rank Analysis (ADQ-Diagrams)

According to equation 7–4, two of the three concentration variables bh_2, bh, and b are linearly independent. It will be shown below that the same generalization applies to absorbances determined at two wavelengths. For this reason, the matrix A (constructed from absorbances ascertained at different wavelengths and pH values) for a two-step acid-base equilibrium must correspond to the rank $s = 2$ (cf. Sec. 3.2.1). This rank assignment can be verified graphically by means of absorption difference quotient diagrams (ADQ-diagrams),[47,49,43,45] as discussed in Sec. 3.2.2.

Equation 7–4 may be used to eliminate the concentration bh_2 from equation 7–6. Making use of the definitions

$$A_{\lambda BH_2} = l \cdot b_0 \cdot \varepsilon_{\lambda BH_2} \tag{7-8a}$$

and

$$\Delta A_\lambda = A_\lambda - A_{\lambda BH_2} \tag{7-8b}$$

it follows that

$$\Delta A_\lambda = q_{\lambda 1} bh + q_{\lambda 2} b \tag{7-9}$$

where $q_{\lambda 1} = l(\varepsilon_{\lambda BH} - \varepsilon_{\lambda BH_2})$
$q_{\lambda 2} = l(\varepsilon_{\lambda B} - \varepsilon_{\lambda BH_2})$

In a two-step titration, ΔA_λ is therefore a function of two linearly independent concentrations. Thus, based on the arguments in Sec. 3.2.2, it is generally to be expected that:

- the corresponding titration spectra will contain no isosbestic points,

- A-diagrams will be non-linear, and

- ADQ-diagrams will be linear (see below).

A-diagrams in this case are prepared starting from equation 7–7. Figure 7–1 illustrates a set of such curves constructed as a function of pK. Molar absorptivities have been selected at random, and it will be noted that all the results are indeed non-linear. For pK differences greater than 3.5–4 the curves tend to follow the boundary lines $\overline{AB}$ and $\overline{BC}$, thereby approximating the course of two non-overlapping one-step protolysis systems.

In an *ADQ-diagram* (cf. Sec. 3.3.2), it is the quotients of absorbance differences at various wavelengths that are plotted against each other.[47,49,43,45] In order to obtain the corresponding analytical expression in the case of a two-step titration, equation 7–8 is applied at three different wavelengths, λ_1, λ_2, and λ_3. The concentrations bh and b are easily eliminated from the resulting three equations, leading to

$$\Delta A_1 = \alpha_1 \cdot \Delta A_2 + \alpha_2 \cdot \Delta A_3 \tag{7-10}$$

where $\alpha_1 = \dfrac{q_{11}q_{32} - q_{12}q_{31}}{q_{21}q_{32} - q_{22}q_{31}} = \dfrac{\begin{vmatrix} q_{11} & q_{12} \\ q_{31} & q_{32} \end{vmatrix}}{|\boldsymbol{Q}|}$ and

$$\alpha_2 = -\dfrac{q_{11}q_{22} - q_{12}q_{21}}{q_{21}q_{32} - q_{22}q_{31}} = -\dfrac{\begin{vmatrix} q_{11} & q_{12} \\ q_{21} & q_{22} \end{vmatrix}}{|\boldsymbol{Q}|}$$

with $|\boldsymbol{Q}| = \begin{vmatrix} q_{21} & q_{22} \\ q_{31} & q_{32} \end{vmatrix} \neq 0$

(For simplicity, ΔA_{λ_1}, $q_{\lambda_1 1}$, etc., have been replaced in the above expressions by ΔA_1, q_{11}, etc.)

7.2 Graphic Matrix-Rank Analysis (ADQ-Diagrams)

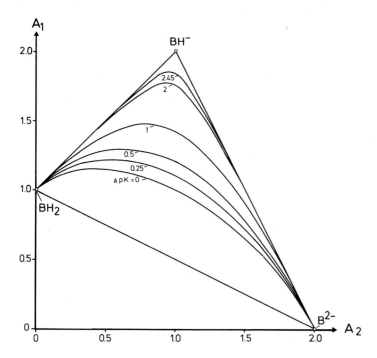

Fig. 7–1. A-diagrams for a two-step acid-base equilibrium computed as a function of ΔpK [$l = 1$ cm, $b_0 = 10^{-3}$ mol/L; $\varepsilon_{1BH_2} = 1000$, $\varepsilon_{1BH} = 2000$, $\varepsilon_{1B} = 0$, $\varepsilon_{2BH_2} = 0$, $\varepsilon_{2BH} = 1000$, $\varepsilon_{2B} = 2000$ (cm^{-1}M^{-1})].

The coefficients α_1 and α_2 are functions only of molar absorptivities, and are therefore constant (the pathlength l cancels). According to equation 7–4, the concentrations bh_2, bh, and b are linearly related to one another. Equation 7–10 shows the same to be true of the ΔA_{λ_i} values at the three wavelengths. This relationship can be verified graphically by rearranging equation 7–10 into the form

$$\frac{\Delta A_1}{\Delta A_2} = \alpha_1 + \alpha_2 \frac{\Delta A_3}{\Delta A_2} \tag{7-11}$$

Thus, if the two quotients $\Delta A_1/\Delta A_2$ and $\Delta A_3/\Delta A_2$ are plotted against each other in the form of an ADQ-diagram, equation 7–11 states that all points will fall on a straight line. In Sec. 7.8 it will be shown that this relationship is valid regardless of the choice of reference point $A_{\lambda i}$. In equation 7–8b, ΔA_λ has been referenced against the point BH$_2$, but B^{2-} or any other point would serve equally well.

Equation 7–10 may be rewritten as a relationship in terms of A_λ at three wavelengths:

$$A_1 = \alpha_1 A_2 + \alpha_2 A_3 + \alpha_3 \tag{7-12}$$

where $\alpha_3 = A_{1BH_2} - \alpha_1 A_{2BH_2} - \alpha_2 A_{3BH_2}$

Thus, equations 7–12 and 7–10 show that values of both ΔA_λ and A_λ at three wavelengths are linearly related so long as all points fall on a straight line in an ADQ-diagram. It therefore becomes possible to use a two-dimensional representation to test whether the matrix A with elements $A_{\lambda,\mathrm{pH}}$ is of the rank two (*graphic matrix-rank analysis*, cf. Sections 3.3 and 6.2)

Nevertheless, it is important to recognize that *exceptions* may interfere with this type of graphic matrix-rank analysis.[47] For example, it may happen in practice that two rows (or columns) of the matrix A will prove to be linearly dependent upon each other. In this event one of the three determinants in equation 7–10 will equal zero, hence one of the three possible A-diagrams (e.g., A_1 vs. A_2) will be a straight line rather than a curve. This may in turn result in an ADQ-diagram consisting of a straight line parallel to one of the coordinate axes, in which case the validity of equation 7–10 is illusory. Other exceptions include ADQ-diagrams comprised of points (or clusters of points), or of lines passing through the origin (cf. Sec. 7.8).

In summary, the relationships expressed by equations 7–10 and 7–12 are only meaningful if the corresponding ADQ-diagram consists of lines that avoid the origin and are not parallel to an axis.

The application of graphic matrix-rank analysis may best be demonstrated by a specific example, such as the titration system presented by *o-phthalic acid*.[43–46,50,51] Figure 7–2 shows the corresponding titration spectra, which intersect in complicated ways. No isosbestic points (or even quasi-isosbestic points) are observed, so a determination of rank is impossible on the basis of spectra alone.

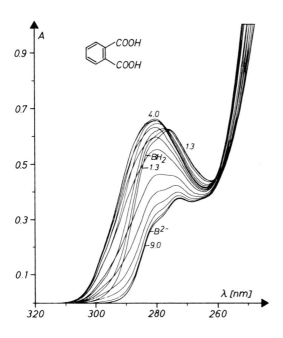

Fig. 7–2. Titration spectra for *o*-phthalic acid ($b_0 = 5 \cdot 10^{-4}$ M; 25.0 °C; pK = 2.45).

7.2 Graphic Matrix-Rank Analysis (ADQ-Diagrams)

Each spectrum is recorded at a specific pH, and it is easy to ascertain at what wavelengths the absorbance varies most. One begins by selecting an appropriate set of such wavelengths (ca. 5–10) distributed as widely as possible over the complete spectral range. From such a set it will normally prove possible to prepare suitable A-diagrams.

Parts (a) and (c) of Fig. 7–3 contain titration curves A_λ vs. pH for o-phthalic acid recorded at 290 and 272 nm, respectively. The first curve has a "bell-shaped" form, whereas the second consists of a set of steps. Each shows two points of inflection. Part (b) of Figure 7–3 shows the A-diagram A_{290} vs. A_{272} for the same system. Here the pH is not explicitly represented, but with the aid of the titration curves each point in the diagram may easily be associated with a particular pH, a matter of some importance in subsequent steps of the evaluation.

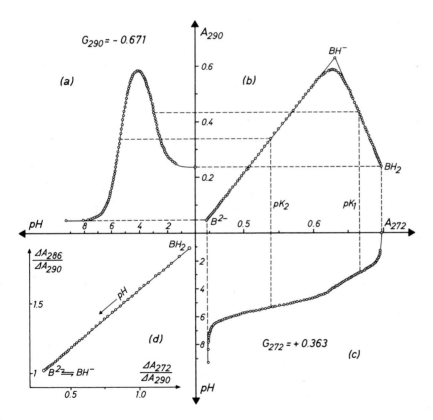

Fig. 7–3. Spectrometric titration curves (a,c), A-diagram (b), and ADQ-diagram (d) for o-phthalic acid. $A_{\lambda BH_2} = l \cdot b_0 \cdot \varepsilon_{BH_2}$ was determined with the aid of equation 7–17. Curves (a) and (c) are based on computations with the values $pK_1 = 2.91$ and $pK_2 = 5.35$. In (d), ΔA_λ is referenced against point B^{2-} (cf. Sections 7.2 and 7.8).

In practice it is wise to construct A-diagrams based on as many wavelength combinations as possible. Additional A-diagrams for *o*-phthalic acid are shown in Fig. 7–4. The fact that all are curved suggests that the corresponding matrix A has the rank two. This conclusion is supported by the ADQ-diagrams shown in Fig. 7–5, which are based on the point BH_2, as well as by those in part (d) of Fig. 7–3 (and another in Fig. 7–28, Sec. 7–8) based on the point B^{2-}. None of the curves passes through the origin, and none is parallel to a coordinate axis. This protolyte therefore exemplifies the spectrometric behavior of a typical two-step acid-base system. Comparison of Fig. 7–4 with Fig. 7–1 permits one to estimate that the pK difference for *o*-phthalic acid lies between 2 and 3.

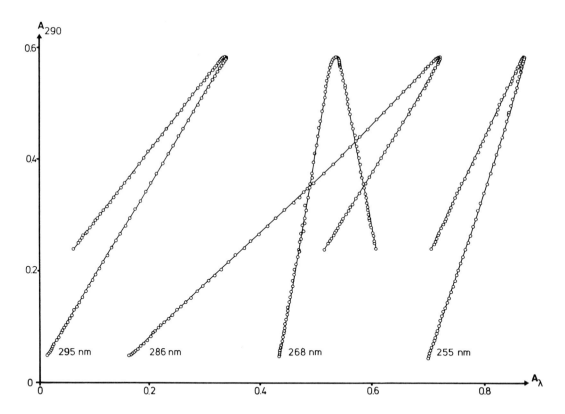

Fig. 7–4. A-diagrams for *o*-phthalic acid (cf. Fig. 7–2).

7.3 The Determination of Absolute pK Values

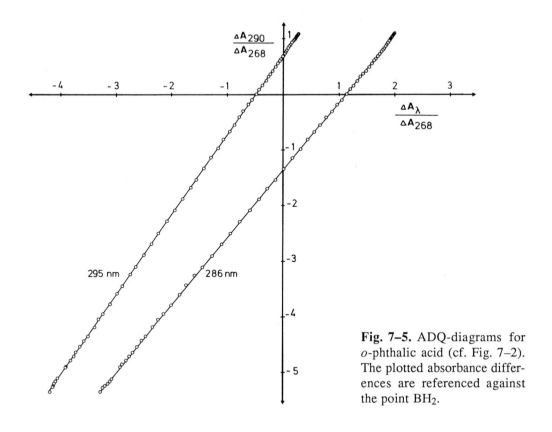

Fig. 7–5. ADQ-diagrams for *o*-phthalic acid (cf. Fig. 7–2). The plotted absorbance differences are referenced against the point BH_2.

7.3 The Determination of Absolute p*K* Values

7.3.1 Inflection Point Analysis (First Approximate Method)

The most familiar method for determining p*K* values for one-step acid-base equilibria relies on the assumption that the point of inflection in a titration curve is the point at which pH = pK_1. However, this approach generally fails in the case of a two-step protolyte unless certain restrictive conditions are fulfilled. As seen in Table 7–1, the *titration curves* for a two-step (non-degenerate) protolytic equilibrium either have a stepped appearance (Type I) or they show a maximum or minimum (Type II). The following quotient may be taken as a measure of the relative spectrometric change associated with two successive titration steps[43,45,50] (for the significance of $q_{\lambda i}$ see equation 7–9):

Table 7–1. Types and degeneracies of two-step spectrometric titration curves (schematic).

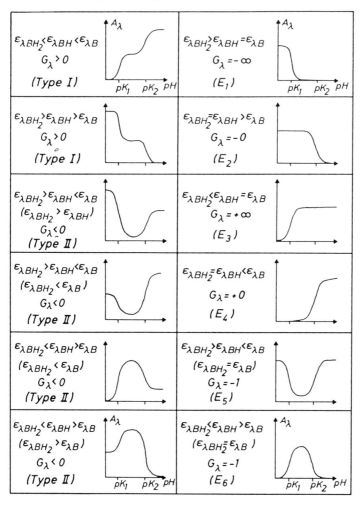

$$G_\lambda = \frac{q_{\lambda 1}}{q_{\lambda 2} - q_{\lambda 1}} = \frac{\varepsilon_{\lambda BH} - \varepsilon_{\lambda BH_2}}{\varepsilon_{\lambda B} - \varepsilon_{\lambda BH}} \qquad (7\text{–}13)$$

For a titration curve of Type I, $G_\lambda > 0$; therefore either

$$\varepsilon_{\lambda BH_2} < \varepsilon_{\lambda BH} < \varepsilon_{\lambda B}$$

or

$$\varepsilon_{\lambda BH_2} > \varepsilon_{\lambda BH} > \varepsilon_{\lambda B}$$

7.3 The Determination of Absolute pK Values

In the case of a Type II curve, $G_\lambda < 0$, which means either that

$$\varepsilon_{\lambda BH_2} < \varepsilon_{\lambda BH} > \varepsilon_{\lambda B} \text{ with } \varepsilon_{\lambda BH_2} \neq \varepsilon_{\lambda B}$$

or

$$\varepsilon_{\lambda BH_2} > \varepsilon_{\lambda BH} < \varepsilon_{\lambda B} \text{ with } \varepsilon_{\lambda BH_2} \neq \varepsilon_{\lambda B}$$

The left portion of Table 7-1 contains schematic illustrations of the various possibilities.[43,45,50] In the event that the pK differences are large (i.e., $\Delta pK > 4$), the overall titration curves can be regarded with sufficient accuracy to be combinations of two successive one-step titration curves (cf. Fig. 7–6). In this *limiting case* the two steps may be examined separately using the methods described in Chapter 6 (cf. especially Sec. 6.4).

If the molar absorptivities of two components are equal, Type I and Type II become "degenerate". This may mean that only a single titration curve will be spectrometrically visible, or it could be that the two steps will cancel each other (cf.

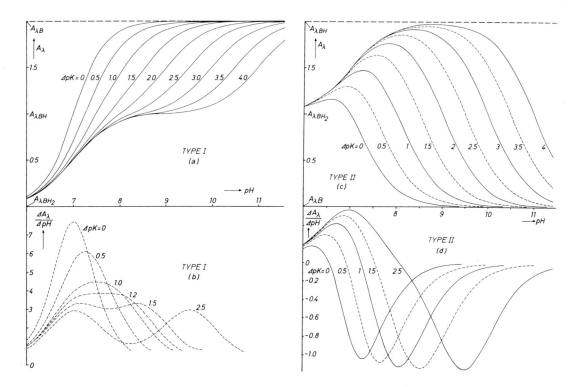

Fig. 7–6. Calculated spectrometric titration curves as a function of ΔpK, together with the corresponding first derivatives ($b_0 = 10^{-4}$ M; $pK_1 = 7.0$; $l = 1$ cm).
(a), (b): Type I [$G_\lambda = 1$; $\varepsilon_{\lambda BH_2} = 0$, $\varepsilon_{\lambda BH} = 10\,000$, $\varepsilon_{\lambda B} = 20\,000$ (cm^{-1}M^{-1})].
(c), (d): Type II [$G_\lambda = 0.5$; $\varepsilon_{\lambda BH_2} = 10\,000$, $\varepsilon_{\lambda BH} = 20\,000$, $\varepsilon_{\lambda B} = 0$ (cm^{-1}M^{-1})].

the right-hand portion of Table 7–1, E_1–E_4). Although the degenerate cases illustrated look rather like single-step titration curves they must not be so evaluated, because they arise from a very different type of analytical expression.

Just as in the case of the more fundamental titration curves in the left-hand half of Table 7–1, analysis of the degenerate cases (E_1–E_4) as one-step systems is only possible to a sufficient degree of accuracy if $\Delta pK > 4$. Only then does the analytical form of the curve correspond to that of a single one-step titration or a sequence of two single-step titrations (cf. Fig. 7–7)

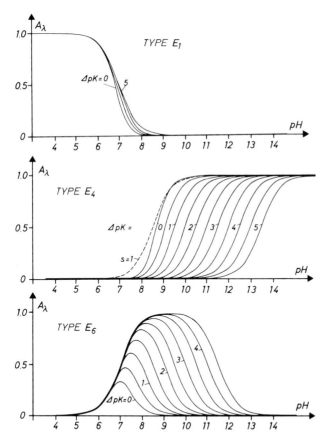

Fig. 7–7. Computed spectrometric titration curves as a function of ΔpK for the degenerate cases E_1, E_4, and E_6 (cf. Table 7–1; $b = 10^{-4}$ M; $pK_1 = 7.0$; $l = 1$ cm).
(a) Type E_1 [$G_\lambda = -\infty$; $\varepsilon_{\lambda BH_2} = 10\,000$, $\varepsilon_{\lambda BH} = \varepsilon_{\lambda B} = 0$ (cm^{-1}M^{-1})].
(b) Type E_4 [$G_\lambda = 0$; $\varepsilon_{\lambda BH_2} = \varepsilon_{\lambda BH} = 0$, $\varepsilon_{\lambda B} = 10\,000$ (cm^{-1}M^{-1})].
(c) Type E_6 [$G_\lambda = -1$; $\varepsilon_{\lambda BH_2} = 0$, $\varepsilon_{\lambda BH} = 10\,000$, $\varepsilon_{\lambda B} = 0$ (cm^{-1}M^{-1})].

7.3 The Determination of Absolute pK Values

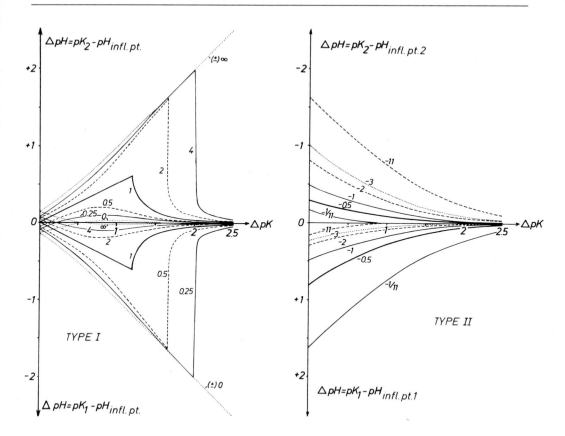

Fig. 7–8. The systematic error ΔpH accompanying an inflection point analysis, plotted as a function of ΔpK and G_λ. Curves for which the initial slope is positive refer to the systematic deviation between pK_2 and the corresponding inflection point, while those for which the initial slope is negative correspond to (pK_1 – pH$_{\text{infl.pt.}}$). (a) Type I, (b) Type II.

The systematic error associated with inflection point analysis is a function not only of ΔpK, but also of the molar absorptivities of all components (and thus of G_λ). In order to estimate this error quantitatively, calculations have been carried out showing the dependence of a two-step titration curve and its first derivative on ΔpK (cf. also Fig. 7–6). The maxima and minima of first derivative curves coincide with inflection points on the corresponding titration curves. Deviations between pK and the points of inflection as determined using first derivative curves have been plotted in Fig. 7–8 as a function of ΔpK.

Titration curves of Type I show only a single inflection point if ΔpK is small (cf. Fig. 7–6). By contrast, Type II curves always show two points of inflection. This reveals itself in Fig. 7–8 by the fact that the Type II error function varies mono-tonically with ΔpK, whereas the Type I error function passes through an extreme.

<small>The case of $G_\lambda = 1$ for Type I curves has been emphasized in Fig. 7–8 by a heavy line. The error $\Delta pH = pK_2 - pH_{\text{inflection point}}$ increases with ΔpK in an essentially linear fashion until a maximum value of ca. 0.6 pH units is reached at $\Delta pK \approx 1.2$ (i.e., at $K_1/K_2 \approx 16$), after which it decreases rapidly. This behavior is consistent with the fact that below $\Delta pK \approx 1.2$ only a single inflection point exists, while above this value there are three (cf. Fig. 7–6). Additional details are discussed in references 43 and 50.</small>

To summarize: error analysis reveals that systematic error is a function of ΔpK and G_λ, where the nature of the relationship is shown in Fig. 7–8. The more precise the determination of one pK value the more uncertain is the other (except in the case of $G_\lambda = \pm 1$). The degenerate cases E_1 through E_4 in Table 7–1 ($G_\lambda = 0$ and $G_\lambda = \pm \infty$) entail particularly large errors, even if $\Delta pK = 0$. If $\Delta pK > 3$ and $0.1 < G_\lambda > 10$, the systematic error is always less than 0.04 pH units; if $\Delta pK > 4$ it is always less than 0.004 pH units. For these reasons, inflection point analysis should be applied to overlapping protolysis equilibria only if the goal is a rough estimate of pK.

Deceptive resemblances between one- and two-step titrations are not limited to titration curves: A-diagrams can also be misleading. In the event that only one equilibrium is spectroscopically observable—as with the degenerate cases E_1–E_4 in Table 7–1—the corresponding A-diagrams will consist only of straight lines. A two-step titration system of this type thus appears "spectroscopically uniform" irrespective of ΔpK;[47–52] that is to say, the spectroscopic behavior of the system is determined by only *one* linearly independent concentration variable.

The A-diagrams derived from degenerate titration systems E_5 and E_6 also consist of straight lines. Nevertheless, since the concentration of ampholyte BH⁻ first rises and then falls during the titration, the change in absorbance reverses itself. This causes the points making up the straight line in the A-diagram to turn back upon themselves.

We may summarize by saying that, strictly speaking, pK values can only be determined on the basis of inflection point analysis for single-step dissociation systems ($s = 1$). Approximation is possible in certain two-step and multi-step cases, but only if the pK differences are sufficiently large. A useful rule of thumb requires that ΔpK be > 3.

7.3.2 Uniform Subregions in an Absorbance Diagram (Second Approximate Method)

Two-step protolysis systems frequently exhibit relatively large pK differences ($\Delta pK > 1.5$). Depending on the magnitude of ΔpK, such systems often possess wide pH-ranges over which only one of the two equilibria plays a significant role. For example, to a satisfactory degree of approximation only the equilibrium

7.3 The Determination of Absolute pK Values

$$BH_2 \leftrightarrows BH^- \;(H^+)$$

is subject to titration in the region pH $\ll$ pK_2, while in the region pH $\gg$ pK_1 only the equilibrium $BH^- \leftrightarrows B^{2-}\;(H^+)$ is affected. In such regions the titration system appears nearly uniform, leading to the expression *uniform subregion*;[42,43,45,50,51] cf. also ref. 53. If the molar absorptivities of the individual components are sufficiently different from one another, it is generally possible to utilize these differences within the confines of uniform subregions to evaluate the overall system as if it consisted of two discrete one-step protolysis equilibria.

Uniform subregions are particularly easily identified with the aid of A-diagrams, as will be seen from the example of *o*-phthalic acid illustrated in Fig. 7–9. If the only titration occurring corresponds to $BH_2 \leftrightarrows BH^-\;(H^+)$ then the arguments presented in Sec. 6.4 assure that all titration points would fall on the line $\overline{BH_2\;BH^-}$. On the other hand, if the only equilibrium involved were $BH^- \leftrightarrows B^{2-}\;(H^+)$ then all points would appear on the line $\overline{BH^-\;B^{2-}}$. In fact, both equilibria exist, so corresponding A-diagrams must be bounded by the two straight lines $\overline{BH_2\;BH^-}$ and $\overline{BH^-\;B^{2-}}$. It follows that linear portions of the A-diagrams must reflect data derived from uniform subregions, in which evaluation by the methods appropriate to single-step acid-base equilibria is permissible.

According to equations 6–12a and 6–12b from Sec. 6.3.3, a single-step protolysis system can be evaluated even if the titration is not carried to completion, and a uniform subregion represents just such an incomplete titration. Therefore all data

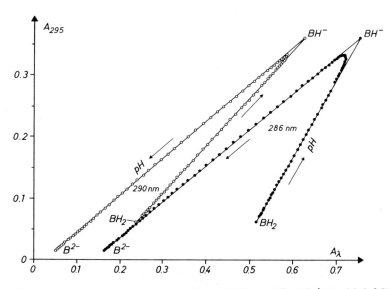

Fig. 7–9. A-diagram for *o*-phthalic acid ($b_0 = 5.8 \cdot 10^{-4}$ M; 20.0 °C). Points BH^- and BH_2 were determined with the aid of equations 7–15 and 7–17.

points from Fig. 7–9 that (within experimental error) fall on the line $\overline{BH_2\,BH^-}$ correspond to the equilibrium $BH_2 \leftrightarrows BH^- \; (H^+)$ and are subject to the following relationship:[42,45,50,51]

$$(A_\lambda - A_{\lambda BH_2})\,10^{-pH} \; := \; \Delta A_\lambda \cdot 10^{-pH} \; = \; -K_1 A_\lambda + K_1 A_{\lambda BH} \tag{7-14}$$

where $A_{\lambda BH} = l\,b_0\,\varepsilon_{\lambda BH}$.

Similarly, points lying on the line $\overline{BH^-\,B^{2-}}$ in Fig. 7–9 can be used to analyze the equilibrium $BH^- \leftrightarrows B^{2-} \; (H^+)$ since

$$(A_\lambda - A_{\lambda B})\,10^{+pH} \; := \; \Delta A'_\lambda \cdot 10^{+pH} \; = \; -\frac{1}{K_2} A_\lambda + \frac{1}{K_2} A_{\lambda BH} \tag{7-15}$$

where $A_{\lambda B} = l\,b_0\,\varepsilon_{\lambda B}$.

Fig. 7–10 is a graph of $\Delta A'_\lambda \cdot 10^{+pH}$ vs. A_λ. Note that the points deviate from linearity at low pH, since here the equilibrium $BH_2 \leftrightarrows BH^- \; (H^+)$ has already begun to make itself felt. These points should be discarded during a least squares analysis. Points that

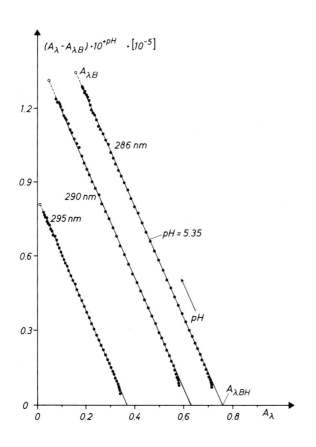

Fig. 7–10. Graph of equation 7–15 for the case of o-phthalic acid ($b_0 = 5.8 \cdot 10^{-4}$ M; 20.0 °C).

7.3 The Determination of Absolute pK Values

lead to small values of $\Delta A'_\lambda$ should also be disregarded. The scatter is certain to be especially large in the event that $\Delta A'_\lambda \approx 0$. In Fig. 7–10 the region corresponding to unreliable data is indicated by a dotted line (o·······o). $A_{\lambda BH}$ values determined on the basis of the two equations 7–14 and 7–15 must be in agreement.

In favorable cases it is also possible to establish a value $A_{\lambda B}$ or $A_{\lambda BH_2}$. For example, if $A_{\lambda BH_2}$ is known, solution of equation 7–14 leads to values for K_1 and $A_{\lambda BH}$. The $A_{\lambda BH}$ obtained in this way can then be used in evaluating the uniform subregion for $BH^- \leftrightarrows B^{2-}$ (H^+) with the aid of the equation

$$(A_\lambda - A_{\lambda BH})\, 10^{-pH} = -K_2 A_\lambda + K_2 A_{\lambda B} \tag{7-16}$$

thereby providing K_2 and $A_{\lambda B}$ as well. On the other hand, if $A_{\lambda B}$ is known, K_2 and $A_{\lambda BH}$ may both be obtained from equation 7–15. Analysis of the uniform subregion for BH^- (H^+) $\leftrightarrows BH_2$ can then be carried out using the relationship

$$(A_\lambda - A_{\lambda BH})\, 10^{+pH} = -\frac{1}{K_1} A_\lambda + \frac{1}{K_1} A_{\lambda BH_2} \tag{7-17}$$

Table 7–2 contains the values of $A_{\lambda BH}$, $A_{\lambda BH_2}$, K_1, and K_2 acquired for o-phthalic acid by this method.

Table 7–2. o-Phthalic acid ($b_0 = 5.8 \cdot 10^{-4}$ M, $l = 1$ cm; 20.0 °C).

a) Determination of $A_{\lambda BH}$ and pK_2 on the basis of equation 7–15.

Wavelength λ (nm)	$A_{\lambda B}$	$A_{\lambda BH}$	pK_2
295	0.012	0.369	5.35
290	0.047	0.628	5.35
286	0.160	0.758	5.35
272	0.447	0.631	5.35

b) Determination of $A_{\lambda BH_2}$ and pK_1 on the basis of equation 7–17.

Wavelength λ (nm)	$A_{\lambda BH}$	$A_{\lambda BH_2}$	pK_1
295	0.369	0.062	2.91
290	0.628	0.238	2.91
286	0.758	0.515	2.91

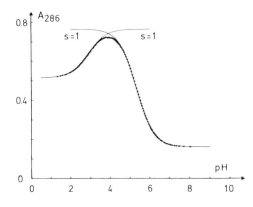

Fig. 7–11. Comparison of simulated single-step titration curves for *o*-phthalic acid (solid line) with the actual two-step curve obtained experimentally (dotted line).

In Sec. 7.3.3 it will be shown that the lines $\overline{BH_2\ BH^-}$ and $\overline{BH^-\ B^{2-}}$ are tangents passing through points corresponding to the species BH_2 and B^{-2}. It therefore follows that points in an A-diagram whose positions correspond within the limits of error to these lines define a uniform subregion.

Approximation on the basis of uniform subregions is distinctly superior to inflection point analysis. The two single-step titration curves constructed in this way for *o*-phthalic acid (based on the values in Table 7–2) are nearly congruent with the experimental titration curve (cf. Fig. 7–11).[42,54] Only in the most strongly overlapping region are deviations apparent. An overall titration curve calculated from the values in Table 7–2 using equation 7–7 shows no deviations whatsoever.

The method described here is valuable, but it remains only an approximation. Satisfactory results are obtained only if uniform subregions are clearly distinguishable within appropriate A-diagrams. Experience has shown that this will normally be the case provided $\Delta pK > 1.5$. Despite its simplicity the method is clearly worthy of consideration, and it often proves sufficient.

7.3.3 Dissociation Diagrams

Preparation of a *dissociation diagram* entails plotting against pH the concentrations bh_2, bh, and b. Often the *relative concentrations* bh_2/b_0, bh/b_0, and b/b_0 are used instead. These will here be abbreviated by the general expression b_i/b_0 (cf. Fig. 7–12). Two points of intersection in such (relative) concentration-pH curves permit the direct determination of pK_1 and pK_2, as may be shown by substitution into the fundamental equation 7–2:[43,55]

$$\text{if } bh_2 = bh, \text{ then } h = K_1 \text{ or } pH = pK_1 \qquad (7\text{–}18a)$$

$$\text{if } bh = b, \text{ then } h = K_2 \text{ or } pH = pK_2 \qquad (7\text{–}18b)$$

7.3 The Determination of Absolute pK Values

The third point of intersection fulfills the relationship

$$\text{if } bh_2 = b, \text{ then } h^2 = K_1 K_2 \text{ or } \text{pH} = \frac{pK_1 + pK_2}{2} \tag{7-18c}$$

In the case of o-phthalic acid the values obtained are as follows:

$$pK_1 = 2.91, \; pK_2 = 5.35, \text{ and } \frac{pK_1 + pK_2}{2} = 4.12$$

In order to construct the necessary diagrams one must know the relevant concentrations. These may often be obtained from molar absorptivities, provided the latter are accurately known with respect to at least two wavelengths. An appropriate starting point for such calculations is equation 7–9. What results is a *system of linear equations* from which, with the aid of equation 7–4, one may readily extract all the required concentrations. Nevertheless, a general examination of multi-component systems[56-69] reveals that satisfactory results may be expected only if the values available for the molar absorptivities are very precise.

The alternative approach discussed below leads to results that are less susceptible to distortion. Fig. 7–13 is an A-diagram illustrating what is known as an *absorbance triangle*, with vertices corresponding to the pure species BH_2, BH^-, and B^{2-}. Coordinates have been selected arbitrarily. The vertices specify the absorbances of the pure species BH_2, BH^-, and B^{2-}, and the inscribed curve corresponds to a typical titration system. A special relationship exists between such a curve and the absorbance triangle, one that can be very useful in carrying out an analysis of the system. The validity of the following assertions will be demonstrated in Sections 7.5 and 7.6:

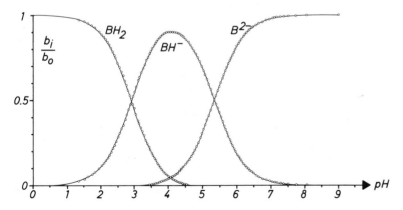

Fig. 7–12. Dissociation curves for o-phthalic acid. Relative concentrations (bh_2/b_0, bh/b_0, and b/b_0—abbreviated collectively as b_i/b_0) were ascertained using distance relationships within the diagram A_{290} vs. A_{272}. The solid lines shown are a result of computation ($pK_1 = 2.91$, $pK_2 = 5.35$).

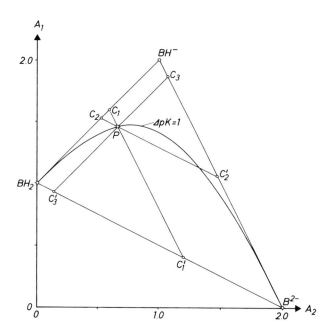

Fig. 7–13. Concentration determinations made with the aid of the *absorbance triangle* in an A-diagram. The concentrations bh_2, bh, and b may all be ascertained on the basis of the distance relationships specified in equation 7–19.

- the plotted curve is a conic section
- point BH⁻ is the pole of the polar $\overline{BH_2\ B^{2-}}$

A *polar* is to be understood in general as a line joining the contact points with a conic section of two tangents to it drawn from a single point, the *pole*. The lines $\overline{BH_2\ BH^-}$ and $\overline{B^{2-}\ BH^-}$ are two such tangents to the curve in Fig. 7–13.[43–45,50,51,54]

The fact that the species in question are linked in a pole–polar relationship opens the possibility of determining the molar absorptivity of the ampholyte BH⁻ solely on the basis of spectrometric data, provided only that the absorbances are known for BH_2 and B^{2-}. No pH data are needed. All that must be done is to construct tangents to the points BH_2 and B^{2-} in an A-diagram. Their point of intersection must be the point BH⁻ (where $A_{\lambda BH} = l\, b_0 \varepsilon_{\lambda BH}$).

Table 7–3 contains absorbances $A_{\lambda BH}$ determined for *o*-phthalic acid using this *tangent method*. The tangents themselves were not actually obtained graphically, but rather by non-linear regression analysis using the program EDIA.[44,45,51]

7.3 The Determination of Absolute pK Values

Table 7–3. o-Phthalic acid: Determination of $A_{\lambda BH}$ strictly on the basis of spectrometric data ($b_0 = 5.8 \cdot 10^{-4}$ M, $l = 1$ cm; 20.0 °C).

Wavelength λ (nm)	$A_{\lambda BH_2}$	$A_{\lambda BH}$	$A_{\lambda B}$
305	0.0843	...	0.060
295	0.0617	0.376	0.012
290	0.2383	0.633	0.047
286	0.5148	0.763	0.160
272	0.6979	0.633	0.447
268	0.6063	0.534	0.432
255	0.7050	0.892	0.700
std. dev. (s)	0.0005	0.001	0.0005

In Sec. 7.4.2 it is demonstrated that the curves constituting an A-diagram are conic sections associated with the function (see p. 99)

$$F = A_1^2 + \alpha A_2^2 + \beta A_1 A_2 + \gamma A_1 + \delta A_2 + \varepsilon = 0$$

If the coefficients α, β, γ, δ, and ε are known as a result of non-linear regression analysis, tangents can be calculated at every point. From the relationship[70,71]

$$\frac{\partial F}{\partial A_1} + \frac{\partial F}{\partial A_2}\left(\frac{dA_2}{dA_1}\right) = (2A_1 + \beta A_2 + \gamma) + (2\alpha A_2 + \beta A_1 + \delta)\left(\frac{dA_2}{dA_1}\right) = 0$$

the following expression is obtained for the slope $\frac{dA_2}{dA_1}$ of the required tangent:

$$\frac{dA_2}{dA_1} = -\frac{\frac{\partial F}{\partial A_1}}{\frac{\partial F}{\partial A_2}} = -\left(\frac{2A_1 + \beta A_2 + \gamma}{2\alpha A_2 + \beta A_1 + \delta}\right)$$

This in turn provides the equation for the tangent through the point (A_{10}, A_{20}):

$$\frac{A_2 - A_{20}}{A_1 - A_{10}} = -\left(\frac{2A_{10} + \beta A_{20} + \gamma}{2\alpha A_{20} + \beta A_{10} + \delta}\right)$$

Accordingly, if A_{BH_2} and A_B are substituted for $A_{\lambda 0}$ (with $\lambda = 1,2$), then the tangents through the points BH$_2$ and B^{2-} can be calculated from the coefficients of the conic section. The advantage of this method is that all data points are considered in establishing the tangents.

In many cases it will be found that a graphic approach to tangent construction suffices (as is true, for example, with o-phthalic acid). Nevertheless, if a graphic solution proves problematic one will usually meet with success by performing a least squares fit of a parabola through several data points and then taking its derivative to obtain the slope of the tangent.

Due to the pioneering work of J. W. Gibbs[72] and H. W. B. Roozeboom,[73] it has become common practice to utilize triangular coordinates in the depiction of ternary systems.[74] The concentrations of individual components can be elucidated from such a "Gibbs triangle" simply by taking ratios of distances. An analogous procedure is also applicable to the *absorbance triangle,* which is simply the result of *an affine transformation of a Gibbs (concentration) triangle* (cf. Sections 7.5.4 and 7.6). For example, the concentrations in Fig. 7–13 can be established from the following distance relationships (additional relationships are described in Sec. 7.5.4):[43,45,50,51,54]

$$bh_2 = \frac{\overline{C_1\ BH^-}}{\overline{BH_2\ BH^-}} b_0 \ , \quad bh = \frac{\overline{C_2\ BH_2}}{\overline{BH_2\ BH^-}} b_0 \ ,$$

(7–19)

$$\text{and} \quad b = \frac{\overline{C_3\ BH^-}}{\overline{BH^-\ BH^{2-}}} b_0$$

The relative concentrations represented in Fig. 7–12 were calculated from the A-diagram A_{290} vs. A_{272} using just such a distance relationship.

In principle, a single pH measurement would suffice for the determination of absolute pK values, provided it were not made near the acidic or basic titration limits. That is, K_1 and K_2 can be calculated from equation 7–2 using the measured pH and concentrations obtained on the basis of equation 7–19, assuming the corresponding pH falls within the region in which the two equilibria overlap. However, limitations on the accuracy of measurement may be such that only one of the two pK values is accessible in this way (e.g., because at the pH in question the system is essentially governed by only one of the equilibria). In this case the spectrometrically determined ratio K_1/K_2 may still permit calculation of the second dissociation constant, although it will be shown in Sec. 7.4.2 that this approach is valid only if the ratio $K_1/K_2 < 1000$.

Other methods may also be used to analyze such titration systems. Those described below are a result of further exploitation of relationships associated with the Gibbs triangle.

7.3.4 Bisection of the Sides of an Absorbance Triangle

Figures 7–13 and 7–14 both illustrate the same absorbance triangle. In Fig. 7–14, however, the data points have been incorporated not into lines parallel to the sides of the triangle (as in Fig. 7–13), but rather into *rays* that intersect the vertices. Each of these rays defines the geometric location of points that conform to a specific and constant concentration relationship. Thus, along the line $\overline{B^{2-}\ P_1}$ the ratio bh_2/bh is everywhere constant. Similarly, the lines $\overline{BH_2\ P_2}$ and $\overline{BH^-\ P_3}$ define constancy of bh/b and bh_2/b, respectively. This conclusion follows directly from the general definition of the Gibbs triangle (cf. also Sec. 7.5.4).

7.3 The Determination of Absolute pK Values

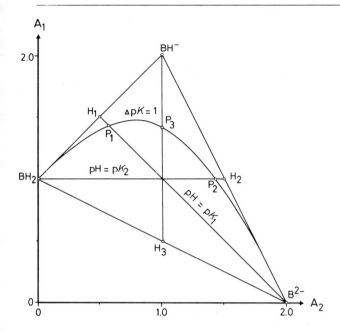

Fig. 7–14. Determination of pK_1 and pK_2 from an A-diagram using the *bisection of sides method*. The line $\overline{B^{2-}H_1}$ intersects the experimental curve at the point P_1, where $pH = pK_1$. Similarly, the line $\overline{BH_2 H_2}$ intersects the curve at P_2, where $pH = pK_2$. At the point P_3, the intersection of the experimental curve with the line $\overline{BH^- H_3}$, the relationship $pH = (pK_1 + pK_2)/2$ is applicable.

In summary, the rays shown in Fig. 7–14, which bisect the sides of the triangle at the midpoints H_1, H_2, and H_3, encompass all the points at which two of the concentrations are equal. However, consideration of equations 7–18a, b, and c shows that these three lines bisecting the sides of the triangle also intersect the experimental curve at the points P_1, P_2, and P_3, the corresponding pH values of which must fulfill the following relationship:

$$pH(P_1) = pK_1, \quad pH(P_2) = pK_2, \quad pH(P_3) = \frac{pK_1 + pK_2}{2} \tag{7-20}$$

These pH values may be read directly from the titration curves A_λ vs. pH on the basis of which the A-diagram was constructed (cf. Fig. 7–3). This *bisection of sides method*[43,45,50,51,54,75] is probably the simplest general approach to the spectrometric determination of pK values; it utilizes only graphic techniques, and it is both accurate and rapid. The following results[43] are obtained for *o*-phthalic acid from the graph of A_{290} vs. A_{272} (Fig. 7–3):

$$pK_1 = 2.92, \quad pK_2 = 5.35, \quad \text{and} \quad \frac{pK_1 + pK_2}{2} = 4.14$$

It is worth noting in this context that the method of "complementary tristimulus colorimetry"[76–78] leads to similar results, in that if ΔpK is large, a chromaticity diagram may be used to obtain:[76]

(a) values of pK_1 and pK_2 (from the midpoints of appropriate line segments), and

(b) a spectroscopic characterization of the ampholyte on the basis of the point at which these two segments intersect.

7.3.5 Pencils of Rays in an Absorbance Triangle

The above-described method of bisection of sides can be further generalized by reference to Fig. 7–15. Here, absorbances for BH_2, BH^-, and $\underline{B^{2-}}$ have been chosen at random, and the triangle has been so oriented that the side $\overline{BH_2\,B^{2-}}$ rests on the A_2 axis. Two curves are shown in the absorbance triangle, each corresponding to the same pK_1. The curves are intersected by two *pencils of rays* emerging from the vertices BH_2 and B^{2-}. By means of these rays the two-step protolysis system may be divided into two single-step equilibria. The process is discussed below, and what results is equivalent to accomplishing separate, isolated titrations of the two equilibrium steps.[75]

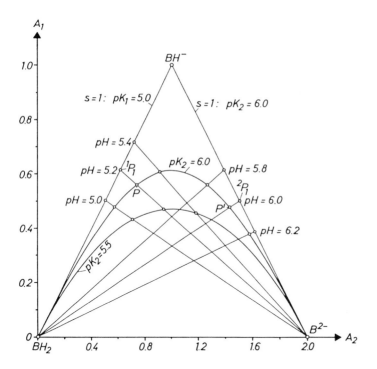

Fig. 7–15. Absorbance curves calculated as a function of pK_2 ($pK_1 = 5.0$ = constant; $pK_2 = 5.5$ and 6.0; the points shown for BH_2, BH^-, and B^{2-} were chosen arbitrarily). For a given value of pH (e.g., pH = 5.2) and a given pK_1, the points for all values of pK_2 within the range $pK_1 \leq pK_2 < \infty$ fall on the line $\overline{P\,B^{2-}} = {}^1P_1\,P$. In the same manner, for a given value of pH (e.g., pH = 6.0) and a given pK_2, the points for all values of pK_1 within the range $-\infty < pK_1 \leq pK_2$ fall on the line $\overline{P'\,BH_2} = {}^2P_1\,P'$. Rays emanating from the points BH_2 and B^{2-} bisect the triangle sides at ${}^1P_1$ and ${}^2P_1$, permitting evaluation of the system by a technique analogous to that used for a one-step acid-base system (the *pencil of rays method*).

7.3 The Determination of Absolute pK Values

For each line within a given pencil of rays, the ratio of concentrations for two of the components remains constant (cf. Sec. 7.3.4). Thus, all points on the lines $\overline{B^{2-}P}$ and $\overline{BH_2 P'}$ represent constant values of the ratios bh/bh_2 and b/bh, respectively. It follows from equation 7–2 that each line also corresponds to a specific pH. For example, the line passing through the point B^{2-} and bisecting the line $\overline{BH_2\,BH^-}$ in Fig. 7–15 is associated with the value pH = 5 (= pK_1). The hypothetical experimental curves shown must also intersect this line at pH 5. The point of intersection with the latter will fall nearer the side $\overline{BH_2\,BH^-}$ the smaller is the value of K_2, and thus the larger is pK_2 (in Fig. 7–15 such points of intersection are shown for pK_2 = 5.5 and 6.0). In the limiting case of K_2 = 0 the point of intersection would fall on the line $\overline{BH_2\,BH^-}$. It is this observation that makes it possible to simplify the two-step titration system by eliminating the effects of the equilibrium $BH^- \leftrightarrows B^{2-}$ (H^+).

The starting point is the pencil of rays emanating from the point B^{2-}, which is used to project the experimental curve onto triangular side $\overline{BH_2\,BH^-}$. All points (e.g., P and P') of a two-step titration system (s = 2) are transformed by such a projection into the corresponding points of the "isolated" (first) equilibrium $BH_2 \leftrightarrows BH^-$ (H^+). Projected points may be designated as in Fig. 7–15 by a symbol like "$^1P_1$", where the superscript defines which equilibrium has been isolated (in this case the first), while the subscript supplies the rank s of the reduced titration system (also 1 in this example).

The second equilibrium, $BH^- \leftrightarrows B^{2-}$ (H^+) , can be separated from the two-step titration in a similar way. All that is required is to project the points (e.g., P and P') onto the side $\overline{BH^-\,B^{2-}}$ using rays emerging from BH_2. This produces a set of points designated $^2P_1$, where "2" stands for the second equilibrium step and "1" specifies its reduced rank (cf. Fig. 7–15).

The resulting sets of points $^1P_1$ and $^2P_1$ thus define two isolated single-step equilibria. The associated pH values correspond to those of the original data points (such as P and P') ascertained for the combined two-step system, and the various point systems are linked by appropriate pencils of rays.

Absorbances and pH values for the points $^1P_1$ and $^2P_1$ may be established using the methods applied to single-step protolysis equilibria. Equations 7–14 through 7–17 of Sec. 7.3.2 are particularly convenient for this purpose. Such an evaluation in the case of o-phthalic acid leads to essentially the same results as those reported in Table 7–2 (Sec. 7.3.2).

The evaluation method discussed in the previous section (Sec. 7.3.4), involving bisection of the sides of a triangle, serves to locate those points on an A-diagram where the relationships $bh_2 = bh$ and $bh = b$ are valid. By contrast, the present method uses rays to establish points on the triangular sides $\overline{BH_2\,BH^-}$ and $\overline{BH^-\,B^{2-}}$ where two isolated single-step equilibria would share the same concentration ratios bh/bh_2 and b/bh as the overall two-step system.

> Analogously to the equilibria $BH_2 \rightleftarrows BH^-$ (H^+) and $BH^- \rightleftarrows B^{2-}$ (H^+), the equilibrium $BH_2 \rightleftarrows B^{2-}$ ($2H^+$) may also be examined in isolation. All the points along the lines of the pencil of rays (omitted from Fig. 7–15) emanating from BH^- and passing through the points P

and P′ of the two-step system correspond to a constant ratio b/bh_2 (= K_1K_2/h^2). The points of intersection of these rays with the triangular side $BH_2 \, B^{2-}$ may be determined in a manner analogous to that involving equations 7–14 and 7–15; i.e.:

$$(A_\lambda - A_{\lambda BH_2}) \, 10^{-2pH} = -K_1K_2A_\lambda + K_1K_2A_{\lambda B}$$

or

$$(A_\lambda - A_{\lambda B}) \, 10^{+2pH} = -\frac{1}{K_1K_2} A_\lambda + \frac{1}{K_1K_2} A_{\lambda BH_2}$$

The validity of the pencil of rays method is completely general. No limitations restrict its applicability with respect to determining values for pK_1 and pK_2.

7.3.6 Linear and Non-linear Regression Methods

Spectrometric analyses of acid-base equilibria have long been the subject of discussion in the literature. The usual approach to ascertaining pK values and molar absorptivities of protolytes involves mathematical processing of data collected at n different points. In some cases iterative techniques are employed, whereas others entail the solution of sets of simultaneous equations;[79–97,38,42,48] cf. also Sec. 6.3.4. The following procedure is appropriate for the investigation of two-step acid-base equilibria. One begins by reformulating equation 7–7:[80,93]

$$(A_\lambda - l b_0 \varepsilon_{\lambda BH}) h = x_1 \cdot l b_0 - x_2 A_\lambda + x_3 \left(l b_0 \varepsilon_{\lambda B} - A_\lambda \right) \frac{1}{h} \quad (7\text{–}21a)$$

where $x_1 = \varepsilon_{\lambda BH} K_1$, $x_2 = K_1$, and $x_3 = K_1 \cdot K_2$

Three pairs of A_λ-pH values suffice for establishing the three unknowns x_1, x_2, and x_3, and therefore K_1, K_2, and $A_{\lambda BH}$, provided one knows the absorbances $A_{\lambda BH_2} = lb_0\varepsilon_{\lambda BH_2}$ and $A_{\lambda B} = lb_0\varepsilon_{\lambda B}$. If more than three A_λ-pH pairs are available the minimum requirements for a solution are exceeded, in which case it is advantageous to subject the entire set of data to *linear regression analysis* based on the method of least squares.

In some cases this approach fails, because $A_{\lambda BH_2}$ and $A_{\lambda B}$ are not known with sufficient accuracy. Equation 7–21a should then be rewritten as

$$A_\lambda = x_1 \cdot l b_0 - x_2 \frac{A_\lambda}{h} + x_3 \cdot l b_0 \cdot \frac{1}{h} - x_4 \frac{A_\lambda}{h^2} + x_5 \cdot l b_0 \cdot \frac{1}{h^2} \quad (7\text{–}21b)$$

where $x_1 = \varepsilon_{\lambda BH_2}$, $x_2 = K_1$, $x_3 = \varepsilon_{\lambda BH} \cdot K_1$, $x_4 = K_1K_2$, and $x_5 = \varepsilon_{\lambda B} \cdot K_1K_2$

Now at least five sets of A_λ-pH data are required to define the five unknowns. Equation 7–21b shows that linear regression makes it possible—in principle—to extract all the constants characterizing a titration system. Nevertheless, this method is rarely employed, since it leads to poor results whenever the matrix of normalized equations is essentially singular (i.e., it is "ill-conditioned"). The effects are particularly serious if all absorbances are calculated on the basis of measurements obtained at a single wavelength.

7.3 The Determination of Absolute pK Values

Today it is more common to apply iterative methods of solution. Thus, a set of molar absorptivities ($\varepsilon_{\lambda BH_2}$, $\varepsilon_{\lambda BH}$, and $\varepsilon_{\lambda B}$) and dissociation constants (K_1 and K_2) is introduced into equation 7–7 and then continuously varied until the calculated titration curve shows optimum agreement with experimental results (*non-linear regression* or "*curve fitting analysis*"). The resulting "best" curve then serves to define the desired set of constants. A major advantage of this method is that molar absorptivities need not be known in advance.

Non-linear regression methods were actually applied to two-step spectrometric titrations at a relatively early date.[80,98] A number of sophisticated computer programs have more recently become available for general application.[38,98–109] In many cases appropriate modifications permit programs designed primarily for use with metal-complex equilibria to be used for studying multi-step acid-base equilibria, since the formal treatments in the two cases are very similar.[38,110–117,107,108]

One well-known program for the solution of non-linear equations is contained in the SPSS 8 statistical package[118,119] available at many computer centers. The subprogram SPSS NONLINEAR operates at the discretion of the user either with the *Gauss-Newton algorithm*[120] or according to the *Marquardt method*.[121,122] First partial derivatives of absorbances with respect to the desired constants may either be supplied in explicit form or they may be determined numerically.

Figure 7–16 shows a set of titration curves for *o*-phthalic acid at several wavelengths. Data points are represented by circles, while the lines were determined on the basis of equation 7–7 with the aid of the program SPSS NONLINEAR. The experimental and calculated curves are virtually congruent. The corresponding results are summarized in 7–4.[46] Each titration curve A_λ vs. pH was separately evaluated.

Table 7–4. Evaluation of individual titration curves for *o*-phthalic acid using the program SPSS-NONLINEAR [($BH_2 \rightleftarrows BH^- (H^+)$, $BH^- \rightleftarrows B^{2-} (H^+)$, where $pK_i = -\log K_i$ ($i = 1,2$), $b_0 = 5.8 \cdot 10^{-4}$ M, $l = 1$ cm; 20.0 °C].

Wavelength λ (nm)	$A_{\lambda BH_2}$ calc.	exp.	$A_{\lambda BH}$ calc.	$A_{\lambda B}$ calc.	exp.	pK_1	pK_2
295	0.0659	0.062	0.3787	0.0130	0.013	2.93	5.33
290	0.2388	0.238	0.6413	0.0489	0.048	2.93	5.33
286	0.5149	0.515	0.7655	0.1610	0.161	2.92	5.33
272	0.6975	0.698	0.6315	0.4474	0.447	2.94	5.34
268	0.6062	0.606	0.5327	0.4328	0.432	2.93	5.35
255	0.7051	0.705	0.8921	0.7009	0.701	2.91	5.33

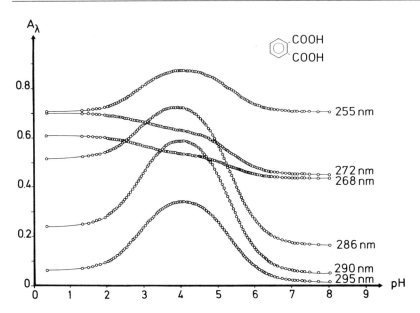

Fig. 7–16. Titration curves A_λ vs. pH for *o*-phthalic acid. Circles indicate actual data points, while the solid lines represent computed curves obtained from non-linear regression analysis.

SPSS NONLINEAR also permits the simultaneous evaluation of all A_λ-pH data. Results of such an analysis are presented in Table 7–5. The standard deviations for all the calculated absorbances are less than ± 0.001 and are thus of the same order of magnitude as the uncertainty of spectrometric measurement. The corresponding calculated values for pK are:

Table 7–5. Combined evaluation of a complete set of titration curves for *o*-phthalic acid using the program SPSS-NONLINEAR and equation 7–7.

Wave-length λ (nm)	$A_{\lambda BH_2}$ calc. (± 0.001)	exp.	$A_{\lambda BH}$ calc. (± 0.001)	$A_{\lambda B}$ calc. (± 0.001)	exp.	pK_1	pK_2
295	0.0612	0.062	0.3762	0.0127	0.013		
290	0.2379	0.238	0.6379	0.0481	0.048		
286	0.5147	0.515	0.7636	0.1611	0.161	2.92 ± 0.01	5.34 ± 0.01
272	0.6981	0.698	0.6320	0.4477	0.447		
268	0.6064	0.606	0.5332	0.4330	0.432		
255	0.7059	0.705	0.8922	0.7010	0.701		

$pK_1 = 2.91_6 \pm 0.005$

$pK_2 = 5.33_6 \pm 0.005$

Comparison of the curves obtained from separate and simultaneous evaluations show little difference; i.e., the results are self-consistent.

Subjection of the same data set to the program TIFIT (described in detail in Chapter 10) leads to the following results:

$pK_1 = 2.91_5 \pm 0.005$

$pK_2 = 5.33_6 \pm 0.005$

Literature values of pK_1 and pK_2 for *o*-phthalic acid vary within the following ranges, depending upon titration conditions:[123,124]

$pK_1 = 2.89 - 3.0$

$pK_2 = 5.28 - 5.51$

The pK values reported above should not be regarded as precise, since no correction was made for the effect of ionic strength. The reason the data appear precise is that the titration was conducted under nitrogen and involved a total of 113 pH measurements.

7.4 Determining the Quotient K_1/K_2 ($\Delta pK = pK_2 - pK_1$)

As Sec. 7.3.3 has shown, it is always possible to derive a complete profile of concentrations, and thus the ratio K_1/K_2, solely from spectrometric data. No knowledge is required of the pH—a poorly defined quantity from the standpoint of thermodynamics—so the precision of the results depends only on the accuracy of spectrometric data.

Nevertheless, the concept "pH" still warrants explicit consideration. Current practice favors the definition pH = $- \log a_{H_3O^+}$. Since the activity $a_{H_3O^+}$ is dependent upon all ions present in solution it cannot be determined precisely regardless how it is measured or how much mathematical skill and effort are applied to the problem. For this reason, R. G. Bates[125,126] introduced a new pH scale convention that has received considerable attention (cf., for example, Standard DIN 19 266) despite the associated measurement problems. The Bates scale is an attempt to closely approximate the hypothetical thermodynamic pH scale by requiring that data be extrapolated to infinite dilution.

Two-step and multi-step titrations commonly present other less fundamental pH problems, however, especially with respect to the need for precise pH measurement over wide ranges of acidity. It is generally recommended that one employ an extensive set of standardized buffers, and that all measured pH values be corrected using statistically determined electrode parameters.[127] The degree of effort required is further increased if one is to take into account systematic errors common to glass electrodes operating in extreme regions of the pH scale (e.g., alkalinity errors, concentration-dependent diffusion potentials, etc.;[4,6] cf. also Chapter 17).

The difficulties associated with pH measurement can be avoided if no need exists to determine absolute pK values for the protolytes in question.[43] For example, one might relate the two pK values for a diprotic acid to the pK of some external standard (cf. Sec. 11.3), in which case the problem becomes one of determining two pK *differences*. An even simpler approach is to be content with establishing the *quotient* of the two dissociation constants, which results in the relative quantity K_1/K_2. Such a quotient is quite sufficient for characterizing many systems, especially those involving non-aqueous solutions, in which the measurement of pH can be particularly problematic. The following discussion is devoted to techniques for the graphic and iterative determination of K_1/K_2.

7.4.1 Distance Relationships Within an Absorbance Triangle

The pair of equations 7–2 which serves to define the systems in question may be combined to produce the desired quotient K_1/K_2:

$$\frac{K_1}{K_2} = \frac{bh}{bh_2} \cdot \frac{bh}{b} \tag{7-22}$$

In Sec. 7.3.4 it was shown that bisection of the sides of an absorbance triangle serves to locate geometrically all those points at which two of the relevant concentrations are equal. In Fig. 7–17, for example, $bh_2 = bh$ along the entire bisecting line $\overline{H_1 B^{2-}}$; similarly, the lines $\overline{H_2 BH_2}$ and $\overline{H_3 BH^-}$ identify regions where $bh = b$ and $bh_2 = b$, respectively. Equation 7–22 establishes an important set of relationships applicable to these lines:[43]

$$\text{for } bh_2 = bh: \quad \frac{K_1}{K_2} = \frac{bh}{b} = \frac{bh_2}{b} \tag{7-23a}$$

$$\text{for } bh = b: \quad \frac{K_1}{K_2} = \frac{bh}{bh_2} = \frac{b}{bh_2} \tag{7-23b}$$

$$\text{for } bh_2 = b: \quad \sqrt{\frac{K_1}{K_2}} = \frac{bh}{bh_2} = \frac{bh}{b} \tag{7-23c}$$

7.4 Determining the Quotient K_1/K_2 ($\Delta pK = pK_2 - pK_1$)

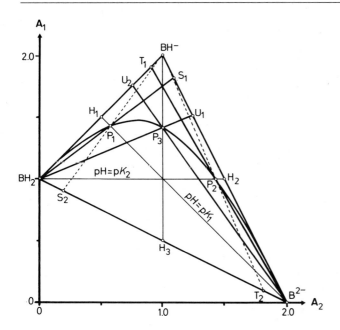

Fig. 7–17. Determination of the pK difference $\Delta pK = pK_2 - pK_1$ using distance relationships within an A-diagram. The side-bisecting lines $\overline{H_1 B^{2-}}$, $\overline{H_2 BH_2}$, and $\overline{H_3 BH^-}$ intersect the curve at the points P_1, P_2, and P_3. With the aid of these points it is possible to construct three pairs of auxilliary lines: $\overline{BH_2 S_1}$ and $\overline{BH^- S_2}$ (using P_1), $\overline{B^{2-} T_1}$ and $\overline{BH^- T_2}$ (using P_2), and $\overline{B^{2-} U_2}$ and $\overline{BH_2 U_1}$ (using P_3). These divide the sides of the triangle into segments from which, with the aid of equations 7–24 a–f, ΔpK may be calculated.

The points in Fig. 7–17 designated P_1, P_2, and P_3 satisfy the three conditions 7–23a, 7–23b, and 7–23c, respectively. Point P_1 is the point of intersection of the bisecting line $\overline{H_1 B^{2-}}$ with the experimental curve. If the concentration ratio bh/b at this point is known, then according to equation 7–23a so is K_1/K_2. Section 7.3.5 shows that this concentration quotient can be obtained with the aid of the auxiliary line $\overline{BH_2 P_1}$, a line that intersects the side $\overline{BH^- B^{2-}}$ at point S_1 (cf. Fig. 7–17). Thus[43] (cf. also Sec. 7.5.4):

$$\frac{K_1}{K_2} = \frac{\overline{S_1 B^{2-}}}{\overline{S_1 BH^-}} = \frac{bh}{b} \tag{7–24a}$$

In an analogous fashion, using the auxiliary line $\overline{BH^- P_1}$ and the point S_2 (cf. Fig. 7–17):

$$\frac{K_1}{K_2} = \frac{\overline{S_2 B^{2-}}}{\overline{S_2 BH_2}} = \frac{bh_2}{b} \tag{7–24b}$$

Similarly, the bisector $\overline{H_2\,BH_2}$ together with P_2 furnishes the points T_1 and T_2 (cf. Fig. 7–17), and thus the relationships

$$\frac{K_1}{K_2} = \frac{\overline{T_1\;BH_2}}{T_1\;BH^-} = \frac{bh}{bh_2} \tag{7-24c}$$

and

$$\frac{K_1}{K_2} = \frac{\overline{T_2\;BH_2}}{T_2\;B^{2-}} = \frac{b}{bh_2} \tag{7-24d}$$

Finally, the third bisector $\overline{H_3\,BH^-}$ and the point P_3 lead to the points U_1 and U_2 (Fig. 7–17), and therefore

$$\sqrt{\frac{K_1}{K_2}} = \frac{\overline{U_1\;B^{2-}}}{U_1\;BH^-} = \frac{bh}{b} \tag{7-24e}$$

and

$$\sqrt{\frac{K_1}{K_2}} = \frac{\overline{U_2\;BH_2}}{U_2\;BH^-} = \frac{bh}{bh_2} \tag{7-24f}$$

A total of six equations for the ratio K_1/K_2 can be derived from the three bisectors of the triangle sides, and none contains the quantity pH. This graphic procedure generally leads to good results provided $\Delta pK < 2$. With larger pK differences the relative error becomes large. The case of o-phthalic acid produces the value[50] $\Delta pK \approx 2.5$.

According to equation 7–22, every point in an A-diagram is also a potential source of the ratio K_1/K_2, since the quotients bh/bh_2 and bh/b may be established using rays emerging from the vertices of the absorbance triangle and passing through the data points. This procedure results in effective utilization of all the information an A-diagram can produce. Thus, the diagram A_{290} vs. A_{268} for o-phthalic acid provides the value[43] $\Delta pK = 2.47 \pm 0.03$ (a mean based on seven data points).

7.4.2 Non-linear Regression Analysis

The ratio of the two dissociation constants

$$\frac{K_1}{K_2} = \kappa \tag{7-3}$$

may also be arrived at by an iterative process,[44] although extensive calculations are required. In contrast to the graphic method of the preceding section, this arithmetic approach is also applicable to precise determination of relatively large pK differences ($\Delta pK < 3$; i.e., $\kappa < 1000$). The procedure described below assumes prior knowledge

7.4 Determining the Quotient K_1/K_2 ($\Delta pK = pK_2 - pK_1$)

of the limiting titration values $A_{\lambda BH_2}$ and $A_{\lambda B}$. As with other methods, its use is restricted to overlapping titration equilibria.

The curve constituting an A-diagram is a conic section (cf. Sections 7.5.1 and 7.6). The form of the absorbance triangle by which it is circumscribed is determined exclusively by the molar absorptivities of the protolytes (BH_2, BH^-, and B^{2-}). By contrast, the curve itself is dependent on κ as well.

The general equation for the conic section comprising the diagram A_1 vs. A_2 is as follows:

$$F = A_1^2 + \alpha A_2^2 + \beta A_1 A_2 + \gamma A_1 + \delta A_2 + \varepsilon = 0 \qquad (7\text{–}25)$$

The five coefficients α, β, γ, δ, and ε unambiguously specify the conic section (the coefficient of A_1^2 is normalized to 1), which is either an ellipse or a hyperbola. (The special case $\kappa = 4$ specifying a parabola is of theoretical interest only. In practice no example has ever been reported that precisely fulfills this relationship.)

The cross term $\beta A_1 A_2$ in equation 7–25 means that the conic section is rotated within the coordinate system. Moreover, the linear terms γA_1 and δA_2 cause it to be linearly displaced parallel to the coordinate axes. The center C is a center of symmetry; thus, replacement (relative to C) of the conic section coordinates ($\overline{A_1}, \overline{A_2}$) by the coordinates ($-\overline{A_1}, -\overline{A_2}$) leaves the equation of the curve unchanged. In order to determine the coordinates of the center C the origin of the "old" coordinate system (A_1, A_2) is displaced parallel to its axes to the center (A_{1C}, A_{2C}). The following relationships then exist between the "new" ($\overline{A_1}, \overline{A_2}$) and "old" coordinate systems:

$$A_1 = \overline{A_1} + A_{1C} \quad \text{and} \quad A_2 = \overline{A_2} + A_{2C} \qquad (7\text{–}26)$$

Introduction of these relationships into equation 7–25 leads to:

$$(\overline{A_1} + A_{1C})^2 + \alpha(\overline{A_2} + A_{2C})^2 + \beta(\overline{A_1} + A_{1C})(\overline{A_2} + A_{2C}) +$$
$$+ \gamma(\overline{A_1} + A_{1C}) + \delta(\overline{A_2} + A_{2C}) + \varepsilon = 0$$

or, after multiplication and rearrangement:

$$\overline{A_1}^2 + \alpha \overline{A_2}^2 + \beta \overline{A_1}\overline{A_2} + (2A_{1C} + \beta A_{2C} + \gamma) \overline{A_1} +$$
$$+ (2\alpha A_{2C} + \beta A_{1C} + \delta) \overline{A_2} + f = 0 \qquad (7\text{–}27)$$

where

$$f = A_{1C}^2 + A_{2C}^2 + \beta A_{1C} A_{2C} + \gamma A_{1C} + \delta A_{1C} + \varepsilon$$

Since the equation is to remain unchanged after replacement of ($\overline{A_1}, \overline{A_2}$) by ($-\overline{A_1}, -\overline{A_2}$) the linear terms must disappear. It therefore follows that

$$2A_{1C} + \beta A_{2C} + \gamma = 0$$

and

$$2\alpha A_{2C} + \beta A_{1C} + \delta = 0$$

This leads to the following coordinates for the center C:[44]

$$A_{1C} = \frac{\beta\delta - 2\alpha\gamma}{4\alpha - \beta^2} \quad \text{and} \quad A_{2C} = \frac{\beta\gamma - 2\delta}{4\alpha - \beta^2} \tag{7-28}$$

Thus, the points A_{1C} and A_{2C} may be calculated provided one knows the coefficients of the conic section. Equation 7–6 also establishes the following condition applicable to C:

$$A_{\lambda C} = l\,(\varepsilon_{\lambda BH_2} \cdot bh_{2C} + \varepsilon_{\lambda BH} \cdot bh_C + \varepsilon_{\lambda B} \cdot b_C).$$

But since $b_0 = bh_{2C} + bh_C + b_C$:

$$A_{\lambda C} = l\,[(\varepsilon_{\lambda BH_2} - \varepsilon_{\lambda BH})\,bh_{2C} + (\varepsilon_{\lambda B} - \varepsilon_{\lambda BH})\,b_C + \varepsilon_{\lambda BH} \cdot b_0] \tag{7-29}$$

We show in Sec. 7.5.1 that there exists the further relationship $bh_{2C} = b_C$. It therefore follows from equation 7–29 (after multiplication by b_0 and rearrangement) that:

$$\frac{b_0}{b_C} = \frac{A_{\lambda BH_2} + A_{\lambda B} - 2A_{\lambda BH}}{A_{\lambda C} - A_{\lambda BH}} \tag{7-30}$$

Here it is to be noted that we have used the identities $A_{\lambda BH_2} = lb_0\varepsilon_{\lambda BH_2}$, $A_{\lambda BH} = lb_0\varepsilon_{\lambda BH}$, and $A_{\lambda B} = lb_0\varepsilon_{\lambda B}$ to further simplify the results. The value of κ may now be calculated from the quotient b_0/b_C given the following relationship, the origin of which is to be found in Sec. 7.5.1:

$$\kappa = \frac{K_1}{K_2} = 4 - \frac{-2b_0}{b_C} \tag{7-45b}$$

In order to solve equation 7–30 for b_0/b_C it is necessary that one know the values for $A_{\lambda BH_2}$, $A_{\lambda BH}$, and $A_{\lambda B}$. $A_{\lambda BH}$ can be determined from the coefficients of the conic section, just as can $A_{\lambda C}$, since the point BH⁻ is the pole of the polar $\overline{BH_2\,B^{2-}}$ (cf. Sec. 7.3.3). The coefficients of the conic section are advantageously determined with the aid of non-linear regression analysis. The process entails searching for that curve which best conforms to the data points, and the corresponding coefficients are taken to be the desired constants.

The goal of the procedure described below is to calculate the value of κ using absorbances measured at all wavelengths, as well as to determine the standard deviation of κ. One begins by examining the absorbances at two wavelengths λ_i ($i = 1, 2$) for n different values of pH. From this, one constructs an A-diagram for the data A_{1m} and A_{2m} (where $m = 1, 2, \ldots n$). If all data were precise, each point (A_{1m}, A_{2m}) would fulfill the conditions of equation 7–25:

$$F_m = A_{1m}^2 + \alpha A_{2m}^2 + \beta A_{1m} A_{2m} + \gamma A_{1m} + \delta A_{2m} + \varepsilon = 0 \tag{7-31}$$

However, since the data are subject to error, each titration point $\begin{pmatrix} A_{1m} \\ A_{2m} \end{pmatrix}$ must be

7.4 Determining the Quotient K_1/K_2 ($\Delta pK = pK_2 - pK_1$)

assigned a correction vector $\begin{pmatrix} a_{1m} \\ a_{2m} \end{pmatrix}$.

The value of each correction vector should be very small, and the following conditions should apply:

(1) every point $\begin{pmatrix} A_{1m} + a_{1m} \\ A_{2m} + a_{2m} \end{pmatrix}$ must fall on the curve one is attempting to "fit" (cf. equation 7–25)

(2) the expression $(a_{1m}^2 + a_{2m}^2)$ is to be minimized for all the vectors implicit in (1); i.e., each vector $\begin{pmatrix} a_{1m} \\ a_{2m} \end{pmatrix}$ is the shortest possible correction vector.

If, in equation 7–31, $(A_{1m} + a_{1m})$ and $(A_{2m} + a_{2m})$ are substituted for A_{1m} and A_{2m}, respectively, and the second order terms involving a_{1m} and a_{2m} are ignored (i.e., $a_{1m}^2 = a_{2m}^2 = a_{1m}a_{2m} = 0$), it follows that:

$$F_m = a_{1m} s_m + a_{2m} t_m = 0 \tag{7-32}$$

where $s_m = 2A_{1m} + \beta A_{2m} + \gamma$ and $t_m = 2\alpha A_{2m} + \beta A_{1m} + \delta$.

One thus obtains the following expression for a_{2m}:

$$a_{2m} = -\left(\frac{F_m + a_{1m} s_m}{t_m}\right) \tag{7-33}$$

This in turn gives, for $(a_{1m}^2 + a_{2m}^2)$:

$$f = a_{1m}^2 + a_{2m}^2 = a_{1m}^2 + \frac{(F_m + a_{1m} s_m)^2}{t_m^2}$$

Condition (2) above specifies that f is to be minimized; therefore:

$$\frac{\partial f}{\partial a_{1m}} = 0 = 2a_{1m} + 2s_m \frac{(F_m + a_{1m} s_m)}{t_m^2}$$

which in turn leads to

$$a_{1m} = -\left(\frac{s_m}{s_m^2 + t_m^2}\right) F_m \tag{7-34}$$

For a_{2m} it follows from equation 7–33 that

$$a_{2m} = -\left(\frac{t_m}{s_m^2 + t_m^2}\right) F_m \tag{7-35}$$

Accordingly, equations 7–34 and 7–35 permit the following expression to be derived for $(a_{1m}^2 + a_{2m}^2)$:[44]

$$\left(a_{1m}^2 + a_{2m}^2\right) = -\frac{F_m^2}{s_m^2 + t_m^2} - \frac{F_m^2}{\left(2A_{1m} + \beta A_{2m} + \gamma\right)^2 + \left(2\alpha A_{2m} + \beta A_{1m} + \delta\right)^2} \tag{7-36}$$

This equation produces a good approximation to the square of the shortest distance from the data point

$$\begin{pmatrix} A_{1m} \\ A_{2m} \end{pmatrix}$$

to the fitted curve only if the correction vector

$$\begin{pmatrix} a_{1m} \\ a_{2m} \end{pmatrix}$$

is small. The function Q has been introduced as a measure of the adequacy of the fit provided by the least squares curve:

$$Q = \sum_{m=1}^{n} \left(a_{1m}^2 + a_{1m}^2\right) \tag{7-37}$$

The coefficients α, β, γ, δ, and ε that produce the smallest value of Q correspond to the best fit. For this reason, minimization of Q requires that all the partial derivatives involving α, β, γ, δ, and ε be zero. In principle, the resulting five equations permit solution in terms of all five of the required coefficients. However, because of equation 7–36, these equations contain fourth degree terms, a circumstance that seriously complicates the task. The difficulties can be circumvented by non-linear regression (curve fitting) analysis. Provided the denominator of equation 7–36 is constant, the coefficients may be determined from normalized linear equations, since F_m is linearly dependent on the coefficients (cf. equation 7–31). The procedure starts from equations 7–36 and 7–37:

$$Q = \sum_{m=1}^{n} w_m F_m^2 \quad \text{with} \quad w_m = \frac{1}{s_m^2 + t_m^2} \tag{7-38}$$

In order to determine the value of Q at constant values of w_m, partial derivatives of equation 7–38 are taken with respect to the individual coefficients, and these are subsequently set equal to zero. For example, considering $\partial Q/\partial \alpha$ on the basis of equations 7–38 and 7–31:

$$\frac{\partial Q}{\partial \alpha} = \sum_{m=1}^{n} w_m F_m \frac{\partial F_m}{\partial \alpha} = 2 \sum_{m=1}^{n} w_m F_m A_{2m}^2 = 0$$

7.4 Determining the Quotient K_1/K_2 ($\Delta pK = pK_2 - pK_1$)

$$\frac{1}{2}\frac{\partial Q}{\partial \alpha} = [wA_1^2 A_2^2] + \alpha[wA_2^4] + \beta[wA_1 A_2^3] + \gamma[wA_1 A_2^2] + \delta[wA_2^3] + \varepsilon[wA_2^2] = 0$$

The brackets in the above expression represent the usual abbreviation for the Gaussian error summation:

$$[wA_1^i A_2^j] = \sum_{m=1}^{n} w_m A_{1m}^i A_{2m}^j \quad \text{with } i, j = 0, 1, \ldots, 4$$

A similar differentiation of Q with respect to the coefficients β, γ, δ, and ε produces the following normal equations (here given in matrix form):

$$\begin{pmatrix} [wA_2^4] & [wA_1 A_2^3] & [wA_1 A_2^2] & [wA_2^3] & [wA_2^2] \\ [wA_1 A_2^3] & [wA_1^2 A_2^2] & [wA_1^2 A_2] & [wA_2 A_2^2] & [wA_1 A_2] \\ [wA_1 A_2^2] & [wA_1^2 A_2] & [wA_1^2] & [wA_1 A_2] & [wA_1] \\ [wA_2^3] & [wA_1 A_2^2] & [wA_1 A_2] & [wA_2^2] & [wA_2] \\ [wA_2^2] & [wA_1 A_2] & [wA_1] & [wA_2] & [w] \end{pmatrix} \cdot \begin{pmatrix} \alpha \\ \beta \\ \gamma \\ \delta \\ \varepsilon \end{pmatrix} = - \begin{pmatrix} [wA_1^2 A_2^2] \\ [wA_1^3 A_2] \\ [wA_1^3] \\ [wA_1 A_2] \\ [wA_1^2] \end{pmatrix}$$

(7–39)

To solve this equation, a first iteration is performed with all quantities w_m set to one. This leads to a first approximation for the coefficients α, β, γ, δ, and ε. Using these it is possible to calculate values for the individual quantities w_m (cf. equations 7–38 and 7–32), and the next iteration can be executed. The process is continued until constant values result for the desired coefficients (usually after ca. 5–10 steps).[44]

In practice it is normal to examine the titration system at several (L) wavelengths, with L typically equal to about 10. This permits the construction of $L(L-1)/2$ different A-diagrams. Each such diagram can be subjected to the above treatment to provide a series of values for κ. In order to distinguish the resulting κ values from one another the subscripts l and k are introduced. Thus, κ_{lk} refers to a constant derived from data acquired at the wavelengths λ_l and λ_k.

In the case of o-phthalic acid it quickly becomes apparent that establishment of a mean value for κ_{lk} presents certain difficulties (cf. Table 7–3 in Sec. 7.3.3). Roughly half the values do indeed cluster about a mean, but several clearly show serious deviations. A few are nearly two orders of magnitude removed from the principal cluster, and some are even negative, a result which is physically meaningless. Apparently it is the case that small errors have a very major effect on certain wavelength combinations (cf. Table 7–3). Thus, the matrix of secular equations may prove quasi-singular, or tangents to the points BH_2 and B^{2-} may be nearly parallel. Alternatively, the denominator in equations 7–28 and 7–30 may approach zero. Such behavior is easily recognized with the aid of the Monte Carlo method.[128]

The program EDIA[44] is specifically designed to deal with situations of this type. The approach is as follows:

(1) Values κ_{lk} are first determined for all possible wavelength combinations (as described above) using the coefficients $(\alpha, \beta, \gamma, \delta, \varepsilon)_{lk}$, and these in turn provide data A_{lm}, A_{km} ($m = 1 \ldots n$). It is preferable that the coefficients be determined by a modification of Newton's method,[44] in which a wavelength combination is discarded if the correction vector cannot be minimized after five iterations, the assumption being that Newton's method is here inapplicable.

(2) An "error-free" data set (A'_{lm}, A'_{km}) is then constructed from the coefficients $(\alpha, \beta, \gamma, \delta, \varepsilon)_{lk}$. The corresponding points all lie directly on the conic section in close proximity to the experimental data points (A_{lm}, A_{kn}).

(3) The standard deviation is calculated based on the distances of the "error-free" points from the actual data points.

(4) The "error-free" points are next displaced from their calculated positions on the basis of a set of normally distributed random numbers, the displacements being of such a magnitude that the distribution obtained has the same standard deviation as that calculated in (3). The resulting simulated experimental data are then used to calculate a new value κ_{lk}.

If this Monte Carlo method is repeated several times, each time using a new set of random numbers, the standard deviation of the final set of simulated data can be interpreted as an estimate of the standard deviation of κ_{lk}. Such results normally become relatively stable after a minimum of ten iterations.

The procedure described makes it possible to distinguish between "good" and "bad" data. That is, one assumes that a given wavelength combination will produce unsatisfactory results if (a) the matrix of secular equations is quasi-singular, (b) Newton's method fails to cause convergence, or (c) the experimental or simulated data lead to a negative value for κ_{lk}.

With o-phthalic acid it is found that 11 of the 21 possible κ_{lk} values show substantially greater standard deviations than do the others. Table 7–3 (Sec. 7.3.3) reveals that at the wavelength $\lambda_l = 305$ nm there is little change in absorbance as a function of pH. The wavelength λ_7 is equally problematic, as is the combination λ_5/λ_6.[44] Apparently the results at these wavelengths are too seriously affected by random fluctuations. Which κ_{lk} values should be accepted for incorporation into the final analysis is a matter left to the judgment of the user of the program.

The final determination of a pK difference (log κ) is carried out on the mean of the "acceptable" κ_{lk} values. The corresponding standard error S (mean error in the mean) and the confidence interval are established in the usual ways.[129]

The above set of computations in the case of o-phthalic acid (7 wavelengths, 113 pH measurements, 30 Monte Carlo simulations) required ca. 1 minute with a high-speed computer. Approximately 90% of that time was consumed by the Monte Carlo simulations. Final results were as follows:

$\Delta pK = pK_2 - pK_1 = \log \kappa = 2.43$
Standard error: $S = 0.01$
Confidence interval (95%) $\Delta pK = \pm 2.2 S$

7.5 The Characteristic Concentration Diagram

The *characteristic concentration diagram* of a system is a graphic presentation of the concentrations of the corresponding linearly independent species. In the case of a protolysis system this amounts to a plot of concentrations as a function of pH, and the preparation of such a diagram is analogous to the preparation of an A-diagram. A complete portrayal of a two-step acid base equilibrium requires three diagrams: bh_2 vs. b, bh_2 vs. bh, and bh vs. b.*

In practice it is A-diagrams themselves that are of greatest relevance, but there are certain important relationships that can be derived most simply from characteristic concentration diagrams.[47] It will be shown in Sec. 7.6 that an A-diagram results from an *affine transformation* of a concentration diagram, and therefore that the relation-ships we are about to establish apply in both types of diagram because they are invariant with respect to affine transformation.

In the section that follows we will examine three different aspects of the diagram bh_2 vs. b:

(1) characteristics related to conic sections (second-order curves),

(2) poles and polars of these conic sections, and

(3) the unique geometric properties of the "concentration triangle".

---

* The need for three characteristic concentration diagrams may be avoided by the use of barycentric coordinates (cf. Sec. 1.4.5).[130] We refrain from that practice here because we are concerned with the relationship between standard concentration diagrams and A-diagrams.

7.5.1 Second-Order Curves (Conic Sections)

In equation 7–3, the following definition was established for κ:

$$\kappa = \frac{K_1}{K_2}$$

It follows from equation 7–2 that

$$\kappa = \frac{bh^2}{bh_2 \cdot b} \tag{7-22}$$

If this relationship is solved for bh and the result is introduced into equation 7–4, rearrangement leads to the following expression:

$$bh_2^2 + b^2 + (2 - \kappa)bh_2 \cdot b - b_0(bh_2 - b) + b_0^2 = 0 \tag{7-40}$$

This is the equation of a second-order curve (a conic section). Depending on the magnitude of κ it may represent any one of the standard types of conic section (see Fig. 7–18):*

$0 < \kappa < 4$ ellipse
$\kappa = 4$ parabola
$\kappa > 4$ hyperbola

In the special case $\kappa = 2$ the concentration diagram consists of a circle. Further transformation into an A-diagram generally results in an ellipse.

As a result of the cross term $(2 - \kappa)\, bh_2 \cdot b$ in equation 7–40, this particular conic section is rotated with respect to the coordinate system. Moreover, its center C is displaced from the origin because of the linear terms $(-2b_0 b)$ and $(-2b_0 bh_2)$. The coordinates of C will be finite in the case of an ellipse or hyperbola, but infinite for a parabola. The actual position of C relative to the origin may be determined using a procedure like that described in Sec. 7.4.2. If, in analogy to equation 7–26, one develops the relationships

$$bh_2 = \overline{bh_2} + bh_{2C} \quad \text{and} \quad b = \overline{b} + b_C \tag{7-41}$$

and introduces these into equation 7–41, the result after rearranging terms is

$$\overline{bh_2}^2 + \overline{b}^2 + (2-\kappa)\,\overline{bh_2}\,\overline{b} + [2bh_{2C} + (2-\kappa)b_C - 2b_0]\,\overline{bh_2} +$$
$$+ [2b_C + (2-\kappa)\,bh_{2C} - 2b_0]\,\overline{b} + f = 0 \tag{7-42}$$

where $f = bh_{2C}^2 + b_C^2 + (2-\kappa)\,bh_{2C}\,b_C - 2b_0(bh_{2C} + b_C) + b_0^2$

---

* Since the absolute concentration is of relatively little interest in a characteristic concentration diagram for a multi-step acid base equilibrium, the coordinate axes of Fig. 7–18 are based on unit concentrations, which may be taken to represent "reduced values".

7.5 The Characteristic Concentration Diagram

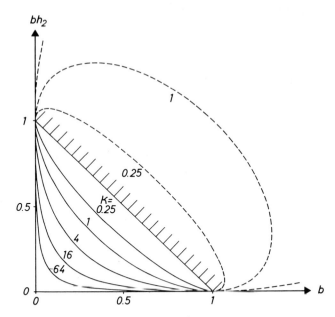

Fig. 7–18. Computed curves for the characteristic concentration diagram bh_2 vs. b. Results are shown as a function of ΔpK and κ, respectively.

Since the linear terms must disappear, the following relationships must also be valid:

$$2bh_{2C} + (2 - \kappa) b_C - 2b_0 = 0 \tag{7-43a}$$

and

$$2b_C + (2 - \kappa) bh_{2C} - 2b_0 = 0 \tag{7-43b}$$

Subtraction of the two equations leads to

$$\kappa(bh_{2C} - b_C) = 0 \quad \text{or} \quad bh_{2C} = b_C \tag{7-44}$$

Together with equation 7–43a this gives the following relationship for the coordinates of C:[44]

$$bh_{2C} = b_C = \frac{2b_0}{4 - \kappa} \tag{7-45a}$$

Assuming C is now known, κ is unambiguously established:

$$\kappa = 4 - \frac{2b_0}{b_C} \tag{7-45b}$$

7.5.2 Poles and Polars of Conic Sections

Fig. 7–19 is a diagram of bh_2 vs. b. The conic section commences with BH_2 (i.e., at low pH) and ends at high pH in the point B^{2-}. These two points have already been equated with the titration limits, where certain concentrations are defined:

very low pH (BH_2): $bh_2 = b_0$, $b = bh = 0$

very high pH (B^{2-}): $bh_2 = bh = 0$, $b = b_0$

The point BH^- ($b = bh_2 = 0, bh = b_0$) coincides with the origin. The *concentration triangle*, with vertices BH_2, BH^-, and B^{2-}, incorporates that region which corresponds to physical reality. Outside the triangle at least one of the concentrations is negative.

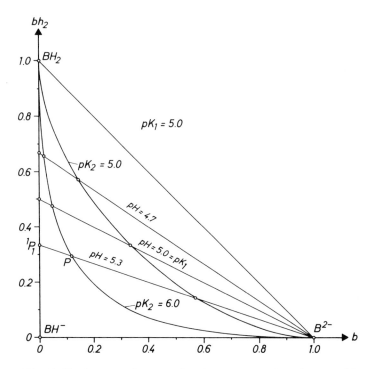

Fig. 7–19. Computed curves for the concentration diagram bh_2 vs. b ($pK_1 = 5.0$ = constant; $pK_2 = 5.0$ and 6.0) Point BH^- is the pole of the polar $\overline{BH_2\,B^{2-}}$. At constant pH (e.g., pH = 5.3) all points corresponding to a pK_2 value in the range $pK_1 \leq pK_2 < \infty$ fall on the straight line $\overline{P\,B^{2-}} = \overline{^1P_1\,P}$. Points $^1P_1$ correspond to those of the isolated first equilibrium $BH_2 \leftrightarrows BH^-$ (H^+) (the *pencil of rays method*).

The point BH⁻ is the pole of the polar $\overline{BH_2\ B^{2-}}$ (for the definitions of pole and polar see Section 7.3.3). This assertion may be readily demonstrated. For the conic section

$$A x^2 + B xy + C y^2 + D x + E y + F = 0$$

the general equation for a polar with the pole (x_1, y_1) is:[131]

$$(2A x_1 + B y_1 + D) x + (2 C y_1 + B x_1 + E) y + D x_1 + E y_1 + 2F = 0 \tag{7-46}$$

Since according to equation 7–40 $A = C = 1$, $B = 2 - \kappa$, $D = E = -2b_0$, and $F = b_0^2$, it is apparent from the diagram bh_2 vs. b (cf. also equation 7–40) that for the polar having the pole BH⁻ (0,0):

$$-2 b_0 b - 2 b_0 bh_2 + 2 b_0^2 = 0 \quad \text{or} \quad bh_2 = -b + b_0 \tag{7-47}$$

This polar intersects the conic section at the points BH_2 and B^{2-}, providing confirmation for the fact that BH⁻ is the pole of the polar $\overline{BH_2\ B^{2-}}$. Another way of expressing the same thing is to say that the tangents to the points BH_2 and B^{2-} intersect at the point BH⁻. This result is independent of the value of κ and is valid for all conic sections.

> The same result is obtained by determining the slopes of the tangents to BH_2 and B^{2-}. This requires taking the first derivatives of bh_2 and b with respect to h in equation 7–5. Thus, for the differential quotient dbh_2/db:
>
> $$\frac{dbh_2}{db} = \frac{\frac{dbh_2}{dh}}{\frac{db}{dh}} = -\frac{2K_2 h + h^2}{K_1 K_2 + 2K_2 h} \tag{7-48}$$
>
> The limiting value for h ($h \to \infty$) is reached at the point BH_2. It thus follows for the slope of the tangent to BH_2 that
>
> $$\left.\frac{dbh_2}{db}\right|_{h \to \infty} \to -\infty \tag{7-49}$$
>
> The point B fulfills the boundary condition $h \to 0$. Therefore the slope here must have the value:
>
> $$\left.\frac{dbh_2}{db}\right|_{h \to 0} \to 0 \tag{7-50}$$
>
> The tangents thus coincide with the coordinate axes.

7.5.3 Pencils of Rays, and Lines Bisecting the Sides of the Triangle

Fig. 7–19 in Sec. 7.5.2 illustrates a pencil of rays connecting various points P on the curve with the titration limit B^{2-}. The equations for these lines can be developed from equation 7–4 by substitution based on the relationship $bh = (K_1 bh_2)/h$ followed by solution for bh_2:

$$bh_2 = \frac{(b_0 - b)h}{K_1 + h} \tag{7-51}$$

If h and K_1 are held constant in equation 7–5 but K_2 is varied, then equation 7–51 states that all possible concentrations bh_2 and b will fall on a single straight line. If, on the other hand, both h and K_2 are treated as variable, then each successive value chosen for h generates a new line. These lines in turn constitute a pencil of rays.

All the points $^1P_1$ within the pencil of rays (see Sec. 7.3.5 for an explanation of the symbol $^1P_1$) lie along the bh_2 axis, in which case $K_2 = 0$ (i.e., $b = 0$). According to equation 7–51, the coordinates of $^1P_1$ (b, bh_2) are thus:

$$^1P_1 \left(0, \frac{b_0 h}{K_1 + h}\right) \tag{7-52}$$

The concentrations associated with $^1P_1$ are equivalent to those for a one-step acid-base system (cf. equation 6–4). Consequently, the points $^1P_1$ indicate the titration points one would obtain if the first equilibrium $BH_2 \leftrightarrows BH^-$ (H^+) could be titrated separately. To determine such points $^1P_1$ one need only construct the lines $\overline{P\,B^{2-}}$ and then extend them until they intersect the bh_2 axis.[45,51,75] The pH values applicable to the points $^1P_1$ are simply those of the actual two-step titration system (i.e., at P). Thus, the isolated equilibrium $BH_2 \leftrightarrows BH^-$ (H^+) is amenable to study by the familiar methods applied to single-step equilibria.

Fig. 7–20 depicts the *side-bisecting line* $\overline{H_1\,B^{2-}}$, a distinctive member of the pencil of lines passing through the point B^{2-} and illustrated in Fig. 7–19. The equation of this side-bisecting line is obtained directly from equation 7–51 by introduction of the relationship $h = K_1$:

$$bh_2 = \frac{1}{2}(b_0 - b) \tag{7-53}$$

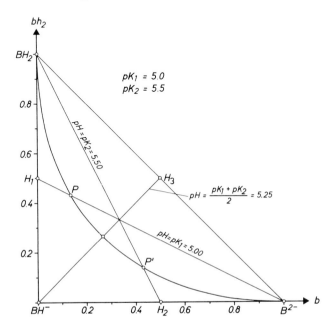

Fig. 7–20. Illustration of the *method of bisecting lines*: the lines $\overline{H_1\,B^{2-}}$, $\overline{BH_2\,H_2}$, and $\overline{BH^-\,H_3}$ bisect the sides of the concentration triangle and intersect the curve at the three points where pH = pK_1, pH = pK_2, and pH = (pK_1 + pK_2)/2, respectively.

7.5 The Characteristic Concentration Diagram

The side-bisecting line $\overline{H_1 \, B^{2-}}$ thus intersects the experimental curve in the concentration diagram at that point (P in Fig. 7–20) at which

$$pH = pK_1 \tag{7-54}$$

We may conclude that the side-bisection method is simply a special case of the general "pencil of rays" method.[75]

Analogous relationships apply with respect to the remaining two sides of the triangle as well. If one replaces the concentrations bh in Equation 7–4 by the expression $bh = (b \cdot h)/K_2$ and then rearranges terms, the result is

$$bh_2 = b_0 - \frac{K_2 + h}{K_2} b \tag{7-55}$$

Provided h and K_2 are constants, this equation describes a straight line. As seen in Fig. 7–20, the line in question passes through the points BH_2 and P'. If K_1 is varied, then all the possible concentrations bh_2 and b lie along this line, which approaches the b axis in the limit of $K_1 \to \infty$. It follows that the titration points $^2P_1$ of the second isolated equilibrium $BH^- \leftrightarrows B^{2-}$ (H^+) may be obtained by a procedure analogous to that described above: a pencil of rays is constructed from the point BH_2 and allowed to intersect the b axis (Fig. 7–20 shows only one such line, $\overline{BH_2 \, H_2}$). This statement may be generalized by noting from equation 7–55 that for any set of coordinates $^2P_1 (b, bh_2)$ with $bh_2 = 0$:

$$^2P_1 \left(\frac{b_0 K_2}{K_2 + h}, \, 0 \right) \tag{7-56}$$

In the case $h = K_2$ (i.e., pH = pK_2) the point $^2P_1$ coincides with the midpoint H_2 of the line $\overline{BH^- B^{2-}}$ in Fig. 7–20 ($b = b_0/2$). The side-bisector $\overline{BH_2 \, H_2}$ thus intersects the concentration curve at the point where the pH is identical to pK_2.

Finally, the concentration triangle may be used to construct a pencil of rays converging in the point BH^- (Fig. 7–20 shows only one of the corresponding lines, $\overline{BH^- \, H_3}$). These rays intersect the triangle side $\overline{BH_2 \, B^{2-}}$ at points that would apply to the equilibrium

$$BH_2 \leftrightarrows B^{2-} \, (2H^+)$$

if it could be studied in isolation. The following relationship is generally valid for these lines through the origin:

$$bh_2 = \frac{h^2}{K_1 K_2} \cdot b \tag{7-57}$$

If h is held constant while K_1 and K_2 are varied in such a way that the product $K_1 \cdot K_2$ remains constant, then according to equation 7–5 all points in a diagram of bh_2 vs. b will fall on a straight line. This line intersects the triangle side $\overline{BH_2 \, B^{2-}}$ at the point where $bh = 0$. The side-bisecting line $\overline{BH^- \, H_3}$ in Fig. 7–20 fulfills the condition $bh_2 = b$. It therefore follows from equation 7–57 that

$$h^2 = K_1 K_2 \tag{7-58}$$

It must also be true that the line $\overline{BH^- \, H_3}$ intersects the concentration curve at a point where this relationship is valid.

7.5.4 Distance Relationships

J. W. Gibbs and H. W. B. Roozeboom have shown that it is often useful to describe three-component chemical systems, such as two-step acid-base equilibria, with the aid of an equilateral triangle.[72-74] Figure 7–21 shows an example of such a "Gibbs triangle" with the vertices BH_2, BH^-, and B^{2-}. Each vertex corresponds to the maximum concentration (b_0) of one of the three pure components. The example shown differs from the usual Gibbs triangle in that only two of its sides are of equal length (i.e., it is an isosceles triangle).

In order to determine the actual concentrations corresponding to a given point within the triangle one constructs three straight lines through the point in question and parallel to the sides of the triangle. The following relationships apply with reference to the points where these lines intersect the sides of the triangle (i.e., the length of each of the sides of the triangle of Fig. 7–21 is regarded as proportional to the starting concentration b_0):

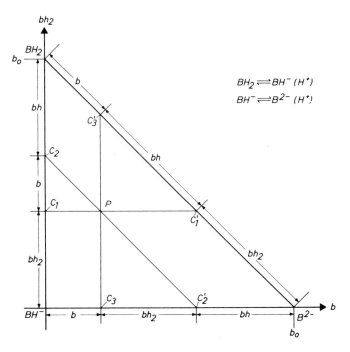

Fig. 7–21. Example of a concentration diagram in the form of a Gibbs triangle with vertices BH_2, BH^-, and B^{2-} (only one data point P has been entered in the diagram).

7.5 The Characteristic Concentration Diagram

$$bh_2 = \frac{\overline{C_1 \; BH^-}}{\overline{BH_2 \; BH^-}} \cdot b_0 = \frac{\overline{C_1' \; B^{2-}}}{\overline{BH_2 \; B^{2-}}} \cdot b_0 = \frac{\overline{C_3 \; C_2'}}{\overline{BH^- \; B^{2-}}} \cdot b_0$$

$$bh = \frac{\overline{C_2 \; BH_2}}{\overline{BH_2 \; BH^-}} \cdot b_0 = \frac{\overline{C_3' \; C_1'}}{\overline{BH_2 \; B^{2-}}} \cdot b_0 = \frac{\overline{C_2' \; B^{2-}}}{\overline{BH^- \; B^{2-}}} \cdot b_0 \qquad (7\text{--}59)$$

$$b = \frac{\overline{C_1 \; C_2}}{\overline{BH_2 \; BH^-}} \cdot b_0 = \frac{\overline{C_3' \; BH_2}}{\overline{BH_2 \; BH^-}} \cdot b_0 = \frac{\overline{C_3 \; BH^-}}{\overline{BH^- \; B^{2-}}} \cdot b_0$$

Thus, concentrations may be deduced simply by measuring an appropriate set of distances, provided b_0 is known.

As was demonstrated in Sec. 7.4.1, the *ratio* K_1/K_2 may also be established from distance ratios. The Gibbs triangle makes this particularly easy to accomplish graphically. Thus, from equation 7–23a:

$$\text{for } bh_2 = bh, \quad \frac{K_1}{K_2} = \frac{bh}{b} = \frac{bh_2}{b}$$

In order to determine K_1/K_2, it is necessary that the line $\overline{BH_2 \; P}$ in Fig. 7–22 be examined. This line establishes the geometric locations of all points with a constant ratio bh/b. This conclusion follows directly from equation 7–55 in the preceding section, which shows that all concentrations consistent with a given pair of h and K_2 values (but different K_1 values) fall on a single line; equation 7–2 in turn assures that the ratio bh/b is also constant along this line.

Similarly, all points along the line $\overline{B^{2-} \; P}$ in Fig. 7–22 share a constant value for the ratio bh_2/bh. One now begins by constructing the side-bisecting line $\overline{H_1 \; B^{2-}}$, which intersects the concentration curve of interest at the point P, where $bh_2/bh = 1$ or $bh_2 = bh$. Once the location of the point P has been determined in this way it immediately becomes possible to establish the ratio K_1/K_2: all that remains is to construct the line $\overline{BH_2 \; P}$, extending it until it intersects the triangle side $\overline{BH^- \; B^{2-}}$ at the point S_1, to which the following relationship applies (cf. Fig. 7–22):

$$\frac{bh}{b} = \frac{\overline{S_1 \; B^{2-}}}{\overline{S_1 \; BH^-}} = \frac{K_1}{K_2}$$

This expression is identical to Equation 7–24a in Sec. 7.4.1. The validity of equations 7–24b through 7–24f may be demonstrated in an analogous way.

As a further example based on Fig. 7–22 consider the line $\overline{BH_2 \; P'} = \overline{BH_2 \; H_2}$, which fulfills the criterion $bh/b = 1$. Consistent with equation 7–23b the following relationship must be valid:

$$\frac{K_1}{K_2} = \frac{bh}{bh_2}$$

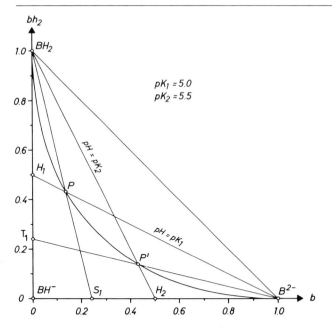

Fig. 7–22. Determination of K_1/K_2 using distance relationships within a concentration triangle. The points T_1 and S_1 are established with the aid of the bisecting lines $\overline{BH_2\,H_2}$ and $\overline{H_1\,B^{2-}}$. This permits one to calculate K_1/K_2 from such relationships as:

$$\frac{K_1}{K_2} = \frac{\overline{S_1\,B^{2-}}}{\overline{S_1\,BH^-}} = \frac{\overline{T_1\,BH_2}}{\overline{T_1\,BH^-}}$$

The required concentration relationship bh/bh_2 may be determined with the aid of the line $\overline{B^{2-}\,P'}$, which intersects the line $\overline{BH_2\,BH^-}$ at T_1, leading to:

$$\frac{bh}{bh_2} = \frac{\overline{T_1\,BH_2}}{\overline{T_1\,BH^-}} = \frac{K_1}{K_2}$$

This expression is identical to equation 7–24c.

7.6 Geometric Relationships Between Concentration Diagrams and Absorbance Diagrams

In order to construct an absorbance difference (AD) diagram, one plots for a series of pH values the differences in the measured absorbance at two wavelengths λ_i ($i = 1, 2$). These absorbance differences may be described (as in equation 7–9) by expressions such as:

$$\Delta A_1 = q_{11} \cdot bh + q_{12} \cdot b, \quad \Delta A_2 = q_{21} \cdot bh + q_{22} \cdot b$$

Matrix notation provides a more compact representation of the above two statements:

$$\begin{pmatrix} \Delta A_1 \\ \Delta A_2 \end{pmatrix} = \begin{pmatrix} q_{11} & q_{12} \\ q_{21} & q_{22} \end{pmatrix} \begin{pmatrix} bh \\ b \end{pmatrix} \tag{7–60}$$

A *vector representation* of a concentration diagram (e.g., *bh* vs. *b*) makes use of what are called *position vectors* to indicate the location of a particular data point with respect to the origin of the coordinate system. Such a vector is explicitly represented in the following way:

$$\vec{c} = bh\,\vec{e}_1 + b\,\vec{e}_2 \tag{7–61a}$$

Here $\vec{e}_1$ and $\vec{e}_2$ are the corresponding *basis vectors*, and *bh* and *b* are the coordinates of the vector $\vec{c}$. Since basis vectors generally have lengths and directions that are unambiguous these need not be further specified. In this book, all vectors will be represented by boldface lowercase letters, and coordinates will be supplied in the usual form of a one-column matrix. Following these conventions, equation 7–61a would be written:

$$\mathbf{c} = \begin{pmatrix} bh \\ b \end{pmatrix} \tag{7–61b}$$

Multi-column matrices will be indicated by boldface uppercase letters; that is, equation 7–60 may be abbreviated:

$$\mathbf{a}' = \mathbf{Q}\,\mathbf{c} \tag{7–62}$$

where $\mathbf{a}' = \begin{pmatrix} \Delta A_1 \\ \Delta A_2 \end{pmatrix}$ and $\mathbf{Q} = \begin{pmatrix} q_{11} & q_{12} \\ q_{21} & q_{22} \end{pmatrix}$

Using the relationships $\Delta A_\lambda = A_\lambda - A_{\lambda BH_2}$ and $A_{\lambda BH_2} = l \cdot b_0 \cdot \varepsilon_{\lambda BH_2}$ (cf. equations 7–8a and 7–8b), it follows that:

$$\begin{pmatrix} A_1 \\ A_2 \end{pmatrix} = \begin{pmatrix} q_{11} & q_{12} \\ q_{21} & q_{22} \end{pmatrix} \begin{pmatrix} bh \\ b \end{pmatrix} + \begin{pmatrix} A_{1BH_2} \\ A_{2BH_2} \end{pmatrix}$$

or

$$\mathbf{a} = \mathbf{Q}\mathbf{c} + \mathbf{a}_0 \tag{7–63}$$

where $\mathbf{a} = \begin{pmatrix} A_1 \\ A_1 \end{pmatrix}$ and $\mathbf{a_0} = \begin{pmatrix} A_{1BH_2} \\ A_{2BH_2} \end{pmatrix}$

Note that the following relationship is obtained by combining equations 7–62 and 7–63:

$$\mathbf{a} = \mathbf{a'} + \mathbf{a_0} \tag{7-64}$$

From this it follows that an A diagram is simply an AD diagram that has been displaced according to the (constant) vector $\mathbf{a_0}$.[51] In other words, absorbance difference diagrams and absorbance diagrams may be interconverted through simple parallel displacements, which means that relationships existing within the one are also present in the other.

Equation 7–63 (or 7–62) permits the transformation of vectors $\mathbf{c}$ within concentration space into vectors $\mathbf{a}$ (absorbance space) or $\mathbf{a'}$ (absorbance difference space). Since the relationships linking these vectors are linear, the transformations themselves may be classified as *affine* or *linear conversions*.[132,133]

Affine conversions play an important role in the discussion that follows. They are characterized by a number of what might be termed "conservation laws", the utilization of which will prove advantageous. For example, as shown in Fig. 7–23, parallel projection (followed by a similarity transformation[71]) can be used to convert a concentration diagram into an absorbance diagram.[51] Here the "Gibbs concentration triangle" occupies a plane that might be described as the "concentration plane". An absorbance diagram results when all points in the concentration plane are projected by means of parallel rays into a second plane, the absorbance plane. In the process, the original vectors are displaced, rotated, and perhaps stretched or compressed.

The following geometric properties are conserved in the course of any affine transformation:[132,133]

(1) straight lines are transformed into straight lines; thus

   (a) tangents remain tangents, and
   (b) asymptotes remain asymptotes,

(2) parallel lines are transformed into parallel lines,

(3) distance relationships present within straight lines are preserved,

(4) points of intersection are conserved,

(5) planes are transformed into planes, and

(6) conic sections remain conic sections: circles and ellipses become ellipses, hyperbolas remain hyperbolas, and parabolas remain parabolas.

7.7 Concentration Difference Quotient (CDQ) Diagrams

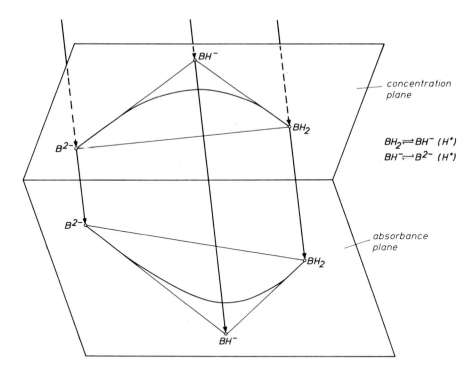

Fig. 7–23. Transformation of a (two-dimensional) concentration diagram into an absorbance triangle (schematic illustration of an affine transformation).

These laws of conservation provide the necessary assurance that all the relationships discussed in the context of concentration diagrams remain valid for A-diagrams, rendering further discussion of the latter superfluous.

7.7 Concentration Difference Quotient (CDQ) Diagrams

The mass balance requirement for a two-step protolysis system is as follows:

$$b_0 = bh_2 + bh + b \tag{7-4}$$

This is also the equation of a plane, that plane in which all titration points are to be found. A corresponding three-dimensional representation would be based upon the three concentrations bh_2, bh, and b. If one chooses any particular titration point P' (bh_2', bh', b'), then parallel displacement of the original coordinate axes into this point P' generates a new coordinate system, through whose origin the titration curve passes. If these coordinate are now given the designations Δbh_2, Δbh, and Δb then the following relationships link the new and old coordinate systems (in analogy to equations 7–26 and 7–41):

$$\Delta bh_2 = bh_2 - bh'_2$$
$$\Delta bh = bh - bh' \qquad (7\text{–}65)$$
$$\Delta b = b - b'$$

Since point P′ also satisfies equation 7–4, the preceding three equations may be added to give:

$$\Delta bh_2 + \Delta bh + \Delta b = 0 \qquad (7\text{–}66)$$

This in turn leads to

$$\frac{\Delta bh_2}{\Delta b} = -\frac{\Delta bh}{\Delta b} - 1 \qquad (7\text{–}67)$$

In what is known as a *concentration difference quotient (CDQ) diagram*,[47] the quotients $\Delta bh_2/\Delta b$ and $\Delta bh/\Delta b$ are plotted against one another to produce a curve. Equation 7–67 reveals that the resulting points will lie on a straight line provided the concentrations of the three components are linearly related, a condition specified by equation 7–4. It is thus possible to use a two-dimensional representation to verify the existence of the required linear relationship. Such CDQ-diagrams may also be transformed into ADQ-diagrams, a fact which is of some practical significance (cf. the following section, Sec. 7.8).

Any titration points may be utilized in forming the differences required for a CDQ-diagram. Nonetheless, it is usually preferable to choose the particular points BH_2 (where $bh_2 = b_0$) and B^{2-} (where $b = b_0$). In the case of the point BH_2 (i.e., $bh'_2 = b_0$ and $bh' = b' = 0$) equation 7–67 is transformed into the relationship

$$\frac{bh_2 - b_0}{b} = -\frac{bh}{b} - 1 \qquad (7\text{–}68)$$

Analogously, for B^{2-} (where $bh'_2 = bh' = 0$ and $b' = b_0$):

$$\frac{bh_2}{b - b_0} = -\frac{bh}{b - b_0} - 1 \qquad (7\text{–}69)$$

The quotients of such concentration differences provide the slopes of the lines joining the point of reference in the appropriate concentration diagram with the individual titration points.[51] For example, the quotient $\Delta bh_2/\Delta b$ in equation 7–67 provides the slope of the line $\overline{P'P}$ in Fig. 7–24a (point P has been chosen at random). Analogously, the lines shown in Fig. 7–24b have the slopes $\Delta bh/\Delta b$. In the cases described by equations 7–68 and 7–69, these lines pass through the points BH_2 and B^{2-}, respectively.

7.7 Concentration Difference Quotient (CDQ) Diagrams

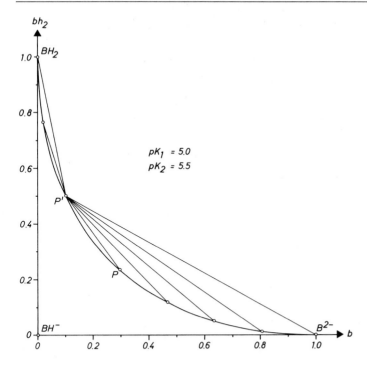

Fig. 7–24a. Characteristic concentration diagram bh_2 vs. b. Lines passing through the point P′ have the slope $\Delta bh_2/\Delta b$.

The denominator in a concentration-difference quotient may occasionally assume the value zero, in which case the expression becomes indeterminate. Such a situation may often be countered by simple geometric determination of the required slope from the appropriate concentration diagram. For example, at the point BH_2 in Fig. 7–24b the concentration b is zero. If one permits the point P in this diagram to migrate along the curve in the direction of the point BH_2 (= P′) it will be seen that the slope of the line $P\,BH_2$ increasingly approaches that of the bh axis. At the limit of $b \to 0$ the slope becomes infinite and P is identical to BH_2. The quotient bh/b in equation 7–68 then assumes its limiting value:

$$\lim_{b \to 0} \frac{bh}{b} \to +\infty$$

If this value is introduced into equation 7–68 it follows for the quotient $(bh_2 - b_0)/b$ that:

$$\lim_{b \to 0} \frac{bh_2 - b_0}{b} \to -\infty$$

Similarly, in the limiting case of $b \to b_0$ one obtains for the quotients in equation 7–68 (using Fig. 7–24b) the following values:

$$\lim_{b \to b_0} \frac{bh}{b} \to 0 \quad \text{and} \quad \lim_{b \to b_0} \frac{bh_2 - b_0}{b} \to -1$$

The location of the desired points in the CDQ-diagram is thus established. If the point corresponding to $b \to 0$ in a CDQ-diagram is designated as BH_2—a form previously reserved for the limiting value of the concentration—and that for $b \to b_0$ as B^{2-}, then all titration points will fall on the line joining these two points [cf. illustration (a) in Fig. 7–25].

Besides the above geometric method for determining otherwise inaccessible concentration difference quotients, there also exist other possible approaches, including direct examination of limiting values or use of the Bernoulli-De L'Hôpital rule.[70,71]

In the case that $P' = B^{2-}$ the quotients of equation 7–69 have the values:

$$\lim_{b \to 0} \frac{bh}{b - b_0} = 0 \quad \text{and} \quad \lim_{b \to b_0} \frac{bh}{b - b_0} = -1$$

Introduction of these results into equation 7–69 leads to the following:

$$\lim_{b \to 0} \frac{bh_2}{b - b_0} = -1 \quad \text{and} \quad \lim_{b \to b_0} \frac{bh_2}{b - b_0} = 0$$

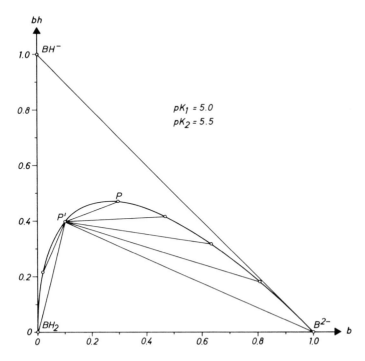

Fig. 7–24b. Characteristic concentration diagram bh vs. b. Lines passing through the point P' have the slope $\Delta bh/\Delta b$.

7.8 Geometric Relationships Between ADQ-, CDQ-, and A-Diagrams

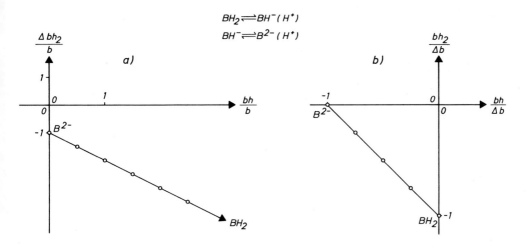

Fig. 7–25. Concentration difference quotient (CDQ) diagrams (schematic representations). The concentration differences in (a) are referenced against the point BH_2 (cf. equation 7–68); those in (b) relate to the point B^{2-} (cf. equation 7–69)

If the points associated with these coordinates are designated in the usual way as BH_2 and B^{2-}, what results is the CDQ-diagram labeled (b) in Fig. 7–25.[51]

As is apparent from Fig. 7–25a, it is possible to choose a reference point such that points in a CDQ-diagram will have the value infinity. The same is true with ADQ-diagrams, since data points here may also fall on line segments that are either finite or infinite in length (see the following section).

7.8 Geometric Relationships Between ADQ-, CDQ-, and A-Diagrams

According to equation 7–6 the following general statement applies to A_λ:

$$A_\lambda = l(\varepsilon_{\lambda BH_2} \cdot bh_2 + \varepsilon_{\lambda BH} \cdot bh + \varepsilon_{\lambda B} \cdot b)$$

Therefore, if one selects at random some data point P' as a reference for the subsequent construction of differences, the following assertion must be valid:

$$A_\lambda' = l(\varepsilon_{\lambda BH_2} \cdot bh_2' + \varepsilon_{\lambda BH} \cdot bh' + \varepsilon_{\lambda B} \cdot b')$$

Upon subtracting the second equation from the first one obtains:

$$\Delta A_\lambda = A_\lambda - A_\lambda' = l(\varepsilon_{\lambda BH_2} \cdot \Delta bh_2 + \varepsilon_{\lambda BH} \cdot \Delta bh + \varepsilon_{\lambda B} \cdot \Delta b) \tag{7-70}$$

where

$$\Delta bh_2 = bh_2 - bh_2', \quad \Delta bh = bh - bh', \quad \text{and} \quad \Delta b = b - b'$$

Consideration of equation 7–66 shows further that

$$\Delta A_\lambda = q_{\lambda 1} \cdot \Delta bh + q_{\lambda 2} \cdot \Delta b \tag{7-71}$$

in which (cf. equation 7–9)

$$q_{\lambda 1} = l(\varepsilon_{\lambda BH} - \varepsilon_{\lambda BH_2}) \quad \text{and} \quad q_{\lambda 2} = l(\varepsilon_{\lambda B} - \varepsilon_{\lambda BH_2})$$

Setting up three such equations for the three wavelengths λ_i ($i = 1, 2, 3$) and eliminating the quantities Δbh and Δb leads in a formal way to the result already obtained as equation 7–10:

$$\Delta A_\lambda = a_1 \Delta A_2 + a_2 \Delta A_3 \tag{7-72}$$

where the quantities a_1 and a_2 retain their previously defined identities. Division by ΔA_2 then leads to the following result:

$$\frac{\Delta A_1}{\Delta A_2} = \alpha_1 + \alpha_2 \frac{\Delta A_3}{\Delta A_2} \tag{7-73}$$

Equation 7–73 is the general equation for an *absorbance difference quotient* (*ADQ*) *diagram*. In this case the absorbance differences are based on a titration point P' chosen at random. By contrast, equation 7–11 in Sec. 7.2 was derived exclusively with reference to the point BH_2 ($bh_2 = b_0$). Equation 7–11 differs from equation 7–73 only in that here the individual points making up the corresponding ADQ-diagram are a function of which reference values are chosen. Nevertheless, they all lie on the same straight line.

An ADQ-diagram may be generated by carrying out a *projective transformation* with the aid of a CDQ-diagram (followed by a similarity transformation).[51] This assertion follows from the corresponding transformation equations. The first of these is obtained by dividing equation 7–70 by Δb, thereby forming the quotient $\Delta A_1/\Delta A_2$:

$$\frac{\Delta A_1}{\Delta A_2} = \frac{\varepsilon_{1BH_2} \dfrac{\Delta bh_2}{\Delta b} + \varepsilon_{1BH} \dfrac{\Delta bh}{\Delta b} + \varepsilon_{1B}}{\varepsilon_{2BH_2} \dfrac{\Delta bh_2}{\Delta b} + \varepsilon_{2BH} \dfrac{\Delta bh}{\Delta b} + \varepsilon_{2B}} \tag{7-74}$$

A similar expression may be obtained for the quotient $\Delta A_3/\Delta A_2$. As the equations show, ADQ-coordinates are dependent upon complex functions of $\Delta bh_2/\Delta b$ and $\Delta bh/\Delta b$, although the appropriate denominators are identical for both $\Delta A_1/\Delta A_2$ and $\Delta A_3/\Delta A_2$. These relationships then constitute the transformation equations for the required projective transformation.[132,133]

In Fig. 7–26 the line $\overline{BH_2 B^{2-}}$ of the CDQ-diagram lies in the "CDQ-plane". The point BH_2 itself is assumed to be located at infinity (cf. Fig. 7–25a). A pencil of rays is

7.8 Geometric Relationships Between ADQ-, CDQ-, and A-Diagrams

now constructed from some (central) point Z such that all points on this CDQ-line are included. The points where these rays penetrate the ADQ-plane once again constitute a straight line ($\overline{BH_2 B^{2-}}$). Such a transformation based on a central projection is always characterized by certain important features, including the following:

(1) Straight lines are transformed into straight lines; and

(2) angle and distance relationships undergo modification, but quotients of line segments are unaffected.

As can be seen in Fig. 7–26, the point BH_2, though at infinity on the CDQ-line, is transformed into the corresponding "vanishing point" V along the ADQ-line, whereby $\overline{ZV}$ is parallel to the CDQ-line.* Thus, the vanishing point itself (V = BH_2) need not lie at infinity.

The reverse situation may also obtain. A point that is finite along the CDQ-line may be infinitely removed when projected onto the ADQ-line. This occurs, for example, if the line passing through Z and the corresponding CDQ-point happens to be parallel to the ADQ-plane. The question of whether or not all points in an ADQ-diagram will lie on a line of finite length depends on the position of Z.

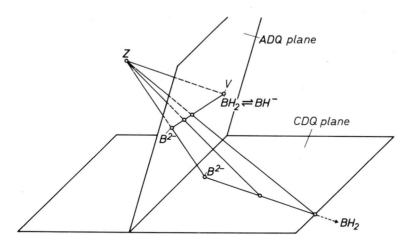

Fig. 7–26. Construction of a (two-dimensional) ADQ-diagram by projective transformation of a CDQ-diagram (schematic illustration based on the CDQ-curve in Fig.7–25a).

---

* All titration points would converge at the point BH_2 on the CDQ- or ADQ-lines if only the equilibrium $BH_2 \leftrightarrows BH^-$ (H^+) were titrated and if differences were calculated using BH_2 as reference.

In practice, ADQ-diagrams are of greatest value when they are based on the points BH_2 and B^{2-}. Specific geometric relationships exist between such ADQ-diagrams and the corresponding three-dimensional A-diagrams, and with the help of these relationships it is possible to conduct a complete analysis of a three-step protolysis system using only spectrometric data (cf. Sec. 8.3.6 and ref. 134 and 135).

Just as is the case with CDQ- and concentration diagrams, certain important geometric relationships exist between ADQ- and A-diagrams.[51] Fig. 7–27 will help to illustrate this point. The figure is comprised of a three-dimensional A-diagram constructed on the basis of the same wavelengths used for the corresponding ADQ-diagram. The appropriate equation for a two-step titration system is analogous to equation 7–63:

$$\begin{pmatrix} A_1 \\ A_2 \\ A_3 \end{pmatrix} = \begin{pmatrix} q_{11} & q_{12} \\ q_{21} & q_{22} \\ q_{31} & q_{32} \end{pmatrix} \begin{pmatrix} bh \\ b \end{pmatrix} + \begin{pmatrix} A_{1BH_2} \\ A_{2BH_2} \\ A_{3BH_2} \end{pmatrix} \qquad (7\text{–}75)$$

or (cf. equation 7–64):

$$\mathbf{a} = \mathbf{Q}\mathbf{c} + \mathbf{a}_0 = \mathbf{a}' + \mathbf{a}_0 \qquad (7\text{–}63)$$

The position vector **a** extends from the origin of the coordinate system toward the point P on the curve (cf. Fig. 7–27). By contrast, the vector **a'** is directed from the point BH_2 to the point P. This is referred to as the *direction vector*. According to equation 7–75:

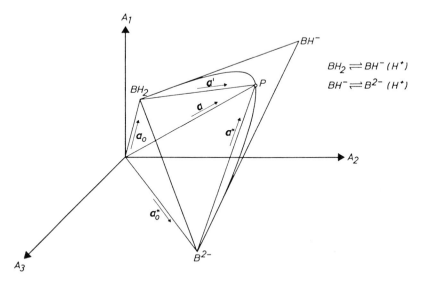

Fig. 7–27. Three-dimensional A-diagram corresponding to a two-step acid-base quilibrium system (schematic). All data points P are in a single plane. Definitions of the vectors: $\mathbf{a}_0 = \overrightarrow{O\,BH_2}$; $\mathbf{a}' = \overrightarrow{BH_2\,P}$; $\mathbf{a} = \overrightarrow{O\,P}$; $\mathbf{a}_0^* = \overrightarrow{O\,B^{2-}}$; $\mathbf{a}^* = \overrightarrow{B^{2-}\,P}$ (O represents the coordinate origin).

7.8 Geometric Relationships Between ADQ-, CDQ-, and A-Diagrams

$$\mathbf{a}' = \mathbf{a} - \mathbf{a_0} = \begin{pmatrix} A_1 - A_{1BH_2} \\ A_2 - A_{2BH_2} \\ A_3 - A_{3BH_2} \end{pmatrix} = \begin{pmatrix} \Delta A_1 \\ \Delta A_2 \\ \Delta A_3 \end{pmatrix} = \mathbf{Qc} \quad (7\text{–}76)$$

ΔA_λ is here referenced against the point BH_2, which is to say that in equation 7–70 $A_\lambda' = A_{\lambda BH_2}$.

Equation 7–72 describes the plane containing the direction vectors $\mathbf{a}'$ (cf. Fig. 7–27). These vectors constitute a pencil of rays emanating from the point BH_2 whose directions are related to the coordinates of the ADQ-diagram.[51] If the vectors are projected in the direction of the A_3 and A_1 axes into the planes defined respectively by the A_1-A_2 and A_3-A_2 axes, then the resulting slopes $\Delta A_1/\Delta A_2$ and $\Delta A_3/\Delta A_2$ of the projected vectors specify the coordinates of the ADQ-diagram.

It follows from these considerations that all points lying on the tangent to BH_2 in a three-dimensional A-diagram coallesce to a single point in an ADQ-diagram. Only points in an A-diagram that deviate from this tangent are depicted separately in an ADQ-diagram. The following relationships are thus established between a (two-dimensional) ADQ-diagram and a three-dimensional A-diagram:[51, 135]

(1) each point in an ADQ-diagram corresponds to a line in an A-diagram, and

(2) each line in an ADQ-diagram corresponds to a plane in an A-diagram.

An ADQ-diagram by its very nature contains less information than the corresponding three-dimensional A-diagram. For example, it is impossible to reconstruct a curve in absorbance space solely on the basis of coordinates taken from an ADQ-diagram. The most that can be established in this way is the direction of the vectors $\mathbf{a}'$.

As an alternative to use of equation 7–75, absorbances can also be described as a function of bh_2 and bh:

$$\begin{pmatrix} A_1 \\ A_2 \\ A_3 \end{pmatrix} = \begin{pmatrix} \bar{q}_{11} \bar{q}_{12} \\ \bar{q}_{21} \bar{q}_{22} \\ \bar{q}_{31} \bar{q}_{32} \end{pmatrix} \begin{pmatrix} bh_2 \\ bh \end{pmatrix} + \begin{pmatrix} A_{1B} \\ A_{2B} \\ A_{3B} \end{pmatrix} \quad (7\text{–}77a)$$

where

$\bar{q}_{\lambda 1} = l(\varepsilon_{\lambda BH_2} - \varepsilon_{\lambda B})$, $\bar{q}_{\lambda 2} = l(\varepsilon_{\lambda BH} - \varepsilon_{\lambda B})$, and $A_{\lambda B} = l \cdot b_0 \cdot \varepsilon_{\lambda B}$ (λ_i, $i = 1, 2, 3$)

which may be abbreviated:

$$\mathbf{a} = \mathbf{a}^* + \mathbf{a_0}^* \quad (7\text{–}77b)$$

The position vector $\mathbf{a_0}^*$ is directed from the origin toward the point B^{2-} on the curve, while the direction vector $\mathbf{a}^*$ proceeds from B^{2-} toward P (cf. Fig. 7–27). Equation 7–77b can be further elaborated in terms of $\mathbf{a}^*$ as follows:

$$\mathbf{a}^* = \mathbf{a} - \mathbf{a_0}^* = \begin{pmatrix} A_1 - A_{1B} \\ A_2 - A_{2B} \\ A_3 - A_{3B} \end{pmatrix} = \begin{pmatrix} \Delta A_1^* \\ \Delta A_2^* \\ \Delta A_3^* \end{pmatrix} \quad (7\text{–}78)$$

Here the quantity ΔA_λ^* is referenced against the point B^{2-}; that is, consistent with equation 7–70, $A_\lambda' = A_{\lambda B}$. ADQ coordinates may be determined from the coordinates of $\mathbf{a}^*$ in a process analogous to that described above, after which an ADQ-diagram may be constructed based on the point B^{2-}.

Figure 7–28 shows such an ADQ-diagram based on B^{2-} for *o*-phthalic acid. All data points that lie on the tangent to B^{2-} in the A-diagram (cf. Fig. 7–27) should of course be coincident in the ADQ-diagram, but slight errors in measurement are here seen to have a particularly significant effect on the results. Experience has shown that considerable scatter is always to be expected from titration points along the absorbance curve in the vicinity of the reference point. It is therefore advisable that one substitute for these points the slopes of the relevant tangents.

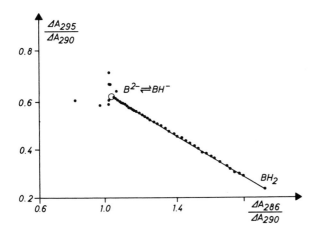

Fig. 7–28. ADQ-diagram for *o*-phthalic acid. Absorbance differences are here referenced against the point B^{2-}.

7.9 Summary and Tabular Overview

Table 7–6 constitutes a review of the methods described above for spectrometric analysis of two-step acid-base equilibria. In each case, the results of primary interest are values for pK_1 and pK_2, as well as the ratio K_1/K_2.

Analysis based on *inflection points* is best avoided for overlapping systems. The only importance of this approach is as a means of obtaining rough estimates of the desired constants to serve as starting points for non-linear regression analysis.

A better approximation method involves evaluation of *uniform subregions* identified from the sides of an absorbance triangle. Usually such sides will be found to contain sufficient data points for the purpose, at least so long as ΔpK is not smaller than ~ 1.5.

7.9 Summary and Tabular Overview

Table 7-6. Summary of the methods for evaluating two-step acid-base equilibria.

Determination	Method	Reference	Limitations*
Confirmation of the presence of two linearly independent equilibria	ADQ-diagrams	Sec. 7.2	No point clusters, no lines through the origin, no lines parallel to an axis, all curves linear
Molar absorptivity of the ampholyte	1) Analysis of uniform subregions from an A-diagram	Sec. 7.3.2	Provides an approximation ($\Delta pK > 1.5$)
	2) Tangent method	Sec. 7.3.3	
	3) Pencil of rays method	Sec. 7.3.5	
	4) Regression analysis	Sec. 7.3.6	
Concentrations	1) Distance relationships in an absorbance triangle	Sec. 7.3.3	
	2) Solution of a system of linear equations (cf. equation 7–63)	Sec. 7.3.6	

Table 7-6 (Continued)

Determination	Method	Reference	Limitations*
pK values	1) Analysis of titration curve inflection points	Sec. 7.3.1	Approximation method ($\Delta pK > 3$)
	2) Evaluation of uniform subregions in an A-diagram	Sec. 7.3.2	Approximation method ($\Delta pK > 1.5$)
	3) Dissociation diagrams	Sec. 7.3.3	
	4) Line bisector method	Sec. 7.3.4	
	5) Pencil of rays method	Sec. 7.3.5	
	6) Regression analysis	Sec. 7.3.6	
pK differences $\Delta pK = pK_2 - pK_1$	1) Distance relationships within an absorbance triangle	Sec. 7.4.1	$\Delta pK < 2$
	2) Non-linear regression	Sec. 7.4.2	$\Delta pK < 3$

* Entries in this column should be regarded only as general guidelines (see, for example, Sec. 7.3.1).

Dissociation diagrams permit the determination of pK values without the need for approximation. However, the method presupposes that one knows the concentrations of BH_2, BH^-, and B^{2-}. This information is advantageously derived from an absorbance triangle.

Lines that *bisect the sides of the absorbance triangle* are the basis for a particularly convenient graphical approach to pK values. The method involved represents a special case of the *pencil of rays method*. Both methods are of general validity.

Numeric evaluation of systems of linear equations (*linear regression analysis*) has not proved to be a popular approach.

The only truly general method is that of *non-linear regression analysis*. No prior knowledge with respect to the titration limits is required, and even though the method entails extensive calculation, numerous computer programs have been developed to lighten the burden (e.g., TITFIT).

The *ratio* K_1/K_2 may be determined even in the absence of pH data, provided an accurate absorbance triangle is available. The quotient K_1/K_2 may be established either graphically (from relationships among rays) or with the aid of iterative optimization procedures (e.g., using the program EDIA).

Literature

1. J. E. Ricci, *Hydrogen Ion Concentration*. Princeton, New Jersey, Princeton University Press 1952.
2. R. G. Bates, *Electrometric pH-Determinations*. J. Wiley & Sons, New York, Chapman & Hall, Ltd., London 1954.
3. E. J. King, *Acid-Base Equilibria*. Pergamon Press, New York 1965.
4. G. Kortüm, *Lehrbuch der Elektrochemie*. Verlag Chemie, Weinheim 1972.
5. G. Kraft, J. Fischer, *Indikationen von Titrationen*. Walter de Gruyter, Berlin, New York 1972.
6. S. Ebel, W. Parzefall, *Experimentelle Einführung in die Potentiometrie*. Verlag Chemie, Weinheim 1975.
7. C. Bliefert, *pH-Wert-Berechnungen*. Verlag Chemie, Weinheim, New York 1978.
8. F. Auerbach, E. Smolczyk, *Z. Physik. Chem.* 110 (1924) 65.
9. J. C. Speakman, *J. Chem. Soc.* 1944, 855.
10. G. Schwarzenbach, A. Willy, R. O. Bach, *Helv. Chim. Acta* 30 (1947) 1303.
11. R. G. Bates, *J. Amer. Chem. Soc.* 70 (1948) 1579.
12. Y. Murakami, K. Nakamura, M. Tokunaga, *Bull. Chem. Soc. Japan* 36 (1963) 669.
13. H. Sigel, *Helv. Chim. Acta* 48 (1965) 1513.
14. L. G. Sillen, *Acta Chem. Scand.* 18 (1964) 1085.
15. L. G. Sillen, *Acta Chem. Scand.* 18 (1964) 1085.
16. N. Ingri, G. Sillen, *Ark. Kemi* 23 (1964) 97.
17. A. R. Emery, *J. Chem. Educ.* 42 (1965) 131.
18. I. G. Sayce, *Talanta* 15 (1968) 1397.

19. R. Arnek, L. G. Sillen, O. Wahlberg, *Ark. Kemi* 31 (1969) 353.
20. P. Branner, L. G. Sillen, R. Whiteker, *Ark. Kemi* 31 (1969) 365.
21. T. Meites, L. Meites, *Talanta* 19 (1972) 1131.
22. P. Gans, A. Vacca, A. Sabatini, *Inorg. Chim. Acta* 18 (1976) 237.
23. H. Stünzi, G. Anderegg, *Helv. Chim. Acta* 59 (1976) 1621.
24. G. Arena, E. Rizzarelli, S. Sammartano, C. Rigano, *Talanta* 26 (1979) 1.
25. A. Ivaska, I.Nagypal, *Talanta* 27 (1980) 721.
26. H. Gampp, M. Maeder, A. D. Zuberbühler, T. Kaden, *Talanta* 27 (1980) 513.
27. F. Gaizer, A. Puskas, *Talanta* 28 (1981) 565.
28. C. A. Chang, B. E. Douglas, *J. Coord. Chem.* 11 (1981) 91.
29. R. J. Motekaitis, A. E. Martell, *Can. J. Chem.* 60 (1982) 168.
30. P. M. May, D. R. Williams, P. W. Linder, R. G. Torrington, *Talanta* 29 (1982) 249.
31. M. C. Garcia, G. Ramis, C. Mongay, *Talanta* 29 (1982) 435.
32. A. Laouenan, E. Suet, *Talanta* 32 (1985) 245.
33. P. M. May, K. Murray, D. R Williams, *Talanta* 32 (1985) 483-489.
34. J. Kostrowicki, A. Liwo, *Computers & Chemistry* 8 (1984) 91 and 8 (1984) 101.
35. R. Fournaise and C. Petifaux, *Talanta* 33 (1986) 499.
36. J. Havel, M. Meloun, *Talanta* 33 (1986) 525.
37. T. Hofman, M. Krzyzanowska, *Talanta* 33 (1986) 851.
38. D. J. Legett, *Computational Methods for the Determination of Formation Constants.* Plenum Press, New York and London 1985.
39. B. J. Thamer, A. F. Voigt, *J. Phys. Chem.* 56 (1952) 225.
40. B. J. Thamer, *J. Phys. Chem.* 59 (1955) 450.
41. K. P. Ang, *J. Phys. Chem.* 62 (1958) 1109.
42. R. Blume, J. Polster, *Z. Naturforsch.* 29 b (1974) 734.
43. R. Blume, H. Lachmann, J. Polster, *Z. Naturforsch.* 30 b (1975) 263.
44. F. Göbber and J. Polster, *Anal. Chem.* 48 (176) 1546.
45. J. Polster, *GIT Fachz. Lab.* 26 (1982) 581.
46. J. Polster, *GIT Fachz. Lab.* 27 (1983) 390.
47. H. Mauser, *Formale Kinetik.* Bertelsmann Universitätsverlag, Düsseldorf 1974.
48. R. Blume, H. Lachmann, H. Mauser, F. Schneider, *Z. Naturforsch.* 29 b (1974) 500.
49. H. Mauser, *Z. Naturforsch.* 23 b (1968) 1025.
50. R. Blume, *Dissertation.* Tübingen 1975.
51. J. Polster, *Habilitationsschrift.* Tübingen 1980.
52. J. Polster, *GIT Fachz. Lab.* 5 (1982) 421.
53. I. Zimmermann, H. W. Zimmermann, *Z. Naturforsch.* 31 c (1976) 656.
54. H. Lachmann, *Habilitationsschrift.* Tübingen 1982.
55. M. Dixo, E. C. Webb, *Enzymes.* Longmans and Green, London (1964), p. 118.
56. J. C. Sternberg, H. S. Stillo, R. H. Schwendeman, *Anal. Chem.* 32 (1960) 84.
57. G. Weber, *Nature* 190 (1961) 27.
58. F. P. Zscheile et al., *Anal. Chem.* 34 (1962) 1776.
59. H. Cerfontain, H. G. Duin, L. Vollbracht, *Anal. Chem.* 35 (1963) 1005; 36 (1964) 1802.
60. I. S. Herschberg, *Z. Anal. Chem.* 205 (1964) 180.
61. N. Ohta, *Anal. Chem.* 45 (1973) 553.
62. J. Sustek, *Anal. Chem.* 46 (1974) 1676.
63. A. Junker, G. Bergmann, *Z. Anal. Chem.* 272 (1974) 267.

64. H. Weitkamp, R. Barth, *Einführung in die quantitative IR-Spektrophotometrie.* Thieme Verlag, Stuttgart (1976), Chap. VIII.
65. D. J. Legett, *Anal. Chem.* 49 (1977) 276.
66. H. H. Perkampu, in: *Ullmanns Encyklopädis der technischen Chemie,* 4. Aufl., Bd.5. Verlag Chemie, Weinheim (1980), p. 279 ff.
67. S. Ebel, E. Glaser, S. Abdulla, U. Steffens, V. Walter, *Z. Anal. Chem.* 313 (1982) 24.
68. K. Weber, *GIT Fachz. Lab.* 27 (1983) 837.
69. K. Weber, *GIT Fachz. Lab.* 27 (1983) 974.
70. H. Netz, *Formeln der Mathematik.* Carl Hanser Verlag, München, Wien 1977.
71. J. Dreszer, *Mathematik Handbuch für Technik und Wissenschaft.* Verlag Harri Deutsch, Zürich, Frankfurt/Main, Thun 1975.
72. J. W. Gibbs, *Trans. Conn. Acad.* 3 (1876) 176.
73. H.W. Roozeboom, *Z. phys. Chem.* 15 (1894) 147.
74. G. Kortüm, H. Lachmann, *Einführung in die chemische Thermodynamik.* Verlag Chemie, Weinheim, Deerfield Beach (Florida), Basel, Vandenhoeck & Ruprecht, Göttingen 1981.
75. J. Polster, *Z. Physik. Chem. N. F.* 97 (1975) 55.
76. C. N. Reilley, E. M. Smith, *Anal. Chem.* 32 (1960) 1233.
77. C. N. Reilley, H. A. Flaschka, S. Laurent, B. Laurent, *Anal. Chem.* 32 (1960) 1218.
78. H. Flaschka, *Talanta* 7 (1960) 99.
79. B. J. Thamer, A. F. Voigt, *J. Phys. Chem.* 56 (1952) 225.
80. B. J. Thamer, *J. Phys. Chem.* 59 (1955) 450.
81. H.-H. Perkampus, T. Rössel, *Z. Elektochem.* 60 (1953) 1140.
82. H.-H. Perkampus, T. Rössel, *Z. Elektochem.* 60 (1956) 1102.
83. H.-H. Perkampus, T. Rössel, *Z. Elektrochem.* 62 (1958) 94.
84. J. H. Baxendale, H. R. Hardy, *Trans. Faraday Soc.* 49 (1953) 1140.
85. D. H. Rosenblatt, *J. Phys. Chem.* 58 (1954) 40-43.
86. R. A. Robinson, A. I. Biggs, *Austr. J. Chem.* 10 (1957) 128.
87. K. P. Ang, *J. Phys. Chem.* 62 (1958) 1109.
88. J. C. Sternberg, H. S. Stillo, R.H.Schwendemann, *Anal. Chem.* 32 (1960) 84.
89. R. M. Wallace, *J. Phys. Chem.* 64 (1960) 899.
90. B. Roth, J. F. Bunnett, *J. Amer. Chem. Soc.* 87 (1965) 334.
91. T. Anfält, D. Jagner, *Anal. Chim. Acta* 47 (1969) 57.
92. J. J. Kankare, *Anal. Chem.* 42 (1970) 1322.
93. G. Heys, H. Kinns, D.D.Perrin, *Analyst* 97 (1972) 52.
94. V. Böhmer, R. Wamsser, H. Kämmerer, *Monatsh. Chem.* 104 (1973) 1315.
95. W. Preetz, G. Schätzel, *Z. Anorg. Allg. Chem.* 423 (1976) 117.
96. A. G. Asuero, M. J. Navas, J. L. Jimenez-Trillo, *Talanta* 33 (1986) 195, 33 (1986) 531 and 33 (1986) 929.
97. H.-H. Perkampus, *UV-VIS-Spektroskopie und ihre Anwendungen.* Springer-Verlag, Berlin 1986.
98. K. Nagano, D. E. Metzler, *J. Amer. Chem. Soc.* 89 (1967) 2891.
99. F. Göbber, H. Lachmann, *Hoppe-Seyler Z. Physiol. Chem.* 359 (1978) 269.
100. R. M. Alcock, F. R. Hartley, D. E. Rogers, *J. Chem. Soc. Dalton Trans.* 1978, 115.
101. M. Meloun, J. Cernak, *Talanta* 26 (1979) 569-573.
102. H. Gampp, M. Maeder, A. D. Zuberbühler, *Talanta* 27 (1980) 1037.
103. M. Meloun, J. Cernak, *Talanta* 31 (1984) 947.
104. M. Meloun, M. Javurek, *Talanta* 31 (1984) 1083.

105. M. Meloun, M. Javurek, *Talanta* 32 (1985) 973.
106. M. Meloun, M. Javurek, A. Hynkova, *Talanta* 33 (1986) 825.
107. H. Gampp, M. Maeder, C. J. Meyer, A. D. Zuberbühler, *Talanta* 32 (1985) 95.
108. H. Gampp, M. Maeder, C. J. Meyer, A. D. Zuberbühler, *Talanta* 32 (1985) 257.
109. I. Bertini, A. Dei, C. Luchinat, R. Monnanni, *Inorg. Chem.* 24 (1985) 301.
110. T. Kaden, A. Zuberbühler, *Talanta* 18 (1971) 61.
111. D. J. Leggett, W. A. E. McBryde, *Anal. Chem.* 47 (1975) 1065.
112. M. Meloun, J. Cernak, *Talanta* 23 (1976) 15.
113. F. Gaizer, A. Puskas, *Talanta* 28 (1981) 925.
114. M. Meloun, M. Javurek, *Talanta* 32 (1985) 1011.
115. J. Havel, M. Meloun, *Talanta* 33 (1986) 435.
116. M. Meloun, M. Javurek, J. Havel, *Talanta* 33 (1986) 513.
117. H. Gampp, M. Maeder, C. J. Meyer, A. Zuberbühler, *Talanta* 32 (1985) 1133.
118. N. H. Nie, C .H. Hull, J. G. Jenkins, K. Steinbrenner, D. H. Bent, *SPSS Statistical Package for the Social Sciences*. New York 1975.
119. N. H. Nie, C.H.Hull, *SPSS 9, Statistik-Programm-System für die Sozialwissenschaften*. Gustav Fischer Verlag, Stuttgart, New York 1983.
120. H. O. Hartley, *Technometrics*. Vol.3 (1961).
121. D. Marquardt, *J. Soc. Ind. Appl. Math.* 11 (1963) 431.
122. P. R. Bevington, *Data Reduction and Error Analysis for the Physical Sciences*. McGraw-Hill, New York 1969.
123. G. Kortüm, W. Vogel, K. Andrussow, *Pure Appl. Chem.* 1 (1960) 190.
124. *Handbook of Chemistry and Physics*. CRC Press, Florida (1979), p. J 198.
125. R. G. Bates, *J. Res. Nat. Bur. Standards* 66 A (2) (1962) 179.
126. R. G. Bates, *Determination of pH, Theory and Practice*. New York, Wiley 1964.
127. S. Ebel, E. Glaser, H. Mohr, *Fresenius Z. Anal. Chem.* 293 (1978) 33.
128. Y. Bard, *Nonlinear Parameter Estimation*. Academic Press, New York, 1974, p. 46.
129. Anal. Chem. 46 (1974) 2258.
130. K.-D. Willamowski, O. E. Rössler, *Z. Naturforsch.* 31 a (1976) 408.
131. H. Wörle, H.-J. Rumpf, *Taschenbuch der Mathematik*. R. Oldenbourg Verlag, München, Wien 1979.
132. H. G. Zachmann, *Mathematik für Chemiker*. Verlag Chemie, Weinheim 1972.
133. B. Baule, *Die Mathematik des Naturforschers und Ingenieurs*, mehrbändig. Hirzel-Verlag, Leipzig 1960.
134. J. Polster, *Z. Physik. Chem. N. F.* 97 (1975) 113.
135. J. Polster, *Z. Physik. Chem. N. F.* 104 (1977) 49.

8 Three-Step Acid-Base Equilibria

8.1 Basic Equations

To date, no methods devoid of approximation have been proposed specifically for the spectrometric investigation of three-step acid-base equilibria. Systems of this magnitude are usually relegated to the larger class of "multi-step" processes and then evaluated by the most universal of methods: regression analysis, conducted on a set of titration curves. Nonetheless, A-diagrams are also useful in the investigation of three-step systems, and they offer the advantage of simplicity without any sacrifice in significance.

In the case of a three-step *unbranched* acid-base system, continuous pH variation involves titration of the following set of equilibria:

$$BH_3 \overset{K_1}{\leftrightarrows} BH_2^- \ (H^+)$$

$$BH_2^- \overset{K_2}{\leftrightarrows} BH^{2-} \ (H^+) \tag{8-1}$$

$$BH^{2-} \overset{K_3}{\leftrightarrows} B^{3-} \ (H^+)$$

As with two-step protolysis systems, the results of interest include:

- the number of linearly independent equilibria,
- the corresponding pK values,
- the quotients K_1/K_2 and K_2/K_3,
- the (molar) absorptivities of each of the components (especially the ampholytes BH_2^- and BH^{2-}), and
- the concentrations.

The approach that will be taken here in presenting various possible methods of analysis is analogous to that used previously in the chapter on two-step acid-base equilibria. We begin by illustrating a series of methods without extensive theoretical justification. Comprehensive discussion is reserved for Sec. 8.5, "The Characteristic Concentration Diagram".

The basis for all analytical treatments of a three-step system is the following set of equations describing three (mixed) dissociation constants:

$$K_1 = a_{H_3O^+} \cdot \frac{bh_2}{bh_3} = h \cdot \frac{bh_2}{bh_3} = 10^{-pH} \cdot \frac{bh_2}{bh_3}$$

$$K_2 = a_{H_3O^+} \cdot \frac{bh}{bh_2} = h \cdot \frac{bh}{bh_2} = 10^{-pH} \cdot \frac{bh}{bh_2} \tag{8-2}$$

$$K_3 = a_{H_3O^+} \cdot \frac{b}{bh} = h \cdot \frac{b}{bh} = 10^{-pH} \cdot \frac{b}{bh}$$

where $a_{H_3O^+} := h = 10^{-pH}$ and

bh_3 = (molar) concentration of BH_3 at equilibrium
(and analogously for bh_2, bh, and b)

Relevant pK values and pK differences are defined as follows:

$$pK_1 := -\log K_1; \quad pK_2 := -\log K_2; \quad pK_3 := -\log K_3 \tag{8-3a}$$

and

$$\Delta_{21} pK := pK_2 - pK_1; \quad \Delta_{32} pK := pK_3 - pK_2;$$
$$\Delta_{31} pK := pK_3 - pK_1 \tag{8-3b}$$

Mass balance requires that:

$$b_0 = bh_3 + bh_2 + bh + b \tag{8-4}$$

With the aid of equations 8–2 and 8–4 it is possible to express the concentrations of individual components as functions of h:

$$bh_3 = \frac{b_0 h^3}{H_3}, \quad bh_2 = \frac{b_0 h^2 K_1}{H_3}, \quad bh = \frac{b_0 h K_1 K_2}{H_3}, \quad \text{and} \quad b = \frac{b_0 K_1 K_2 K_3}{H_3} \tag{8-5}$$

where $H_3 = h^3 + h^2 K_1 + h K_1 K_2 + K_1 K_2 K_3$

Within the range of validity of the Lambert-Beer-Bouguer Law, the following equation expresses the magnitude of the absorbance A_λ:

$$A_\lambda = l(\varepsilon_{\lambda BH_3} \cdot bh_3 + \varepsilon_{\lambda BH_2} \cdot bh_2 + \varepsilon_{\lambda BH} \cdot bh + \varepsilon_{\lambda B} \cdot b) \tag{8-6}$$

where l = pathlength of the cuvette
$\varepsilon_{\lambda BH_3}$ = (molar) absorptivity of BH_3 at the wavelength λ
of measurement (and similarly for $\varepsilon_{\lambda BH_2}$, $\varepsilon_{\lambda BH}$, and $\varepsilon_{\lambda B}$)

From equations 8–5 and 8–6 one obtains the following function for the titration curve A_λ vs. pH:

$$A_\lambda = l \cdot b_0 \cdot \frac{\varepsilon_{\lambda BH_3} \cdot h^3 + \varepsilon_{\lambda BH_2} \cdot h^2 \cdot K_1 + \varepsilon_{\lambda BH} \cdot h \cdot K_1 K_2 + \varepsilon_{\lambda B} \cdot K_1 K_2 K_3}{h^3 + h^2 K_1 + h K_1 K_2 + K_1 K_2 K_3} \qquad (8\text{–}7)$$

Two of the absorptivities, $\varepsilon_{\lambda BH_3}$ and $\varepsilon_{\lambda B}$, can usually be determined directly. In strongly acid medium virtually the only component present is BH_3, assuming its pK is not exceptionally low; i.e, $bh_3 = b_0$. Under relatively strongly basic conditions (provided pK_3 is not too large) only B^{3-} need be considered ($b = b_0$). If the equilibria overlap, however, absorptivity values for the ampholytes ($\varepsilon_{\lambda BH_2}$, $\varepsilon_{\lambda BH}$) are not directly accessible. Nevertheless, they may be estimated rather easily with the aid of A-diagrams (cf. Sec. 8.3.3).

8.2 Graphic Matrix-Rank Analysis (ADQ3-Diagrams)

Generally speaking, the matrix **A** comprised of absorbances measured at various wavelengths and pH values for a three-step titration system shows a rank of three (cf. Sec. 3.2.1) provided each component demonstrates its own characteristic absorbance pattern. The rank of such a matrix is a measure of the number of linearly independent concentrations associated with the system (cf. equation 8–4), and it may be obtained graphically with the aid of what is known as an "ADQ3-diagram".[1-7]

Most three-step titration systems show linearly independent behavior for absorbance differences (or absorbances) at three wavelengths. Thus:

$$\Delta A_1 = \alpha_1 \Delta A_2 + \alpha_2 \Delta A_3 + \alpha_3 \Delta A_4 \qquad (8\text{–}8)$$

where

$$\Delta A_{\lambda_i} = A_{\lambda_i} - A_{\lambda_i BH_3} \text{ and } A_{\lambda_i BH_3} = l b_0 \cdot \varepsilon_{\lambda_i BH_3} \quad (\lambda_i = 1...4)$$

The coefficients α_1, α_2, and α_3 are constants whose values depend only on the absorptivities of the various components. This relationship follows directly from equation 9–8 in Sec. 9.1, in which s (equal to the number of linearly independent equilibria) is assigned the value three.

The relationship expressed by equation 8–8 may be represented two-dimensionally provided one of the coefficients α_i is eliminated. This is achieved with the aid of a single data point, in which case the equation takes the following form:

$$\Delta A_{11} = \alpha_1 \Delta A_{21} + \alpha_2 \Delta A_{31} + \alpha_3 \Delta A_{41} \qquad (8\text{–}9)$$

Note that a second index (here with the value 1) has been introduced in order to distinguish this equation from equation 8–8 (cf. equation 9–22 in Sec. 9.3).

Solving this equation in terms of α_1, substituting the result into equation 8–8, and rearranging terms leads to[6,7]

$$\frac{\Delta A_1 \Delta A_{21} - \Delta A_2 \Delta A_{11}}{\Delta A_3 \Delta A_{21} - \Delta A_2 \Delta A_{31}} = \alpha_2 + \alpha_3 \frac{\Delta A_4 \Delta A_{21} - \Delta A_2 \Delta A_{41}}{\Delta A_3 \Delta A_{21} - \Delta A_2 \Delta A_{31}}$$

Determinant notation may be used to express this relationship in a more compact way (where α_2 and α_3 have been renumbered as α_1 and α_2):

$$\frac{\begin{vmatrix} \Delta A_1 & \Delta A_{11} \\ \Delta A_2 & \Delta A_{21} \end{vmatrix}}{\begin{vmatrix} \Delta A_3 & \Delta A_{31} \\ \Delta A_2 & \Delta A_{21} \end{vmatrix}} = \alpha_1 + \alpha_2 \frac{\begin{vmatrix} \Delta A_4 & \Delta A_{41} \\ \Delta A_2 & \Delta A_{21} \end{vmatrix}}{\begin{vmatrix} \Delta A_3 & \Delta A_{31} \\ \Delta A_2 & \Delta A_{21} \end{vmatrix}} \quad (8\text{--}10)$$

In an *ADQ3-diagram* the two quotients above are plotted against each other, with ΔA_{λ_i} ($\lambda_i = 1...4$) taken to be the variable and $\Delta A_{\lambda_i 1}$ a constant. Relatively small errors in measurement may lead to a great deal of scatter, but the method has nevertheless been shown to be useful provided all data curves are first subjected to smoothing.[4,5]

8.3 Absolute p*K* Determination

8.3.1 Analysis of Inflection Points

Conclusions with respect to inflection points like those drawn previously in the context of two-step titration systems are valid here as well. Thus, inflection points in the titration curve A_λ vs. pH for a three-step protolysis system may lead to reasonable approximations of p*K* values, but only if the p*K* differences are great (i.e., Δ_{21} p*K* > 3 and Δ_{32} p*K* > 3). In the case of smaller differences the inflection point method gives poor results, though even these may be adequate for use as starting points in non-linear optimization (cf. Sec. 8.3.7). Unfortunately it sometimes is questionable whether titration curves of this type can even be resolved in terms of three points of inflection (cf. Fig. 7–6 in Sec. 7.3.1)

8.3.2 Uniform Subregions Within Absorbance Diagrams

In the discussion that follows, the *A-diagram* occupies a particularly central role. The key characteristics of such a representation may be examined either on the basis of a simulation or else directly using experimental data. In the present case the compound 1,2,4-benzenetricarboxylic acid serves as a convenient source of actual data.

1,2,4-Benzenetricarboxylic acid in fact represents a case of a *branched* three-step titration system:

8.3 Absolute pK Determination

The desired macroscopic dissociation constants K_1, K_2, and K_3 are related to the corresponding microscopic constants through the following expressions:[8,9]

$$K_1 = k_1 + k_2 + k_3$$
$$K_1K_2 = k_1k_{12} + k_1k_{13} + k_2k_{23} = k_2k_{21} + k_3k_{31} + k_3k_{32}$$
$$K_1K_2K_3 = k_1k_{12}k_{123} = k_2k_{21}k_{123} = k_2k_{23}k_{231} =$$
$$= k_1k_{13}k_{132} = k_3k_{31}k_{132} = k_3k_{32}k_{231}$$

In order to construct an A-diagram for 1,2,4-benzenetricarboxylic acid it is first necessary that one record the corresponding titration spectra. As is apparent from

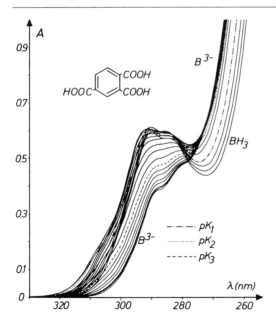

Fig. 8–1. Titration spectra for 1,2,4-benzenetricarboxylic acid ($b_0 = 3.95 \cdot 10^{-4}$ M; 20.0 °C).

Fig. 8–1, the spectra intersect repeatedly, and in ways that are not readily interpretable. No isosbestic points are observed.[10]

However, the spectrometric behavior of the titration is readily understood once an A-diagram has been constructed. Fig. 8–2b shows obvious regions of linearity near the points BH_3 and B^{3-}. These segments may be evaluated to a good approximation as if they resulted from single-step equilibria (cf. Sec. 7.3.2). For example, all points on the line $\overline{BH_3 \, BH_2^-}$ conform well to the relationship (cf. equation 7–14):[3,6,10,11]

$$(A_\lambda - A_{\lambda BH_3}) \cdot 10^{-pH} = -K_1 A_\lambda + K_1 A_{\lambda BH_2} \tag{8–11}$$

where $A_{\lambda BH_3} = lb_0 \varepsilon_{\lambda BH_3}$ and $A_{\lambda BH_2} = lb_0 \varepsilon_{\lambda BH_2}$

The data point shown for BH_2^- in Fig. 8–2 was determined in this way. The overall results achieved are summarized in Table 8–1.[11]

An analogous relationship applies to points along the line $\overline{B^{3-} \, BH^{2-}}$ (cf. equation 7–15):

$$(A_\lambda - A_{\lambda B}) \cdot 10^{+pH} = -\frac{1}{K_3} A_\lambda + \frac{1}{K_3} A_{\lambda BH} \tag{8–12}$$

where $A_{\lambda B} = lb_0 \varepsilon_{\lambda B}$ and $A_{\lambda BH} = lb_0 \varepsilon_{\lambda BH}$

A graph based on this equation is shown in Fig. 8–3. The resulting point for BH^{2-}, derived from $A_{\lambda BH}$, has also been included in Fig. 8–2. The corresponding numerical results are incorporated in Table 8–1.

8.3 Absolute pK Determination

The second equilibrium $BH_2^- \leftrightarrows BH^{2-}$ (H^+) cannot be analyzed in this way because there is no region in which it is essentially the only equilibrium operative (cf. the line $\overline{BH_2^- \, BH^{2-}}$ in Fig. 8–2). Nevertheless, Table 8–1 shows that all molar absorptivities are now known with a reasonable degree of accuracy, so the concentrations of all components may be calculated (cf. Sec. 8.3.3) and the results may be used to obtain an estimate of K_2.

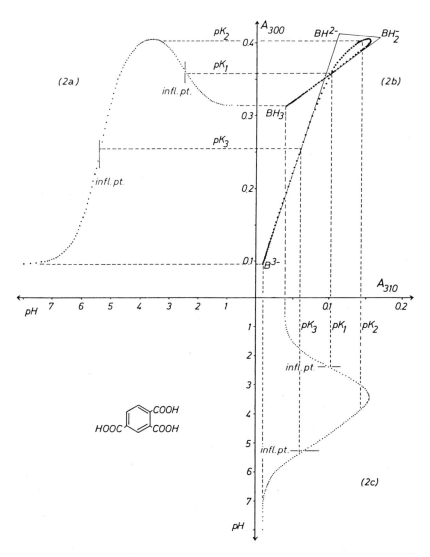

Fig. 8–2. Spectrometric titration curves (a,c) and A-diagram (b) for 1,2,4-benzenetricarboxylic acid ($b_0 = 4.74 \cdot 10^{-4}$ M; 20.0 °C).

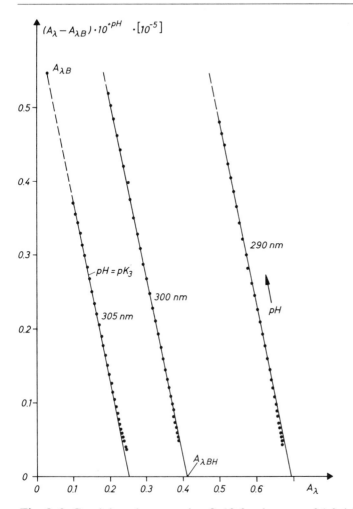

Fig. 8–3. Graph based on equation 8–12 for the case of 1,2,4-benzenetricarboxylic acid.

The fact that the initial and final regions of the A-diagram are so distinctive in this example is a result of relatively large pK differences. As a rule, sufficient data points will be present to define such lines if $\Delta_{21}\text{p}K > 1.5$ and $\Delta_{32}\text{p}K > 1.5$. This assertion has been verified in a large number of cases, and the method will frequently be found applicable.

In the event that one of the pK differences is small while the other is large, the corresponding titration curve consists essentially of a combination of the curves for a one-step titration and a two-step titration. The two-step subsystem may then be evaluated with the aid of the methods presented in Chapter 7.

8.3 Absolute pK Determination

Table 8–1. 1,2,4-Benzenetricarboxylic acid (abbreviated table).
a) Determination of $A_{\lambda BH_2}$ and pK_1 on the basis of equation 8–11.

Wavelength λ (nm)	$A_{\lambda BH_3}$	$A_{\lambda BH_2}$	pK_1
305	0.133	0.280	2.54
300	0.313	0.420	2.56
270	0.537	0.765	2.57

b) Determination of $A_{\lambda BH}$ and pK_3 on the basis of equation 8–12.

Wavelength λ (nm)	$A_{\lambda B}$	$A_{\lambda BH}$	pK_3
305	0.030	0.253	5.38
300	0.096	0.413	5.39
290	0.407	0.694	5.40

8.3.3 Dissociation Diagrams

In the dissociation diagrams b_i vs. pH, the points of intersection of two curves lead to the following relationships (cf. references 11 and 12):

$$
\begin{aligned}
bh_3 &= bh_2: & h &= K_1 \\
bh_2 &= bh: & h &= K_2 \\
bh &= b: & h &= K_3 \\
bh_3 &= bh: & h^2 &= K_1 K_2 \\
bh_2 &= b: & h^2 &= K_2 K_3 \\
bh_3 &= b: & h^3 &= K_1 K_2 K_3
\end{aligned}
\qquad (8\text{–}13)
$$

In order to use this set of relationships to obtain the dissociation constants K_1, K_2, and K_3 it is first necessary that all concentrations be known. In principle these may be established using a system of equations based on equation 8–6 and data taken at three wavelengths, but this requires that very accurate values be available for the corresponding molar absorptivities.

Concentrations may also be obtained through a graphical procedure. Consider the three-dimensional A-diagram shown in Fig. 8–4. Here the experimental data curve lies entirely within an *absorbance tetrahedron*, which is bounded by the points BH_3,

BH_2^-, BH^{2-}, and B^{3-}. These vertices in turn correspond to the molar absorptivities of the pure components (e.g., for BH_3, $A_{\lambda BH_3} = lb_0\varepsilon_{\lambda BH_3}$, etc.). In general, BH_3 and B^{3-} are the only vertices whose positions are known, but the locations of BH_2^- and BH^{2-} must also be available in order to carry out the desired analysis. Fortunately, these may be ascertained by differential geometric means. Thus, the desired vertices obey the following relationships with respect to the experimental curve (cf. Sec. 8.5.1):[6,7,13,14]

(1) The points BH_2^- and BH^{2-} are located on tangents to the curve at the points BH_3 and B^{3-}, respectively.

(2) The points BH_2^- and BH^{2-} lie in the two "osculating planes" within which the curve approaches the points BH_3 and B^{3-}.

In order to determine the locations of the vertices BH_2^- and BH^{2-} it is first necessary that the osculating planes for BH_3 and B^{3-} be extended to reveal their line of mutual intersection, which is the tetrahedral edge $\overline{BH_2^- BH^{2-}}$. This edge in turn intersects tangents constructed through the points BH_3 and B^{3-} at the desired vertices BH_2^- and BH^{2-}. One is thus able to establish the molar absorptivities of BH_2^- and BH^{2-} solely on the basis of spectrometric data.[14]

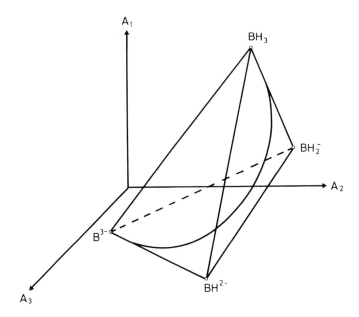

Fig. 8–4. Schematic representation of a three-dimensional A-diagram. The curve corresponding to data points for a three-step acid-base system lies entirely within a tetrahedron (known as the "absorbance tetrahedron") defined by the vertices BH_3, BH_2^-, BH^{2-}, and B^{3-}. The locations of BH_2^- and BH^{2-} are determined by constructing tangents and osculating planes at the points BH_3 and B^{3-} (cf. discussion in the text).

8.3 Absolute pK Determination

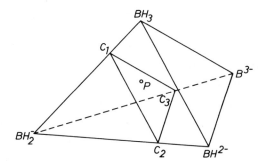

Fig. 8–5. Concentration determination based on distance relationships along an edge of a tetrahedron. For example, equation 8–14 shows that bh_2 may be ascertained by evaluating the expression:

$$bh_2 = \frac{\overline{BH_3\,C_1}}{\overline{BH_3\,BH_2^-}} \cdot b_0$$

The appropriate tetrahedron may be constructed within either an absorbance diagram or a concentration diagram.

The osculating planes may be easily identified with the aid of a two-dimensional ADQ-diagram through a correlation with the corresponding three-dimensional A-diagram (cf. Sec. 8.6).[6,7]

An absorbance tetrahedron constructed as described above can be used to establish concentrations by constructing an appropriate series of planes parallel to the tetrahedral surfaces. Fig. 8–5 shows that such a plane passing through the titration point P and parallel to the tetrahedral face bounded by the vertices BH_3, BH^{2-}, and B^{3-} intersects the sides of the tetrahedron at the points C_1, C_2, and C_3. (It is not important for the present discussion to define the axes in Fig. 8–5.) Analogous to what has already been observed for a Gibbs triangle, the following relationships exist with respect to the concentration bh_2:[3,6,13]

$$bh_2 = \frac{\overline{BH_3\,C_1}}{\overline{BH_3\,BH_2^-}} \cdot b_0 = \frac{\overline{BH^{2-}\,C_2}}{\overline{BH^{2-}\,BH_2^-}} \cdot b_0 = -\frac{\overline{B^{3-}\,C_3}}{\overline{B^{3-}\,BH_2^-}} \cdot b_0 \qquad (8\text{–}14)$$

Concentrations of the other species may be described in a similar way.

8.3.4 Pencils of Rays in an Absorbance Tetrahedron

A three-step acid-base system may be simplified by eliminating either the first or the third equilibrium with the aid of *pencils of rays*.[6,7,15] The equilibria that remain are then evaluated as two-step protolysis systems, furnishing pK values that are identical to those of the original three-step system. The process is illustrated in Fig. 8–6. Eliminating the equilibrium $BH^{2-} \leftrightarrows B^{3-}$ (H^+) requires construction of the pencil of rays emanating from the point B^{3-}. These encompass not only the titration points P of the three-step system, but also the points P_2 associated with the desired two-step system. Furthermore, the points P_2 lie in the osculating plane generated by the experimental curve at the point BH_3 (cf. Sections 8.3.3 and 8.6). Thus, the points P_2 represent the points of intersection of the pencil of rays with the osculating plane for BH_3.

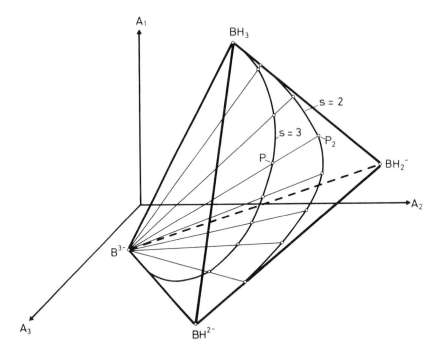

Fig. 8–6. Data points (P) and the points corresponding to the titration limits BH_3 and B^{3-} in a three-step acid-base system permit construction of useful pencils of rays. The lines converging at the point B^{3-} penetrate the osculating plane for BH_3 at the points P_2. These points in turn correspond to the hypothetical titration points of the isolated two-step system $BH_3 \leftrightarrows BH_2^-$ (H^+) and $BH_2^- \leftrightarrows BH^{2-}$ (H^+).

8.3 Absolute pK Determination

The points P_2 constitute the titration curve that would be obtained if one were able to titrate separately the first two equilibria of the three-step system:

$$BH_3 \stackrel{K_1}{\leftrightarrows} BH_2^- \ (H^+)$$
$$BH_2^- \stackrel{K_2}{\leftrightarrows} BH^{2-} \ (H^+) \tag{8-15}$$

This curve may be further examined using the methods described in Chapter 7 to provide values for pK_1 and pK_2.

In order to analyze the isolated equilibria

$$BH_2^- \stackrel{K_2}{\leftrightarrows} BH^{2-} \ (H^+)$$
$$BH^{2-} \stackrel{K_3}{\leftrightarrows} B^{3-} \ (H^+) \tag{8-16}$$

one needs the pencil of rays (omitted from Fig. 8–6) joining the titration points P with the vertex BH_3. These rays intersect the osculating plane for B^{3-} (which also contains the points BH^{2-} and BH_2^-) at the fictitious titration points associated with the system of equations 8–16. Evaluation of this set of points leads to values for pK_2 and pK_3 of the corresponding three-step system.

8.3.5 Pencils of Planes in the Absorbance Tetrahedron

Analogous to the method involving bisecting the side of an absorbance triangle in a two-step protolysis system (cf. Sec. 7.3.4), pK values associated with a three-step system may also be ascertained directly from titration curves A_λ vs. pH provided correlations can be made with the corresponding three-dimensional A-diagrams. Consider the plane defined by the points B^{3-}, BH^{2-}, and H_1 in Fig. 8–7, where H_1 is the midpoint of the line $\overline{BH_3 \ BH^{2-}}$. The point at which the curve penetrates this plane (point P in Fig. 8–7) must fulfill the condition pH = pK_1, so the corresponding pH value taken from the titration curve (A_λ vs. pH) provides a measure of pK_1.[6,7,13] Similarly, the point along the curve that lies in the plane defined by BH_3, B^{3-}, and H_2 provides a value for pK_2, and that in the plane BH_3, BH^{2-}, H_3 gives pK_3 (these are not shown explicitly in Fig. 8–7).

The planes illustrated in Fig. 8–7 are members of a *pencil of planes* useful in examining separately the three equilibria of which the titration system is composed.[6,7,13] Thus, the set of planes illustrated in Fig. 8–8 is constructed around a common line of intersection, $\overline{B^{3-} \ BH^{2-}}$. Any point P of the three-step titration system, taken in combination with this line of intersection, in turn defines one of these planes, and together they comprise a pencil of planes penetrated by the side $\overline{BH_3 \ BH_2^-}$ of the

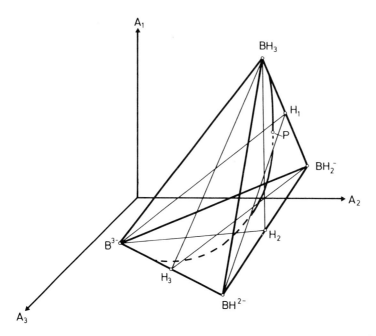

Fig. 8–7. Determination of pK_1, pK_2, and pK_3 from a three-dimensional A-diagram with the aid of selected planes. Data point P, which lies in the plane defined by the points B^{3-}, BH^{2-}, and H_1, fulfills the condition $pH = pK_1$, where H_1 is the point of bisection of the line $\overline{BH_3\,BH_2^-}$. Similarly, points in the planes (BH_3, B^{3-}, H_2) and (BH_3, BH_2^-, H_3) fulfill the conditions $pH = pK_2$ and $pH = pK_3$, respectively.

tetrahedron. This provides the set of points $^1P_1$, where (as in Sec. 7.3.5) the left upper index refers to the sequential number of the equilibrium, while the right lower index signifies the rank s of the reduced titration system. These are the points that would result if it were possible to titrate separately the first equilibrium $BH_3 \leftrightarrows BH_2^-$ (H^+) of the three-step system at pH values corresponding to the points P. Thus, the points $^1P_1$ may be evaluated like titration points from any single-step system (cf. Sec. 6), permitting straightforward assignment of a value to pK_1.

In the same way, the remaining two equilibria may also be evaluated in isolation. Thus, the pencils of planes with common lines of intersection $\overline{BH_3\,B^{3-}}$ and $\overline{BH_3\,BH_2^-}$, taken together with the corresponding tetrahedral edges $\overline{BH_2^-\,BH^{2-}}$ and $\overline{BH^{2-}\,B^{3-}}$, provide the titration points $^2P_1$ and $^3P_1$, respectively, for the second and third equilibria $BH_2^- \leftrightarrows BH^{2-}$ (H^+) and $BH^{2-} \leftrightarrows B^{3-}$ (H^+). These include the bisecting points H_2 and H_3 indicated in Fig. 8–7.

8.3 Absolute pK Determination

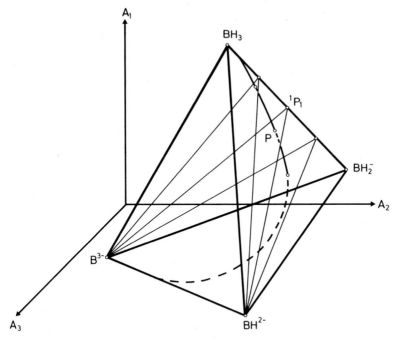

Fig. 8–8. Determination of the points $^1P_1$ for the isolated equilibrium $BH_3 \leftrightarrows BH_2^-$ (H^+) using the pencil of planes generated by the titration points P and the common line of intersection $\overline{B^{3-} \; BH^{2-}}$. The tetrahedral edge $\overline{BH_3 \; BH_2^-}$ penetrates the planes at the points $^1P_1$, which may be evaluated like the points of a single-step titration.

Table 8–2 shows a series of results obtained on the basis of such single-step protolysis equilibria.[13] A corresponding diagram for one of the equilibria, constructed using the relationship

$$(A_\lambda - A_{\lambda BH}) \cdot 10^{+pH} = -\frac{1}{K_2} A_\lambda + \frac{1}{K_2} A_{\lambda BH_2}$$

(cf. equation 8–12) is shown in Fig. 8–9.

Table 8–2. 1,2,4-Benzenetricarboxylic acid: Evaluation with the aid of pencils of planes in a three-dimensional A-diagram ($b_0 = 4.74 \cdot 10^{-4}$ M; 20.0 °C).

Wavelength λ (nm)	$A_{\lambda BH_3}$	$A_{\lambda B}$	$A_{\lambda BH_2}$	$A_{\lambda BH}$	pK_1	pK_2	pK_3
310	0.0380	0.0085	0.187	0.110	2.55	3.87	5.48
300	0.3120	0.0945	0.413	0.402	2.52	3.90	5.43
270	0.5375	0.8415	0.742	0.831	2.55	3.86	5.43

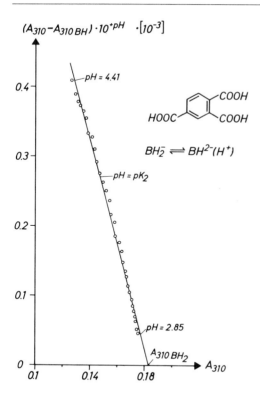

Fig. 8–9. Evaluation of the isolated equilibrium $BH_2^- \leftrightarrows BH^{2-}$ (H^+) for 1,2,4-benzenetricarboxylic acid (cf. text).

8.3.6 Correlation of Two-dimensional A- and ADQ-Diagrams

8.3.6.1 *Determining the Molar Absorptivities of BH_2^- and BH^{2-}*

Three-step acid-base equilibria may be analyzed completely by a two-dimensional geometric method employing two-dimensional A- and ADQ-diagrams.[6,7,16] Figure 8–10 shows the ADQ-diagram for 1,2,4-benzenetricarboxylic acid. The absorbance difference ΔA_λ is here based on the point BH_3 ($\Delta A_\lambda = A_\lambda - A_{\lambda BH_3}$; cf. Sec. 7.8). In the ADQ-diagram, all titration points converge at the single point "BH_3" in the event that only the first equilibrium $BH_3 \leftrightarrows BH_2^-$ (H^+) is spectroscopically observable. If the second equilibrium is also spectroscopically active then the point becomes a curve proceeding in the direction of the line $\overline{BH_3\,BH^{2-}}$. It deviates from this line once the third equilibrium begins to produce an effect. The curve terminates in the point "B^{3-}" where only the protolyte B^{3-} is present.

8.3 Absolute pK Determination

Fig. 8–10 shows the point "BH^{2-}" to lie at the intersection of tangents to the curve at BH$_3$ and B^{3-} The following relationships may be shown to apply at this point (cf. Sec. 8.6):

$$\frac{\Delta A_1}{\Delta A_2} = \frac{q_{12}}{q_{22}} = \frac{\varepsilon_{1BH} - \varepsilon_{1BH_3}}{\varepsilon_{2BH} - \varepsilon_{2BH_3}} = 0.212 \tag{8-17a}$$

and

$$\frac{\Delta A_3}{\Delta A_2} = \frac{q_{32}}{q_{22}} = \frac{\varepsilon_{3BH} - \varepsilon_{3BH_3}}{\varepsilon_{2BH} - \varepsilon_{2BH_3}} = 0.336 \tag{8-17b}$$

Given the further relationship

$$\Delta A_{\lambda_i} = A_{\lambda_i} - A_{\lambda_i BH_3} \quad (\lambda_i = 1, 2, 3)$$

it follows that

$$A_1 = \frac{q_{12}}{q_{22}} A_2 + (A_{1BH_3} - \frac{q_{12}}{q_{22}} A_{2BH_3}) \tag{8-18a}$$

and

$$A_1 = \frac{q_{32}}{q_{22}} A_2 + (A_{3BH_3} - \frac{q_{32}}{q_{22}} A_{2BH_3}) \tag{8-18b}$$

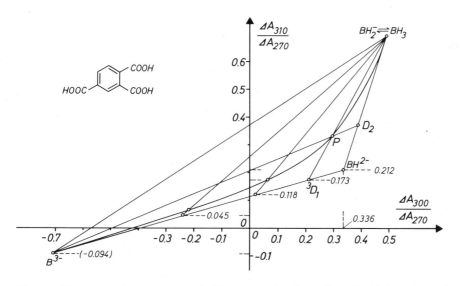

Fig. 8–10. ADQ-Diagram for 1,2,4-benzenetricarboxylic acid ($\Delta A_\lambda = A_\lambda - A_{\lambda BH_3}$). Data points P (only five of which are shown) fall on the indicated curve. The points $^3D_1$ provide the ADQ-coordinates of the isolated equilibrium BH^{2-} ⇆ B^{3-} (H$^+$), while the points D$_2$ provide those of the isolated pair of equilibria BH$_3$ ⇆ BH$_2^-$ (H$^+$) and BH$_2^-$ ⇆ BH^{2-}(H$^+$).

Equations (8–18a) and (8–18b) produce in the diagrams A_1 vs. A_2 and A_3 vs. A_2, respectively, straight lines containing points BH^{2-}. These lines may be constructed provided the quotients q_{12}/q_{22} (= 0.212) and q_{32}/q_{22} (= 0.336) are known, together with $A_{\lambda_i BH_3}$ (λ_i = 1, 2, 3). The straight line represented by equation (8–18a), passing through the points BH_3 and BH^{2-}, is illustrated in Fig. 8–11. In order to determine unambiguously the location of the latter point, the intersection must be established between this line and the tangent to the point B^{3-}, since this tangent must also contain the point BH^{2-} (cf. Sec. 8.3.3). The point of intersection then satisfies the relationship:

$$A_\lambda(BH) = A_{\lambda BH} = l\, b_0\, \varepsilon_{\lambda BH} \tag{8-19}$$

Thus, proper correlation of two-dimensional A- and ADQ-diagrams provides a method for determining $A_{\lambda BH}$ strictly on the basis of spectroscopic measurements.

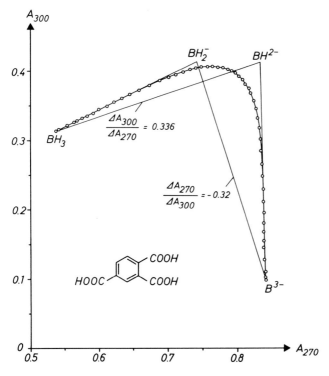

Fig. 8–11. A-Diagram for 1,2,4-benzenetricarboxylic acid. The lines $\overline{BH_3\ BH_2^-}$ and $\overline{B^{3-}\ BH^{2-}}$ are tangents to the points BH_3 and B^{3-}, respectively. The lines $\overline{BH_3\ BH^{2-}}$ and $\overline{B^{3-}\ BH_2^-}$ were constructed with the aid of equations 8–18a and 8–18b on the basis of the points of intersection of the tangents to BH^{2-} and BH_2^- taken from Figures 8–10 and 8–12.

8.3 Absolute pK Determination

In a similarly way, $A_{\lambda BH_2}$ may be determined[16] by constructing an ADQ-diagram based on the point B^{3-}, (cf. Fig. 8–12). The point of intersection of the tangents, BH_2^-, produces a straight line in the corresponding A-diagram, and the point BH_2^- must lie on this line (cf. the line $\overline{B^{3-} BH_2^-}$ in Fig. 8–11). Its point of intersection with the tangent to BH_3 must be the desired point BH_2^-, which satisfies the relationship

$$A_\lambda (BH_2) = A_{\lambda BH_2} = l\, b_0 \varepsilon_{\lambda BH_2} \tag{8-20}$$

The results obtained from applying this procedure to 1,2,4-benzenetricarboxylic acid are given in Table 8–3.

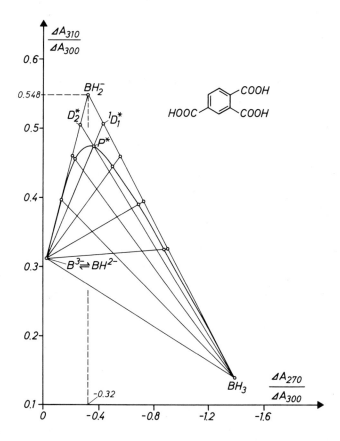

Fig. 8–12. ADQ-Diagram for 1,2,4-benzenetricarboxylic acid ($\Delta A_\lambda = A_\lambda - A_{\lambda B}$). The data points P* (only eight of which are shown) fall on the curve indicated. The points $^1D_1^*$ provide the ADQ-coordinates of the isolated equilibrium $BH_3 \leftrightarrows BH_2^-$ (H$^+$), while the points D_2^* apply to an isolated system comprised of the two equilibria $BH_2^- \leftrightarrows BH^{2-}$ (H$^+$) and $BH^{2-} \leftrightarrows B^{3-}$ (H$^+$).

Table 8–3. 1,2,4-Benzenetricarboxylic acid: Determination of $A_{\lambda BH_2}$ and $A_{\lambda BH}$ through correlation of (two-dimensional) A- and ADQ-diagrams, but without use of pH data ($b_0 = 4.74 \cdot 10^{-4}$ M; 20.0 °C).

Wavelength λ (nm)	$A_{\lambda BH_3}$	$A_{\lambda BH_2}$	$A_{\lambda BH}$	$A_{\lambda B}$
310	0.0380	0.185	0.102	0.0085
300	0.3120	0.412	0.411	0.0945
270	0.5375	0.740	0.839	0.8415

8.3.6.2 Reducing the System by One Stage of Equilibrium

Appropriate use of A- and ADQ-diagrams also makes it possible to reduce a three-step protolysis system to a simpler two-step system:[6]

$$BH_3 \leftrightarrows BH_2^- \ (H^+)$$
$$BH_2^- \leftrightarrows BH^{2-} (H^+) \tag{8–15}$$

Consider the diagram of Fig. 8–10. The two lines $\overline{P\ B^{3-}}$ and $\overline{BH_3\ BH^{2-}}$ intersect in the point D_2, where $\overline{BH_3\ BH^{2-}}$ is the tangent to BH_3. This is the point one would obtain at a pH corresponding to P in the three-step system if it were possible to titrate the reduced system described by equations 8–15. If the pH is now varied, what results is a pencil of rays converging in the point B^{3-} (only one of the corresponding lines is indicated in Fig. 8–10). Their points of intersection (D_2) with the tangent $\overline{BH_3\ BH_2^-}$ provide ADQ-values for the reduced system (cf. Sections 8.6).

The points D_2 in turn generate pencils of rays in the corresponding A-diagrams. As shown by the simulated case in Fig. 8–13, these lines pass through BH_3, and along them lie the desired points for the reduced system. Locations are established for these points by referring to the titration curve for the three-step system (cf. the curve $s = 3$ in Fig. 8–13). The points constituting this curve, taken together with the titration limit B^{3-}, permit construction of another pencil of rays on which the points for system 8–15 must also lie (cf. $\overline{P\ B^{3-}}$ and Sec. 8.3.4). Thus, the desired points are to be sought at the intersections formed by corresponding lines from the two pencils of rays, one originating in the point BH_3 and the other in B^{3-} (cf. the curve $s = 2$ in Fig. 8-13). A single pH value is therefore common to the points on any given pair of lines. The curve corresponding to the reduced system may be evaluated further using the methods introduced in Chapter 7. Consequently, both pK_1 and pK_2 may be ascertained by a two-dimensional approach.

In an analogous way it is also possible to separate the last two equilibria of the three-step system, making use of an ADQ-diagram based on the point B^{3-}.[6]

8.3 Absolute pK Determination

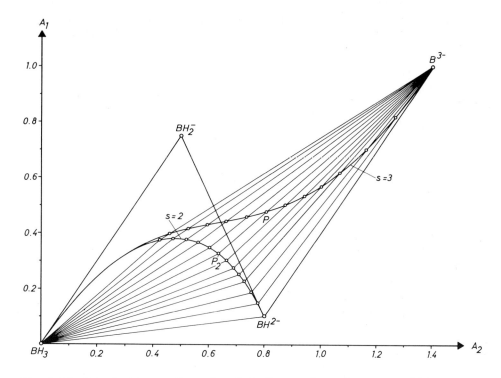

Fig. 8–13. Calculated (two-dimensional) A-diagram for a three-step acid-base system ($s = 3$, data points P, $pK_1 = 6.0$, $pK_2 = 6.5$, $pK_3 = 7.0$). The points P_2 supply the titration points of the isolated pair of equilibria $BH_3 \leftrightarrows BH_2^-$ (H^+) and $BH_2^- \leftrightarrows BH^{2-}$ (H^+) ($s = 2$). The points P_2 occur at the intersections of the pencils of rays emanating from BH_3 and B^{3-}. Intersecting lines correspond in each case to identical values of pH. The pencil of rays associated with BH_3 may be constructed on the basis of the points D_2 in the ADQ-diagram of Fig. 8–10.

8.3.6.3 Determining pK Values from the Three Separate Equilibria

The pK values associated with a three-step system may also be determined sequentially using correlations based on two-dimensional A- and ADQ-diagrams.[6,7,17] The process of establishing pK_3 for the equilibrium

$$BH^{2-} \overset{K_3}{\leftrightarrows} B^{3-} (H^+)$$

begins from a diagram like that shown in Fig. 8–10. The point $^3D_1$ lies at the intersection of the line $\overline{BH_3\,P}$ and the tangent $\overline{B^{3-}BH^{2-}}$ (for an explanation of the symbol $^3D_1$ see Sec. 8.3.6.1). If the points $^3D_1$ corresponding to various titration points are determined, and the lines they generate are introduced into the appropriate A-diagram, there results a pencil of rays emanating from the point BH$_3$ (cf. Fig. 8–14). In order unambiguously to establish the points along these lines that comprise the points ^{3}P$_1$ of the isolated (third) equilibrium it is necessary to seek the points of intersection with the tangent to B^{3-}. The pH values of the points ^{3}P$_1$ are identical to those of the corresponding three-step system. Points ^{3}P$_1$ may be evaluated just like the titration points of a single-step acid-base equilibrium. Table 8–4 contains results obtained by this method for 1,2,4-benzenetricarboxylic acid.[17]

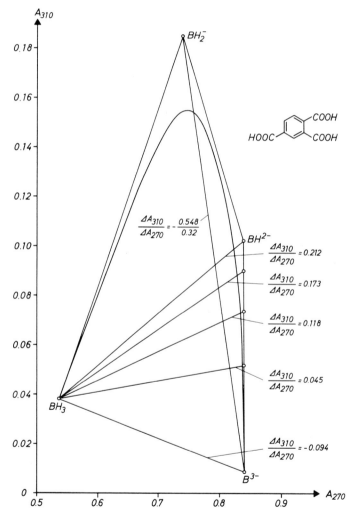

Fig. 8–14. A-Diagram for 1,2,4-benzenetricarboxylic acid. The lines $\overline{\text{BH}_3\,\text{BH}_2^-}$ and $\overline{\text{BH}_3\,\text{BH}^{2-}}$ are tangent to BH$_3$ and B^{3-} (data points defining the absorbance curve have been omitted). Rays emerging from BH$_3$ were determined from the $\Delta A_{310}/\Delta A_{270}$ coordinates of the points ^{3}D$_1$ in Fig. 8–10. The points ^{3}P$_1$ of the isolated equilibrium BH^{2-} ⇆ B^{3-} (H$^+$) fall on the line $\overline{\text{B}^{3-}\,\text{BH}^{2-}}$.

8.3 Absolute pK Determination

Table 8–4. Establishment of a complete set of pK values for 1,2,4-benzenetricarboxylic acid using equations 8–11 and 8–12, together with the relationship

$$(A_\lambda - A_{\lambda BH}) \cdot 10^{+pH} = -\frac{1}{K_2} A_\lambda + \frac{1}{K_2} A_{\lambda BH_2}$$

based on absorbances due to the three isolated equilibria (cf. the discussion in the text).

Wavelength λ (nm)	pK_1	pK_2	pK_3
310	2.56	3.89	5.40
300	2.55	...	5.38
270	2.56	3.92	5.39

In order to determine pK_2 for the equilibrium

$$BH_2^- \overset{K_2}{\leftrightarrows} BH^{2-} (H^+)$$

it is necessary to insert into the appropriate A-diagram the pencil of rays derived from the points D_2 and D_2^* in Figures 8–10 and 8–12, respectively (cf. Fig. 8–15). The pH values corresponding to D_2 and D_2^* are in this case identical.[17] The points of intersection ($^2P_1$) of the two pencils of rays provide the titration points of the isolated second equilibrium and thus pK_2 as well (cf. Table 8–4).

To determine pK_1 of the equilibrium

$$BH_3 \overset{K_1}{\leftrightarrows} BH_2^- (H^+)$$

one begins with the points $^1D_1^*$ in Fig. 8–12. These are the points of intersection between the pencil of rays shown and the tangent to BH_3. The coordinates of $^1D_1^*$ produce a pencil of rays in the A-diagram that converges at the titration limit B^{3-}. (These lines have been omitted from Fig. 8–14 in the interest of clarity.) Intersections of the lines with the tangent to BH_3 represent equivalent titration points for the isolated first equilibrium (cf. Table 8–4).

8.3.7 Linear and Non-linear Regression Analysis

Just as in the case of two-step acid-base equilibria, three-step systems can also be evaluated by solving sets of simultaneous equations. For example, reformulation of equation 8–7 leads to:

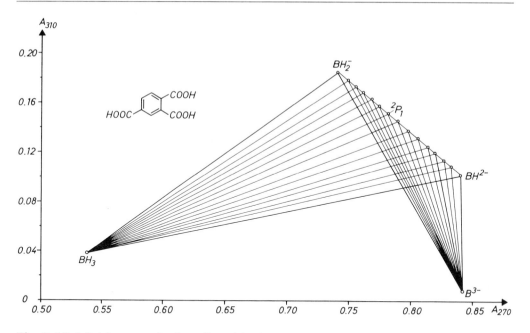

Fig. 8–15. 1,2,4-benzenetricarboxylic acid: determining absorbances for the isolated equilibrium $BH_2^- \leftrightarrows BH^{2-}$ (H^+) using a two-dimensional A-diagram (the absorbance curve for the three-component system has been omitted).
The lines $\overline{BH_3\,BH_2^-}$ and $\overline{B^{3-}\,BH^{2-}}$ are the tangents to BH_3 and B^{3-}. The pencils of rays originating from BH_3 and B^{3-} were constructed on the basis of the ADQ-coordinates of the points D_2 from Fig. 8–10 and D_2^* from Fig. 8–12, respectively (cf. equations 8-18a and 8-18b). The points $^2P_1$ for the isolated equilibrium $BH_2^- \leftrightarrows BH^{2-}$ (H^+), not all of which are shown, fall on the line $\overline{BH_2^-\,BH^{2-}}$.

$$\left(A_\lambda - lb_0\,\varepsilon_{BH_3}\right)h = x_1 \cdot lb_0 - x_2 \cdot A_\lambda + x_3 \cdot lb_0 \cdot \frac{1}{h} - x_4 \cdot \frac{A_\lambda}{h} +$$
$$+ x_5\left(lb_0\,\varepsilon_B - A_\lambda\right)\frac{1}{h^2} \tag{8-21}$$

where

$$x_1 = \varepsilon_{\lambda BH_2}K_1,\quad x_2 = K_1,\quad x_3 = \varepsilon_{\lambda BH}K_1K_2,\quad x_4 = K_1K_2,\quad \text{and} \quad x_5 = K_1K_2K_3$$

In principle, the five unknowns $x_1\ldots x_5$ (and therefore K_1, K_2, K_3, $\varepsilon_{\lambda BH_2}$, and $\varepsilon_{\lambda BH}$) may all be determined from five sets of A_λ-pH measurements, provided the absorbances $A_{\lambda BH_3} = l\,b_0\varepsilon_{\lambda BH_3}$ and $A_{\lambda B} = l\,b_0\varepsilon_{\lambda B}$ are known to a sufficient degree of accuracy. If more than five sets of measurements are available, then the overdetermined system of linear equations may be advantageously solved by *linear regression* (cf. Sec. 7.3.6).

8.3 Absolute pK Determination

It is not necessarily mandatory that $A_{\lambda BH_3}$ and $A_{\lambda B}$ be known in order for such an analysis to be successful. An alternative formulation of equation 8–7 leads to:

$$A_\lambda = x_1 \cdot lb_0 - x_2 \cdot \frac{A_\lambda}{h} + x_3 \cdot lb_0 \cdot \frac{1}{h} - x_4 \cdot \frac{A_\lambda}{h^2} + x_5 \cdot lb_0 \cdot \frac{1}{h^2} -$$
$$- x_6 \cdot \frac{A_\lambda}{h^3} + x_7 \cdot lb_0 \cdot \frac{1}{h^3} \qquad (8\text{--}22)$$

where

$$x_1 = \varepsilon_{\lambda BH_3}, \quad x_2 = K_1, \quad x_3 = \varepsilon_{\lambda BH_2} \cdot K_1, \quad x_4 = K_1 K_2,$$
$$x_5 = \varepsilon_{\lambda BH} \cdot K_1 K_2, \quad x_6 = K_1 K_2 K_3, \text{ and } x_7 = \varepsilon_{\lambda B} \cdot K_1 K_2 K_3$$

This equation should also be amenable to solution by the methods of linear regression, although such an approach is rarely employed today. In general, *non-linear regression* methods based on equation 8–7 are preferred, and a number of powerful computer programs are available for the purpose.[18-30] Often, programs intended for the analysis of metal complex equilibria are suitable after appropriate modification.[31-39,27-29]

In the case of 1,2,4-benzenetricarboxylic acid, application of the program SPSS-NONLINEAR (cf. Sec. 7.3.6) to the titration curves A_λ vs. pH leads to the results presented in Table 8–5 and Fig. 8–16.[40] The circles in Fig. 8–16 indicate measured values; solid lines are calculated curves, which intersect essentially all the data points. Each titration curve was evaluated independently, although the complete set of A_λ-pH data may also be subjected to simultaneous evaluation (cf. Table 8–6), thereby permitting a determination of the standard deviation not only of the calculated absorbances but also of the pK values. The observed standard deviation for the individual absorbances is less than ± 0.002. The following overall results are obtained for the pK values:

$pK_1 = 2.53_0 \pm 0.005$
$pK_2 = 3.98_9 \pm 0.010$
$pK_3 = 5.40_6 \pm 0.005$

Table 8–5. Evaluation of individual titration curves for 1,2,4-benzenetricarboxylic acid using the program SPSS-NONLINEAR and equation 8–7 [($BH_3 \leftrightarrows BH_2^- (H^+)$, $BH_2^- \leftrightarrows BH^{2-} (H^+)$, $BH^{2-} \leftrightarrows B^{3-} (H^+)$, where $pK_i = -\log K_i$ ($i = 1, 2, 3$), $b_0 = 4.74 \cdot 10^{-4}$ M, $l = 1$ cm; 20.0 °C].

Wavelength λ (nm)	$A_{\lambda BH_3}$ calc.	exp.	$A_{\lambda BH_2}$ calc.	$A_{\lambda BH}$ calc.	$A_{\lambda B}$ calc.	exp.	pK_1	pK_2	pK_3
310	0.0381	0.040	0.1863	0.1041	0.0092	0.001	2.55	3.93	5.39
305	0.1303	0.133	0.2782	0.2462	0.0289	0.030	2.54	3.88	5.40
300	0.3118	0.314	0.4194	0.3985	0.0952	0.096	2.56	3.86	5.40
290	0.7100	0.710	0.6811	0.6821	0.4064	0.407	2.39	3.97	5.41
274	0.5273	0.528	0.6137	0.6819	0.6444	0.645	2.54	3.95	5.41
268	0.5766	0.582	0.8819	0.9637	0.9931	0.994	2.53	3.97	5.33

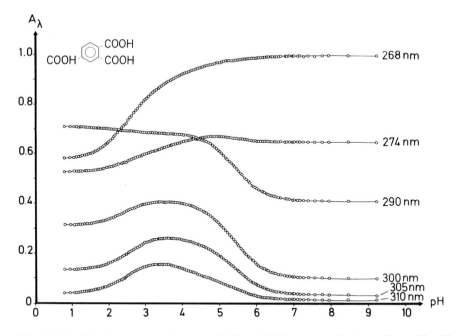

Fig. 8–16. Titration curves A_λ vs. pH for 1,2,4-benzenetricarboxylic acid. The lines are based on calculations made with the aid of SPSS NONLINEAR, while the circles represent data points.

8.3 Absolute pK Determination

Table 8–6. Combined evaluation of a complete set of titration curves for 1,2,4-benzenetricarboxylic acid using the program SPSS-NONLINEAR and equation 8–7 (cf. also Table 8–5). The absorbances reported in the table showed a maximum deviation of ± 0.0015 absorbance units (with $pK_1 = 2.530 \pm 0.005$, $pK_2 = 3.989 \pm 0.010$, and $pK_3 = 5.406 \pm 0.005$).

Wavelength λ (nm)	$A_{\lambda BH_3}$ calc.	exp.	$A_{\lambda BH_2}$ calc.	$A_{\lambda BH}$ calc.	$A_{\lambda B}$ calc.	exp.
310	0.0377	0.040	0.1843	0.1003	0.0092	0.001
305	0.1302	0.133	0.2780	0.2423	0.0289	0.030
300	0.3110	0.314	0.4183	0.3946	0.0953	0.096
290	0.7088	0.710	0.6794	0.6825	0.4064	0.407
274	0.5269	0.528	0.6144	0.6835	0.6442	0.645
268	0.5769	0.582	0.8829	0.9665	0.9933	0.994

A comparison of the titration curves determined independently with those resulting from a simultaneous determination shows that pK values from the former deviate from the average values derived from the latter by about 0.15 pH units. Such behavior is observed whenever the corresponding titration stages lead to only modest changes (< 0.03) in absorbance (cf. Table 8–5). The advantage of a simultaneous evaluation utilizing all the experimental data is therefore obvious.

Performing calculations on the same set of data using the program TIFIT (cf. Chapter 10) provides the following overall result:

$pK_1 = 2.53 \pm 0.01$
$pK_2 = 3.97 \pm 0.01$
$pK_3 = 5.41 \pm 0.01$

Just as in the case of o-phthalic acid, no correction has been made for ionic strength effects on the pH titration of 1,2,4-benzenetricarboxylic acid.[11] The resulting pK values should therefore not be regarded as precise. The high degree of mutual consistency that was achieved may be attributed to the fact that the titration was performed under nitrogen, and to the availability of 129 separate pH measurements.

8.4 The Determination of K_1/K_2 and K_2/K_3 with the Aid of the Absorbance Tetrahedron

Just as with two-step equilibria (cf. Sec. 7.4), important relationships may be established among the dissociation constants in a three-step acid-base system solely on the basis of spectrometric data. According to equation 8–2, the quotient

$$\frac{K_1}{K_2} = \frac{(bh_2)^2}{bh_3 \cdot bh} \qquad (8-23)$$

is dimensionless. Therefore it is subject to determination on the basis of simple distance relationships within an absorbance tetrahedron.[6,13] As will be seen from Fig. 8–17a, the point P on the curve lies in the plane defined by the points B^{3-}, BH^{2-}, and H_1. Here H_1 is the point of bisection of the line $\overline{BH_3\,BH_2^-}$. Like H_1, point P fulfills the condition $bh_3 = bh_2$. Therefore, equation 8–23 may be simplified for any point P to:

$$\left.\frac{K_1}{K_2}\right|_{bh_3\,=\,bh_2} = \frac{bh_2}{bh} \qquad (8-24)$$

The numerical value of the ratio bh_2/bh may be determined from measurements along the edge $\overline{BH_2^-\,BH^{2-}}$ of the tetrahedron. As shown in Fig. 8–17a, this edge penetrates the plane containing the points BH_3, B^{3-}, and P at the point F, dividing the line $\overline{BH_2^-\,BH^{2-}}$ into two segments. The ratio of these two line segments is directly proportional to bh_2/bh, and consequently to K_1/K_2. This leads to the relationship (cf. Sec. 8.5.4):

$$\frac{K_1}{K_2} = \frac{\overline{BH^{2-}\,F}}{\overline{BH_2^-\,F}} \qquad (8-25)$$

An analogous approach permits the relationship K_2/K_3 to be similarly determined. In Fig. 8–17b, H_2 is the bisecting point of the line $\overline{BH^{2-}\,BH_2^-}$. The curve penetrates the plane determined by BH_3, B^{3-}, and H_2 at the point P. This point and the tetrahedral edge $\overline{BH_3\,BH_2^-}$ define another plane, one which is penetrated by the tetrahedral edge $\overline{BH^{2-}\,B^{3-}}$ at the point F, and the following relationship applies with respect to K_2/K_3:

$$\frac{K_2}{K_3} = \frac{\overline{B^{3-}\,F}}{\overline{BH^{2-}\,F}} \qquad (8-26)$$

Additional relationships may be developed as well,[13] and it is even possible to correlate the pH values of points along the curve with the quantities $K_1 \cdot K_2$, $K_2 \cdot K_3$, and $K_1 \cdot K_2 \cdot K_3$. This subject is explored further in Sec. 8.5.4.

8.4 The Determination of K_1/K_2 and K_2/K_3

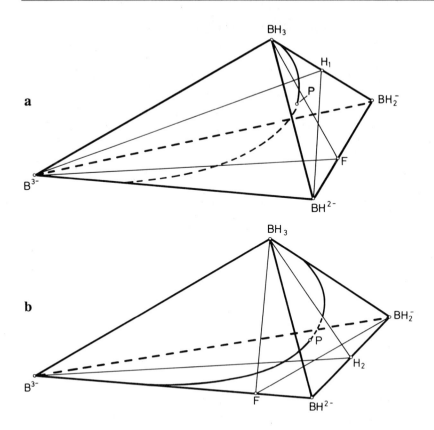

Fig. 8–17. Determination of pK differences on the basis of distance ratios in an absorbance tetrahedron (the nature of the coordinate system has not been specified).
a) The relationship $K_1/K_2 = \overline{BH^{2-} F} / \overline{BH_2^- F}$ may be utilized in the calculation of Δ_{21} pK: = pK_2 − pK_1.
b) Similarly, Δ_{32} pK: = pK_3 − pK_2 may be determined using the relationship $K_2/K_3 = \overline{B^{3-} F} / \overline{BH^{2-} F}$.

The tetrahedra shown in Figures 8–17a and 8–17b may be regarded as existing either in absorbance space or in concentration space, and it is for that reason that axis labels have been omitted. Since absorbance tetrahedra and concentration tetrahedra may be interconverted by affine transformations—leaving distance relationships unaffected—equations 8–25 and 8–26 are equally applicable to both concentration diagrams and absorbance diagrams (cf. also Sec. 7.6).

8.5 The Characteristic Concentration Diagram

This section deals with the most important (differential) geometric characteristics of the curve as it traverses the space defined by the concentration tetrahedron. The relationships discussed also apply to the absorbance tetrahedron, which is simply a concentration tetrahedron that has been distorted through the application of affine transformations. The rules in Sec. 7.6 for affine transformations are equally applicable in three dimensions (although here the role of conic sections is ignored).

8.5.1 Tangents and Osculating Planes

Mass balance for a three-step, unbranched protolysis system requires that:

$$b_0 = bh_3 + bh_2 + bh + b \qquad (8\text{--}4)$$

There are thus three independent concentration variables forming the basis of the characteristic concentration diagram. In the discussion that follows, bh_3, bh, and b represent arbitrary concentrations corresponding to these three variables. Fig. 8–18 shows a typical concentration curve for a three-step system. The curve lies entirely within the so-called *concentration tetrahedron*, three of whose edges correspond to the coordinate axes. The *position vector* $\mathbf{c}$ points from the origin of the coordinate system in the direction of the point P located on the concentration curve. The following relationship may be established, in analogy to equations 7–61a and 7–61b:[41,42]

$$\vec{c} = \mathbf{c} = \begin{pmatrix} bh_3 \\ bh \\ b \end{pmatrix} = bh_3 \vec{e_1} + bh\, \vec{e_2} + b\, \vec{e_3} \qquad (8\text{--}27)$$

A three-dimensional coordinate system is defined by the three mutually orthogonal (basis) vectors $\vec{e_1}$, $\vec{e_2}$, and $\vec{e_3}$. According to equation 8–5, the coordinates bh_3, bh, and b are a function of the parameter h. If one now differentiates with respect to h, the resulting vector $\mathbf{t'}$ indicates the direction of the tangent to the curve:[41,42]

$$\mathbf{t'} = \frac{d\mathbf{c}}{dh} = \begin{pmatrix} \dfrac{d\,bh_3}{dh} \\ \dfrac{d\,bh}{dh} \\ \dfrac{d\,b}{dh} \end{pmatrix} = \frac{d\,bh_3}{dh}\vec{e_1} + \frac{d\,bh}{dh}\vec{e_2} + \frac{d\,b}{dh}\vec{e_3} \qquad (8\text{--}28)$$

The following expression then provides a measure of the length of $\mathbf{t'}$:[41]

8.5 The Characteristic Concentration Diagram

$$|\mathbf{t'}| = \left|\frac{d\mathbf{c}}{dh}\right| = \sqrt{\left(\frac{d\,bh_3}{d\,h}\right)^2 + \left(\frac{d\,bh}{d\,h}\right)^2 + \left(\frac{d\,b}{d\,h}\right)^2} \tag{8-29}$$

If various sets of coordinates determined with the aid of equation 8–5 are now differentiated with respect to h, and the lengths of the vectors $\mathbf{t'}$ are calculated, it becomes apparent that the length varies as one moves along the concentration curve. In order to normalize the vector $\mathbf{t'}$ to a unit length the following quotient may be developed:[41]

$$\mathbf{t} := \frac{\mathbf{t'}}{|\mathbf{t'}|} = \frac{\dfrac{d\mathbf{c}}{dh}}{\left|\dfrac{d\mathbf{c}}{dh}\right|} \tag{8-30}$$

The vector $\mathbf{t}$ is now of unit length and points in the direction of the tangent to the curve.

In the limiting case of $h \to \infty$ only the component BH_3 is present ($bh_3 = b_0$). Therefore, according to equation 8–29:

$$\mathbf{t'}\bigg|_{h \to \infty} = \frac{b_0 K_1}{h^2} \begin{pmatrix} 1 \\ 0 \\ 0 \end{pmatrix} \tag{8-31}$$

For the corresponding vector $\mathbf{t}$ (cf. equation 8–10):

$$\mathbf{t}\bigg|_{h \to \infty} = \begin{pmatrix} 1 \\ 0 \\ 0 \end{pmatrix} = \vec{e_1} \tag{8-32}$$

According to this result, the tangent vector $\mathbf{t}$ constitutes a unit vector along the coordinate axis bh_3![6] Any vector lying along the bh_3-axis must fulfill the following relationship (cf. equation 8–4, where $bh = b = 0$, and Fig. 8-18):

$$b_0 = bh_3 + bh_2 \tag{8-33}$$

Thus for the coordinate origin (where $bh_3 = 0$) it follows that $bh_2 = b_0$. Since this relationship includes the point BH_2^-, it is apparent that BH_2^- must be coincident with the origin. This in turn means that the tangent vector for the point BH_3 supplies a direction for the line along which the point BH_2^- lies!

At the limit $h \to 0$ only the component B^{3-} is present ($b = b_0$). If one now determines with the aid of Fig. 8–18 the vectors $\mathbf{t'}$ and $\mathbf{t}$ for the point B^{3-} it will be found, in analogy to the foregoing case, that

$$\mathbf{t'}\bigg|_{h \to 0} = \frac{b_0}{K_3} \begin{pmatrix} 0 \\ 1 \\ -1 \end{pmatrix} \tag{8-34}$$

and

$$\mathbf{t}\bigg|_{h \to 0} = \frac{1}{\sqrt{2}} \begin{pmatrix} 0 \\ 1 \\ -1 \end{pmatrix} = \frac{1}{\sqrt{2}} \vec{e_2} - \frac{1}{\sqrt{2}} \vec{e_3} \tag{8-35}$$

These vectors indicate the direction of the line $\overline{B^{3-} BH^{2-}}$ and show that the point BH^{2-} lies on the tangent to B^{3-}.[6]

The tangents to BH_3 and B^{3-} show the directions of lines along which the titration points of the equilibria $BH_3 \leftrightarrows BH_2^- (H^+)$ and $BH^{2-} \leftrightarrows B^{3-} (H^+)$ would lie if these could be titrated in isolation. If one inquires into the geometric locations of the titration points of the isolated first two and last two equilibria one encounters yet another property of the curve, a property that can prove very useful in the course of an analysis.

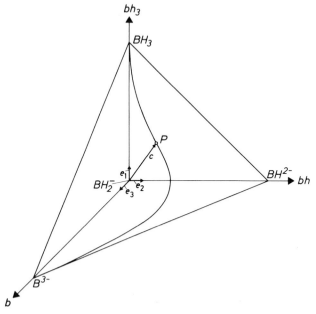

Fig. 8–18. The characteristic concentration diagram for a three-step acid-base system (with $pK_1 = 5.0$, $pK_2 = 5.5$, and $pK_3 = 6.0$). Since in this case absolute concentrations are of relatively little interest, the coordinate axes are scaled in terms of concentration units. As a result, the concentration variables may be regarded as "reduced quantities".

8.5 The Characteristic Concentration Diagram

Consider on the basis of Fig. 8–18 the behavior of the curve in the vicinity of the point BH_3. In particular, note that the curve originates in the plane defined by the coordinate axes bh_3 and bh. Similarly, in the vicinity of the point B^{3-} the curve appears to approach the plane defined by the axes bh and b. In order to verify that these observations are accurate it is necessary to invoke, in addition to the tangent vector, yet another quantity from differential geometry, one associated with the location of a curve in space.

Curvature is a measure of the deviation of a curve from a straight line.[42] Displacing any point along the concentration curve by a small amount leads to a change in the orientation of the spatial tangent. A set of such motions taken together describes a plane whose nature is characteristic of the corresponding segment of the curve. Considering the matter in reverse, imagine the plane defined by three points on the curve, all in relatively close proximity to one another. If two of these points are now caused to approach the third, all motion occurs in a plane passing through the third point and known as the *osculating plane* for that point. The osculating plane for BH_3 is thus identical to the coordinate plane formed by the bh_3 and bh axes (cf. Fig. 8–18).

An osculating plane may be identified with the aid of tangent vectors and what are known as *curvature vectors*. The following relationship defines the curvature vector $\dot{\mathbf{t}}$:[41]

$$\dot{\mathbf{t}} = \frac{\frac{d\mathbf{t}}{dh}}{\left|\frac{d\mathbf{c}}{dh}\right|} \tag{8-36}$$

The curvature vector $\dot{\mathbf{t}}$ is orthogonal to the unit-tangent vector $\mathbf{t}$, a fact that may easily be demonstrated.[41] The tangent vector $\mathbf{t}$ is a unit vector (i.e., $|\mathbf{t}| = 1$), so it follows for a scalar product that

$$\mathbf{t} \cdot \mathbf{t} = 1$$

Differentiating this expression with respect to h leads to:

$$\mathbf{t} \cdot \frac{d\mathbf{t}}{dh} + \frac{d\mathbf{t}}{dh} \cdot \mathbf{t} = 0 \quad \text{or} \quad \mathbf{t} \cdot \frac{d\mathbf{t}}{dh} = 0 \tag{8-37}$$

Thus, $\mathbf{t}$ must be orthogonal to $d\mathbf{t}/dh$, which means it is orthogonal to $\dot{\mathbf{t}}$ (the same is true for the unit curvature vector $\dot{\mathbf{t}}/|\dot{\mathbf{t}}|$).

In order to use equation 8–36 to establish the curvature vector $\dot{\mathbf{t}}$ for the point BH_3 one begins by developing the corresponding tangent vector $\mathbf{t}$ from equation 8–30 with the aid of equation 8–5. Differentiating with respect to h, forming the quotients specified in equation 8–36, dividing the result by $|\dot{\mathbf{t}}|$ in order to normalize the vector $\dot{\mathbf{t}}$, and finally extrapolating to the limiting value $h \to \infty$ provides the following result:

$$\left.\frac{\dot{t}}{|\dot{t}|}\right|_{h \to \infty} = \begin{pmatrix} 0 \\ 1 \\ 0 \end{pmatrix} = \vec{e_2} \tag{8-38}$$

In other words, the (unit) curvature vector for the point BH_3 is directed toward the bh axis (cf. Fig. 8–18). Equation 8–32 indicates that the tangent vector for BH_3 is coincident with the bh_3 axis. Thus, the tangent vector is perpendicular to the curvature vector, a result that is also apparent from equation 8–37. These two vectors define the osculating plane identified previously as the surface containing the coordinate axes bh_3 and bh. It has therefore been demonstrated that in the vicinity of BH_3 the concentration curve approaches the plane containing the points BH_3, BH_2^-, and BH^{2-}.[6]

In order to determine the osculating plane for the point B^{3-} it is necessary to ascertain the (unit) curvature vector for the limiting case of $h \to 0$. Thus:

$$\left.\frac{\dot{t}}{|\dot{t}|}\right|_{h \to 0} = \frac{1}{\sqrt{2}} \begin{pmatrix} 0 \\ -1 \\ -1 \end{pmatrix} = -\frac{1}{\sqrt{2}} \vec{e_2} - \frac{1}{\sqrt{2}} \vec{e_3} \tag{8-39}$$

This unit curvature vector lies in the plane of the coordinate axes bh and b and is perpendicular to the tangent vector of equation 8–35. As a result, BH_2^-, BH^{2-}, and B^{3-} must be points in the osculating plane for B^{3-}. Since according to Fig. 8–18 the points BH_2^- and BH^{2-} belong to the osculating planes for both BH_3 and B^{3-}, the bh axis must be the line along which the two osculating planes intersect, and it is here that the tetrahedral points BH_2^- and BH^{2-} must be sought. Their precise positions may be ascertained by projecting the tangents for BH_3 and B^{3-} until they intersect this line. The result is a highly significant one in that it permits determination of the vertices BH_2^- and BH^{2-} of the absorbance tetrahedron by a differential geometric method based solely on spectrometric data (cf. Sec. 8.3.3).

8.5.2 Pencils of Rays

A three-step protolysis system may be reduced by the equilibrium $BH^{2-} \leftrightarrows B^{3-}$ (H^+) by constructing within the concentration tetrahedron a *pencil of rays* from the limiting point B^{3-} through the titration points P (cf. Fig. 8–19). It is then necessary to establish the points (P_2) at which these lines intersect the osculating plane for BH_3 (cf. Sec. 8.5.1). This is most easily accomplished by expressing the lines $\overline{B^{3-} P}$ in parameter form, which entails introducing the position vectors c_B and c_P (not shown in Fig. 8–19) leading from the coordinate origin to the points B^{3-} and P, respectively. The directional vector ($c_P - c_B$) then indicates the direction from the point B^{3-} to the point P. Using the vectors c_B and ($c_P - c_B$), every point in the line $\overline{B^{3-} P}$ may be described by means of a vector c:

8.5 The Characteristic Concentration Diagram

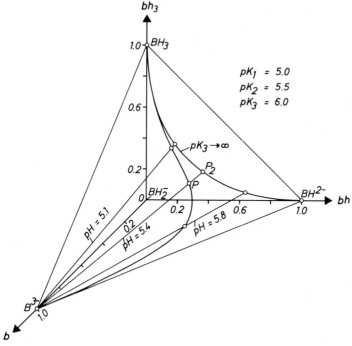

Fig. 8–19. The "pencil of rays" method applied to a three-step acid-base system. The curve associated with the points P represents the three-step system (with $pK_1 = 5.0$, $pK_2 = 5.5$, and $pK_3 = 6.0$). If pK_3 is varied, then at constant pH, pK_1, and pK_2 all points fall on the line $\overline{B^{3-}\ P}$. When pK_3 is large the three-step system is reduced by the equilibrium $BH^{2-} \leftrightarrows B^{3-}\ (H^+)$ (cf. the points P_2). Varying the pH results in a pencil of rays.

$$\mathbf{c} = \begin{pmatrix} bh_3 \\ bh \\ b \end{pmatrix} = \mathbf{c}_B + k(\mathbf{c}_P - \mathbf{c}_B) \quad \text{with} \quad -\infty < k < \infty \tag{8–40}$$

The new parameter k may take any real value (parametric form).[41,42] If the coordinates for B^{3-} and P are now introduced into this equation (cf. equation 8–5), the following expression is obtained:

$$\begin{pmatrix} bh_3 \\ bh \\ b \end{pmatrix} = \begin{pmatrix} 0 \\ 0 \\ b_0 \end{pmatrix} + k \cdot \frac{b_0}{H_3} \begin{pmatrix} h^3 \\ hK_1K_2 \\ K_1K_2K_3 - H_3 \end{pmatrix} \tag{8–41}$$

Solving this for b leads to:

$$b = b_0 + k \cdot \frac{b_0}{H_3} (K_1K_2K_3 - H_3)$$

This in turn may be used to derive an expression for k when $b = 0$. Setting the result into equation 8–41 leads to the position vector $\mathbf{c}_{P_2}$:

$$\mathbf{c}_{P_2} = \frac{b_0}{H_2} \begin{pmatrix} h^2 \\ K_1 K_2 \\ 0 \end{pmatrix} \qquad (8\text{--}42)$$

where $H_2 = K_1 K_2 + K_1 H + h^2$

Thus, the concentrations bh_3 and bh at P_2 have the same functional values as the concentrations of the substances BH_2 and B^{2-} in a two-step acid-base equilibrium (cf. equation 7–5)! That is, the points P_2 provide the titration points one would observe if the first two equilibria could be titrated in isolation.[6]

In an analogous way the first equilibrium $BH_3 \leftrightarrows BH_2^- $ (H^+) may be split apart from the system by constructing the pencil of rays linking the titration points P with BH_3. The rays in this case penetrate the osculating plane of B^{3-} (cf. Sec. 8.5.1) at the points:

$$\begin{pmatrix} bh_3 \\ bh \\ b \end{pmatrix} = \frac{b_0 K_2}{H_2'} \begin{pmatrix} 0 \\ h \\ K_3 \end{pmatrix} \qquad (8\text{--}43)$$

where $H_2' = K_2 K_3 + K_2 h + h^2$

These are the concentrations one would obtain if the last two equilibria could be titrated in isolation.[6]

8.5.3 Pencils of Planes

The equilibria constituting a three-step protolysis system may be evaluated individually with the aid of *pencils of planes*.[6,7,13] To illustrate the method, consider the first equilibrium, $BH_3 \leftrightarrows BH_2^-$ (H^+). One begins by constructing the pencil of planes encompassing the titration points P and the tetrahedral edge $\overline{B^{3-} BH^{2-}}$, which is also the tangent to B^{3-} (cf. Fig. 8–20, in which only one such plane has been indicated). The bh_3 axis penetrates these planes at the points $^1P_1$ (for an explanation of the notation see Fig. 8-20). In order to determine the coordinates of these points it is useful to express the equations of the planes in parametric form.

The following relationships specify the position vectors $\mathbf{c}_B$, $\mathbf{c}_P$, and $\mathbf{c}_{BH}$ leading from the coordinate origin to the points B^{3-}, P, and BH^{2-}, respectively (cf. also equation 8–5):

$$\mathbf{c}_B = \begin{pmatrix} 0 \\ 0 \\ b_0 \end{pmatrix}, \quad \mathbf{c}_P = \frac{b_0}{H_3} \begin{pmatrix} h^3 \\ h K_1 K_2 \\ K_1 K_2 K_3 \end{pmatrix}, \quad \text{and} \quad \mathbf{c}_{BH} = \begin{pmatrix} 0 \\ b_0 \\ 0 \end{pmatrix} \qquad (8\text{--}44)$$

8.5 The Characteristic Concentration Diagram

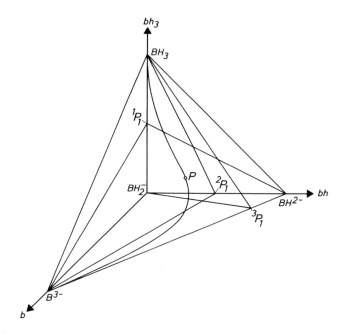

Fig. 8–20. Pencils of planes for a three-step acid-base system. The curve (with the points P) has been calculated for $pK_1 = 5.0$, $pK_2 = 5.5$, and $pK_3 = 6.0$. With the aid of selected planes it is possible to determine the points $^1P_1$, $^2P_1$, and $^3P_1$. Here the lower right index ("1") indicates that an isolated *single-step* equilibrium is involved. The upper left index specifies which of the equilibria this is [i.e., a "1" implies the first equilibrium, $BH_3 \leftrightarrows BH_2^- (H^+)$].

The directional vectors $(c_{BH} - c_B)$ and $(c_P - c_B)$ define the plane containing the points B^{3-}, BH^{2-}, and P. These vectors permit a unique specification for each point in the plane, which is described in parametric notation as follows:[41,42]

$$c = \begin{pmatrix} bh_3 \\ bh \\ b \end{pmatrix} = c_B + k_1(c_{BH} - c_B) + k_2(c_P - c_B) \tag{8-45}$$

where $-\infty < k_1 < \infty$ and $-\infty < k_2 < \infty$

The parameters k_1 and k_2 may assume any real value. The bh_3 axis penetrates this plane at the points $^1P_1$. If coordinates along bh_3 are designated by the symbol σ then the following relationship applies to the position vector $c_{^1P_1}$:

$$c_{^1P_1} = \begin{pmatrix} \sigma \\ 0 \\ 0 \end{pmatrix} \tag{8-46}$$

The coordinate σ may be determined by substituting equations 8–44 and 8–46 into equation 8–45, with $c = c_{1P_1}$. After rearranging, one obtains for the coordinate equations:

$$0 = -\sigma + k_2 \frac{b_0 h^3}{H_3}$$

$$0 = k_1 b_0 + k_2 \frac{b_0 h K_1 K_2}{H_3} \qquad (8\text{–}47)$$

$$-b_0 = -k_1 b_0 + k_2 \frac{b_0}{H_3}(K_1 K_2 K_3 - H_3)$$

Cramer's rule[43] permits solution of this system of equations to provide the desired value for σ, from which c_{1P_1} may be calculated:

$$c_{1P_1} = \frac{b_0 h}{h + K_1} \begin{pmatrix} 1 \\ 0 \\ 0 \end{pmatrix} \qquad (8\text{–}48)$$

From this it is seen that σ has the same functional values as the concentrations in a single-step acid-base equilibrium (cf. equation 6–9). Thus, the pencil of planes approach has made it possible to investigate the first equilibrium in isolation.

In an analogous manner the second equilibrium $BH_2^- \leftrightarrows BH^{2-}$ (H^+) may be isolated by constructing the pencil of planes joining the titration points P and the tetrahedral edge $\overline{BH_3\, B^{3-}}$. The bh axis, which is also the line of intersection of the osculating planes for BH_3 and B^{3-} (cf. Sec. 8.5.1), penetrates the pencil of planes at the points $^2P_1$ (cf. Fig. 8–20). The coordinates of $^2P_1$ may be expressed as follows:

$$c_{2P_1} = \frac{K_2}{h + K_2} \begin{pmatrix} 0 \\ b_0 \\ 0 \end{pmatrix} \qquad (8\text{–}49)$$

Consequently, the points $^2P_1$ specify the titration points of the isolated second equilibrium.

Finally, the third equilibrium $BH^{2-} \leftrightarrows B^{3-}$ (H^+) may be evaluated separately using the pencil of planes incorporating the titration points P and the tetrahedral edge $\overline{BH_3\, BH_2^-}$, which is tangent to BH_3. The tetrahedral edge $\overline{B^{3-}\, BH^{2-}}$ penetrates these planes at the points $^3P_1$ (cf. Fig. 8–20), to which the following relationship applies:

$$c_{3P_1} = \frac{b_0}{h + K_3} \begin{pmatrix} 0 \\ h \\ K_3 \end{pmatrix} \qquad (8\text{–}50)$$

Each plane within a given pencil corresponds to a specific pH value. If the point $^1P_1$ in Fig. 8–20 is taken to coincide with the bisection point H_1 of the line $\overline{BH_3\, BH_2^-}$ (with $bh_3 = b_0/2$), then according to equation 8–48:

$h = K_1$ or $pH = pK_1$

8.5 The Characteristic Concentration Diagram

Thus, the concentration curve penetrates the plane containing the points B^{3-}, BH_2^-, and H_1 at the point where the pH corresponds to the value of pK_1.

Similarly, the bisecting points H_2 ($= {}^2P_1$) and H_3 ($= {}^3P_1$) of the lines $\overline{BH_2^- \; BH^{2-}}$ and $\overline{BH^{2-} \; B^{3-}}$ lead to values for pK_2 and pK_3 (cf. also Sec. 8.3.5).

8.5.4 Determining K_1/K_2 and K_2/K_3 from the Concentration Tetrahedron

As shown in Sec. 8.4, when $bh_3 = bh_2$ then the following relationship holds true for the ratio K_1/K_2:

$$\left. \frac{K_1}{K_2} \right|_{bh_3 \,=\, bh_2} = \frac{bh_2}{bh} \tag{8-24}$$

The point P on the concentration curve of the first diagram in Fig. 8–21 will be seen to fulfill the condition $bh_3 = bh_2$. This point is located where the curve penetrates the plane that contains the tetrahedral edge $\overline{BH^{2-} \; B^{3-}}$ and bisects the line $\overline{BH_3 \; BH_2^-}$ (points of bisection are indicated in Fig. 8–21 by the symbol "$\frac{1}{2}$"). The validity of this conclusion may be demonstrated by constructing a line (not shown in the diagram) through the point P parallel to the bh_3 axis. This line penetrates the surface of the tetrahedron at the points G and H on the tetrahedral faces (BH_2^-, BH^{2-}, B^{3-}) and (BH_3, BH^{2-}, B^{3-}) respectively. It therefore follows (from equation 8–14 in Sec. 8.3.3) for the concentrations bh_3 and bh_2 that

$$bh_3 = \frac{\overline{P\,G}}{\overline{BH_3 \; BH_2^-}} \cdot b_0 \quad \text{and} \quad bh_2 = \frac{\overline{P\,H}}{\overline{BH_3 \; BH_2^-}} \cdot b_0 \tag{8-51}$$

Since all points in the plane illustrated fulfill the condition $\overline{P\,G} = \overline{P\,H}$, then the concentrations bh_3 and bh_2 must everywhere be identical; that is, the ratio of bh_3 to bh_2 is always one.

The ratio bh_2/bh at the point P is also subject to geometric determination. As shown by the first diagram in Fig. 8–21, the plane containing the points BH_3, B^{3-}, and P divides the line $\overline{BH_2^- \; BH^{2-}}$ at the point F. Therefore the following relationship applies to the ratio bh_2/bh and, according to equation 8–24, to the ratio K_1/K_2 as well:

$$\left. \frac{K_1}{K_2} \right|_{bh_3 \,=\, bh_2} = \frac{bh_2}{bh} = \frac{\overline{BH^{2-}\,F}}{\overline{BH_2^-\,F}} \tag{8-25}$$

The ratio K_1/K_2 can thus be determined on the basis of a simple distance relationship.[6,13] Related methods exist for the determination of K_1/K_2 and K_2/K_3 (cf. Fig. 8–21).

The pH values of certain points on the curve are related to the quantities $K_1 \cdot K_2$, $K_2 \cdot K_3$, and $K_1 \cdot K_2 \cdot K_3$.[6,13] The locations of the points fulfilling these relationships are apparent from the last three diagrams in Fig. 8–21 (where $h = 10^{-pH}$, from

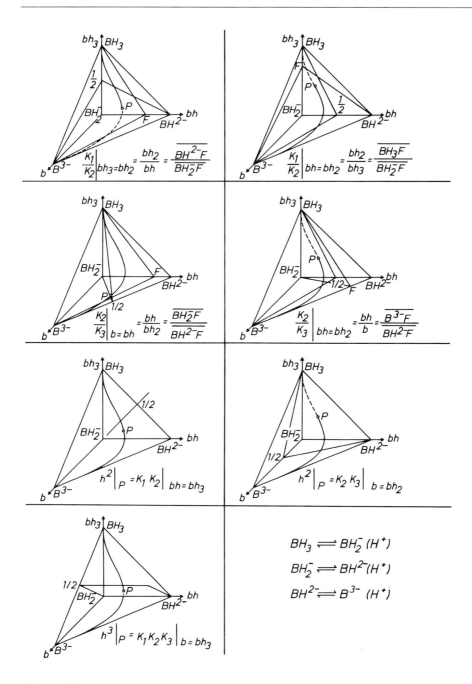

Fig. 8–21. The establishment of a value for K_i/K_j using distance relationships within a concentration tetrahedron (schematic representation). With the aid of selected planes it is also possible to locate points at which the concentrations of two species are identical (e.g., at the point P the concentrations bh and bh_3 are the same: $h^2 = K_1 K_2$).

equation 8–2). All relationships shown in Fig. 8–21 also apply in the absorbance tetrahedron, which may be regarded simply as an affine distortion of the concentration tetrahedron.

8.6 Geometrical Relationships Between A- and ADQ-Diagrams

The results of Sec. 8.5 show that the absorbance tetrahedron possesses the following differential geometric characteristics (cf. Fig. 8–22):

- the vertices BH_2^- and BH^{2-} lie on the tangents to BH_3 and B^{3-}, respectively; and,

- BH_2^- and BH^{2-} are points within the osculating planes for BH_3 and B^{3-}.

As a consequence, it is a straightforward matter to express the equations for the tangents and osculating planes of BH_3 and B^{3-} in an A-diagram. Using parametric representation, the following expression applies to the tangent to BH_3 (cf. equation 8–40):

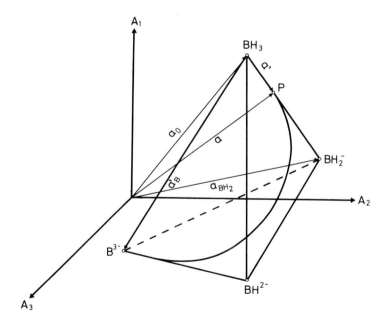

Fig. 8–22. Three-dimensional A-diagram. Each point P on the curve may be represented in the absorbance tetrahedron by a vector $\mathbf{a}$ which fulfills the condition $\mathbf{a} = \mathbf{a_0} + \mathbf{a'}$ (where $\mathbf{a_0} = \overrightarrow{O\,BH_3}$, $\mathbf{a'} = \overrightarrow{BH_3\,P}$, and $\mathbf{a} = \overrightarrow{O\,P}$, "O" being the coordinate origin).

$$\mathbf{a} = \mathbf{a_0} + k(\mathbf{a_{BH_2}} - \mathbf{a_0}) \tag{8-52}$$

where $-\infty < k < \infty$

Thus:

$$\mathbf{a} = \begin{pmatrix} A_1 \\ A_2 \\ A_2 \end{pmatrix}, \quad \mathbf{a_0} = \begin{pmatrix} A_{1BH_3} \\ A_{2BH_3} \\ A_{3BH_3} \end{pmatrix} = l \cdot b_0 \begin{pmatrix} \varepsilon_{1BH_3} \\ \varepsilon_{2BH_3} \\ \varepsilon_{3BH_3} \end{pmatrix}, \quad \text{and}$$

$$\mathbf{a_{BH_2}} = \begin{pmatrix} A_{1BH_2} \\ A_{2BH_2} \\ A_{3BH_2} \end{pmatrix} = l \cdot b_0 \begin{pmatrix} \varepsilon_{1BH_2} \\ \varepsilon_{2BH_2} \\ \varepsilon_{3BH_2} \end{pmatrix}$$

From these results it follows for the directional vector $\mathbf{a'}$ indicating the direction from BH_3 to BH_2^- (cf. Fig. 8–22):

$$\mathbf{a'} = \mathbf{a} - \mathbf{a_0} = k(\mathbf{a_{BH_2}} - \mathbf{a_0}) = \begin{pmatrix} \Delta A_1 \\ \Delta A_2 \\ \Delta A_3 \end{pmatrix} = k \cdot b_0 \begin{pmatrix} q_{11} \\ q_{21} \\ q_{31} \end{pmatrix} \tag{8-53}$$

with

$$q_{\lambda_i 1} = l(\varepsilon_{\lambda_i BH_2} - \varepsilon_{\lambda_i BH_3}) \quad \text{where } \lambda_i = 1, 2, 3$$

As the curve in the absorbance tetrahedron begins to deviate from the tetrahedral edge $\overline{BH_3 \, BH_2^-}$ it passes first along the osculating plane for BH_3. This osculating plane contains the points BH_3, BH_2^-, and BH^{2-}. The plane in which these three points lie fulfills the relationship:[41,42]

$$\begin{vmatrix} \Delta A_1 & b_0 q_{11} & b_0 q_{12} \\ \Delta A_2 & b_0 q_{21} & b_0 q_{22} \\ \Delta A_3 & b_0 q_{31} & b_0 q_{32} \end{vmatrix} = 0 \tag{8-54}$$

As in equation 8-53, $\Delta A_{\lambda_i} = A_{\lambda_i} - A_{\lambda_i BH_3}$ ($\lambda_i = 1, 2, 3$). Furthermore:

$$q_{\lambda_i 1} = l(\varepsilon_{\lambda_i BH_2} - \varepsilon_{\lambda_i BH_3}) \quad \text{and} \quad q_{\lambda_i 2} = l(\varepsilon_{\lambda_i BH} - \varepsilon_{\lambda_i BH_3})$$

Solving this determinant leads to:[6]

$$\Delta A_1 = \alpha_1 \Delta A_2 + \alpha_2 \Delta A_3 \tag{8-55}$$

where

$$\alpha_1 \cdot |\mathbf{Q}| = \begin{vmatrix} q_{11} & q_{12} \\ q_{31} & q_{32} \end{vmatrix}, \quad \alpha_2 \cdot |\mathbf{Q}| = -\begin{vmatrix} q_{11} & q_{12} \\ q_{21} & q_{22} \end{vmatrix}, \quad \text{and} \quad |\mathbf{Q}| = -\begin{vmatrix} q_{21} & q_{22} \\ q_{31} & q_{32} \end{vmatrix} \neq 0$$

This is the equation of the osculating plane for BH_3, expressed in terms of the coordinates of BH_3, BH_2^-, and BH^{2-}.

The curve in the absorbance tetrahedron enters the osculating plane for B^{3-} before it encounters the tetrahedral edge $\overline{B^{3-} BH^{2-}}$. Analogous to equation 8–55, the

8.6 Geometrical Relationships Between A- and ADQ-Diagrams

tetrahedral face containing the points BH_3, BH^{2-}, and B^{3-} may be expressed as:

$$\Delta A_1 = \alpha_1' \Delta A_2 + \alpha_2' \Delta A_3 \tag{8-56}$$

where

$$\alpha_1' \cdot |Q'| = \begin{vmatrix} q_{12} & q_{13} \\ q_{32} & q_{33} \end{vmatrix}, \quad \alpha_2' \cdot |Q'| = -\begin{vmatrix} q_{12} & q_{13} \\ q_{22} & q_{23} \end{vmatrix}, \quad \text{and} \quad |Q'| = \begin{vmatrix} q_{22} & q_{23} \\ q_{32} & q_{33} \end{vmatrix} \neq 0$$

with

$$q_{\lambda_i 2} = l(\varepsilon_{\lambda_i BH} - \varepsilon_{\lambda_i BH_3}) \quad \text{and} \quad q_{\lambda_i 3} = l(\varepsilon_{\lambda_i B} - \varepsilon_{\lambda_i BH_3}) \quad (\lambda_i = 1, 2, 3)$$

The curve ends within the absorbance tetrahedron at the point B^{3-}. The directional vector $\mathbf{a}_B'$, pointing from BH_3 to B^{3-} may be expressed as (cf. Fig. 8–22):

$$\mathbf{a}_B' = \begin{pmatrix} A_{1B} - A_{1BH_3} \\ A_{2B} - A_{2BH_3} \\ A_{3B} - A_{3BH_3} \end{pmatrix} = l \cdot b_0 \begin{pmatrix} \varepsilon_{1B} - \varepsilon_{1BH_3} \\ \varepsilon_{2B} - \varepsilon_{2BH_3} \\ \varepsilon_{3B} - \varepsilon_{3BH_3} \end{pmatrix} \tag{8-57}$$

Equations 8–53 and 8–55 through 8–57 are closely related to an *ADQ-diagram*. It follows from equation 8–53 that all points in an A-diagram falling on the tangent to BH_3 converge to a single point within an ADQ-diagram based on the point BH_3 (cf. point BH_3 in Fig. 8–10, Sec. 8.3.6). Therefore, the ADQ coordinates of this point must be:[6]

$$\frac{\Delta A_1}{\Delta A_2} = \frac{q_{11}}{q_{21}} = \frac{\varepsilon_{1BH_2} - \varepsilon_{1BH_3}}{\varepsilon_{2BH_2} - \varepsilon_{2BH_3}} \quad \text{and}$$

$$\frac{\Delta A_3}{\Delta A_2} = \frac{q_{31}}{q_{21}} = \frac{\varepsilon_{3BH_2} - \varepsilon_{3BH_3}}{\varepsilon_{2BH_2} - \varepsilon_{2BH_3}} \tag{8-58}$$

So long as only the first equilibrium $BH_3 \leftrightarrows BH_2^-$ (H^+) is spectroscopically detectable, only a single point is produced in the ADQ-diagram. If the other two equilibria also affect the spectra, this point is transformed in most cases into a curved line. The tangent to the origin of this curve corresponds to the osculating plane for BH_3 in the A-diagram (cf. the line $\overline{BH_3\ BH^{2-}}$ in Fig. 8–10). This tangent contains all titration points produced while only the first two equilibria are being titrated. According to equation 8–55, the tangent is described by

$$\frac{\Delta A_1}{\Delta A_2} = \alpha_1 + \alpha_2 \frac{\Delta A_3}{\Delta A_2} \tag{8-59}$$

The curve deviates from this line as soon as the corresponding curve in the A-diagram leaves the osculating plane for BH_3. When the curve in the A-diagram approaches the tetrahedral edge $\overline{BH^{2-}\ B^{3-}}$ (cf. Fig. 8–22) only the last equilibrium remains to be titrated. The corresponding points in the ADQ-diagram are then located on the line described, according to equation 8–56, by:

$$\frac{\Delta A_1}{\Delta A_2} = \alpha'_1 + \alpha'_2 \frac{\Delta A_3}{\Delta A_2} \tag{8–60}$$

This line also contains the limiting point on the curve, corresponding to a titration solution consisting only of the protolyte B^{3-}. According to equation 8–57, the coordinates for this limiting point are

$$\frac{\Delta A_1}{\Delta A_2} = \frac{\varepsilon_{1B} - \varepsilon_{1BH_3}}{\varepsilon_{2B} - \varepsilon_{2BH_3}} \quad \text{and} \quad \frac{\Delta A_3}{\Delta A_2} = \frac{\varepsilon_{3B} - \varepsilon_{3BH_3}}{\varepsilon_{2B} - \varepsilon_{2BH_3}} \tag{8–61}$$

The foregoing illustrates the relationship that exists between a three-dimensional A-diagram and a two-dimensional ADQ-diagram. Every point in the ADQ-diagram corresponds to a line in the A-diagram, and every line in the ADQ-diagram represents a plane in the A-diagram (cf. the related discussion in Sec. 7.8).

Equations 8–59 and 8–60 provide expressions for the tangents to the initial and final stages of the curve in an ADQ-diagram. From these relationships it is possible to calculate the point of intersection of the two tangents. Thus:[17]

$$\frac{\Delta A_1}{\Delta A_2} = \frac{q_{12}}{q_{22}} = \frac{\varepsilon_{1BH} - \varepsilon_{1BH_3}}{\varepsilon_{2BH} - \varepsilon_{2BH_3}} \tag{8–17a}$$

and

$$\frac{\Delta A_3}{\Delta A_2} = \frac{q_{32}}{q_{22}} = \frac{\varepsilon_{3BH} - \varepsilon_{3BH_3}}{\varepsilon_{2BH} - \varepsilon_{2BH_3}} \tag{8–17b}$$

The point of intersection provides the ADQ coordinates for the component BH^{2-}. This result is quite logical from the standpoint of the three-dimensional A-diagram. The point at which the two tangents in the ADQ-diagram intersect corresponds in an A-diagram (cf. Fig. 8–22) to the tetrahedral edge along which the two planes containing the points (BH_3, BH^{2-}, B^{3-}) and (BH_3, BH_2^-, BH^{2-}) intersect: this is the edge designated $BH_3 BH^{2-}$. The appropriate ADQ values lead immediately to equations 8–17a and 8–17b. As demonstrated in Sec. 8.3.6, this relationship may be used to determine the molar absorptivity of the protolyte BH^{2-} solely on the basis of spectrometric data.[16]

A similar set of relationships may be shown to apply within an ADQ-diagram based on the point B^{3-}. Thus:

- the ADQ values of the third equilibrium $B^{3-} \leftrightarrows BH^{2-}$ (H^+) are constant, and
- tangents to the limiting points on the curve intersect in the point BH_2^-, the coordinates of which are the ADQ values of the component BH_2^-.

These relationships may also be derived directly as functions of the curve in the ADQ diagram.[16]

The *pencils of rays* shown in the ADQ-diagrams of Figures 8–10 and 8–12, as well as their relevant points of intersection with the tangents, may also be correlated with

8.6 Geometrical Relationships Between A- and ADQ-Diagrams

A-diagrams. For example, the line $\overline{B^{3-}\,P}$ in Fig. 8–10 corresponds to the plane in the A-diagram that contains the points BH_3, B^{3-}, and P (cf. Fig. 8–7 in Sec. 8.3.5). The intersection of this plane with the osculating plane for BH_3 produces the line $\overline{BH_3\,^2P_1}$ (cf. the line $\overline{BH_3\,H_2}$ in Fig. 8–7, with $H_2 = {}^2P_1$). Here ${}^2P_1$ represents the titration points of the isolated second equilibrium (cf. Sec. 8.3.5). The arguments of Sec. 7.3.5 (with respect to the line $\overline{BH_2\,^2P_1}$ in Fig. 7–15) demonstrate that the line $\overline{BH_3\,^2P_1}$ also contains the titration points of the first two isolated equilibria. Proceeding in the opposite fashion, the edge $\overline{BH_3\,^2P_1}$ may be shown to produce the point D_2 in Fig. 8–10. Points D_2 provide not only the ADQ values of the isolated first two equilibria but also those of the isolated second equilibrium (cf. also Fig. 7–15).[17]

Similarly, the points ${}^3D_1$ in Fig. 8–10 represent the intersection of the plane containing the points BH_3, BH^{2-}, and B^{3-} with that containing BH_3, BH_2^-, and P (cf. line $\overline{BH_3\,H_3}$ in Fig. 8–7). Along this line are to be found the titration points (${}^3P_1$) of the isolated third equilibrium (where $H_3 = {}^3P_1$ in Fig. 8–7). These lines produce a pencil of rays emerging from the point BH_3 in Fig. 8–14. Their intersections with the tangent to B^{3-} provide the titration points for the isolated third equilibrium.[17]

An ADQ-diagram based on the point B^{3-} may be interpreted similarly. Thus, the points ${}^1D_1^*$ from Fig. 8–12 of Sec. 8.3.6.1 provide the ADQ coordinates for the isolated first equilibrium $BH_3 \leftrightarrows BH_2^- \,(H^+)$, while the points D_2^* give the points for either the isolated second equilibrium or the last two equilibria considered in isolation.[17]

The geometric properties of pencils of rays in an ADQ-diagram may also be derived in an elementary way,[16] although the required calculations are rather extensive.

Points within an ADQ-diagram lead in the (two-dimensional) A-diagram to pencils of rays, along whose lines are to be found the titration points of the isolated equilibria. In order to specify precisely the locations of the titration points, these lines must be caused to intersect the pencils of rays defined by the curve within the A–diagram itself, together with its limiting values. For example, the line $\overline{P\,B^{3-}}$ in Fig. 8–13 of Sec. 8.3.6.2 contains the titration points of the protolysis system after it has been reduced by the final equilibrium $B^{3-}\,(H^+) \leftrightarrows BH^{2-}$.[15] The validity of this conclusion may be readily demonstrated, and Sec. 9.3 contains an outline of the general approach. In an analogous way it is possible to reduce the three-step system by the first equilibrium $BH_3 \leftrightarrows BH_2^- \,(H^+)$ simply by taking the pencils of rays associated with BH_3 and the points P on the titration curve in an A-diagram and allowing them to intersect with the lines constructed from the corresponding points D_2^* in Fig. 8–12.

Table 8-7. Summary of the various methods for evaluating three-step acid-base equilibria.

Determination	Methods	Section	Limitations*
Verification of the presence of three linearly independent equilibria	ADQ3-Diagrams	8.2	No point clusters No lines parallel to the coordinate axes No curved lines
Molar absorptivities of the two ampholytes	1) analysis of uniform subregions in a (two-dimensional) A-diagram	8.3.2	Approximation (Δ_{21} pK > 1.5; Δ_{32} pK > 1.5)
	2) using tangents and osculating planes in a three-dimensional A-diagram	8.3.3	
	3) using pencils of planes in a three-dimensional A-diagram	8.3.5	
	4) by correlating two-dimensional A- and ADQ-diagrams	8.3.6	
	5) regression analysis	8.3.7	
Concentrations	1) through distance relationships in an absorbance tetrahedron	8.3.3	
	2) solution of a system of linear equations	8.1	

8.7 Summary and Tabular Overview

Table 8-7 (continued)

Determination	Methods	Section	Limitations*
pK-Values	1) inflection point analysis of the titration curves	8.3.1	Approximation (Δ_{21} pK > 3; Δ_{32} pK > 3)
	2) evaluation of uniform subregions	8.3.2	Approximation (Δ_{21} pK > 1.5; Δ_{32} pK > 1.5)
	3) dissociation diagrams	8.3.3	
	4) with the aid of specific planes in an absorbance tetrahedron	8.3.5	
	5) pencil of rays method	8.3.4	
	6) pencil of planes method	8.3.5	
	7) by correlating two-dimensional A- and ADQ-diagrams	8.3.6	
	8) regression analysis	8.3.7	
K_1/K_2 and K_2/K_3	1) using specific distance relationships within an absorbance tetrahedron	8.4	Δ_{21} pK < 2; Δ_{32} pK < 2
	2) from concentrations	8.4	Δ_{21} pK < 2; Δ_{32} pK < 2

* The pK differences listed in this column should be regarded only as approximations.

8.7 Summary and Tabular Overview

Table 8–7 provides a summary of the methods available for studying three-step acid-base equilibria. Of primary interest are the means for establishing values for pK_1, pK_2, and pK_3, together with K_1/K_2 and K_2/K_3.

In the case of overlapping systems, *inflection point analysis* leads to inaccurate results. This approach may only be applied when the pK differences are large ($\Delta_{21}\,pK > 3$, $\Delta_{32}\,pK > 3$).

The evaluation of a *uniform subregions* within an A-diagram generally leads to a better approximation, and the method is applicable to many systems (i.e., whenever $\Delta_{21}\,pK > 1.5$, $\Delta_{32}\,pK > 1.5$).

The pK values may also be ascertained without the need for approximation using *dissociation diagrams*. The requisite concentrations are derived, for example, from absorbance tetrahedra, which in turn may be constructed from *osculating planes* and *tangents*.

A three-step titration system may be dissected into its constituent equilibria with the aid of *pencils of rays* and *pencils of planes*, and each equilibrium may then be evaluated by the methods applicable to single-step protolysis equilibria. This approach leads to values for pK_1, pK_2, and pK_3, and its applicability is general.

Another approach to determining the pK values associated with a titration system involves correlation of *two-dimensional A- and ADQ-diagrams*. The procedure is an easy one, but it is nonetheless significant and of general utility.

Non-linear regression analysis avoids the necessity of establishing the limiting values of the titration curve A_λ vs. pH, but the method requires extensive calculation (using, e.g., the program TIFIT).

Values for the quotients K_1/K_2 and K_2/K_3 need not be dependent on poorly defined pH values. Instead, these ratios may be ascertained with the aid of selected distance relationships within an absorbance tetrahedron, and thus based exclusively on spectrometric data.

Literature

1. H. Mauser, *Formale Kinetik*. Bertelsmann Universitätsverlag, Düsseldorf 1974.
2. H. Mauser, private communication (1974).
3. R. Blume, *Dissertation*. Tübingen 1975.
4. D. Danil, G. Gauglitz, H. Meier, *Photochem. Photobiol.* 26 (1977) 225.
5. G. Gauglitz, *GIT Fachz. Lab.* 3 (1982) 205.
6. J. Polster, *Habilitationsschrift*. Tübingen 1980.
7. J. Polster, *GIT Fachz. Lab.* 26 (1982) 690.
8. T. Ishimitsu, S. Hiroso, H. Sakurai, *Talanta* 24 (1977) 555.

Literature

9. T. Ishimitsu, S. Hirose, H. Sakurai, *Chem. Pharm. Bull.* 26 (1978) 74.
10. R. Blume, J. Polster, *Z. Naturforsch.* 30 b (1975) 358.
11. R. Blume, J. Polster, *Z. Naturforsch.* 29 b (1974) 734.
12. M. Dixon, E. C. Webb, *Enzymes*. Longmands and Green, London (1964), p. 118.
13. J. Polster, *Z. Anal. Chem.* 276 (1975) 353.
14. J. Polster, *Z. Anal. Chem.* 290 (1978) 116.
15. J. Polster, *Z. Physik. Chem. N. F.* 97 (1975) 55.
16. J. Polster, *Z. Physik. Chem. N. F.* 97 (1975) 113.
17. J. Polster, *Z. Physik. Chem. N. F.* 104 (1977) 49.
18. K. Nagano, D. E. Metzler, *J. Amer. Chem. Soc.* 89 (1967) 2891.
19. F. Göbber, H. Lachmann, *Hoppe-Seylers Z. Physiol. Chem.* 359 (1978) 269.
20. R. M. Alcock, F. R. Hartley, D. E. Rogers, *J. Chem. Soc. Dalton Trans.* 1978, 115.
21. M. Meloun, J. Cernak, *Talanta* 26 (1979) 569.
22. H. Gampp, M. Maeder, A. D. Zuberbühler, *Talanta* 27 (1980) 1037.
23. M. Meloun, J. Cernak, *Talanta* 31 (1984) 947.
24. M. Meloun, M. Javurek, *Talanta* 31 (1984) 1083.
25. M. Meloun, M. Javurek, *Talanta* 32 (1985) 973.
26. M. Meloun, M. Javurek, A. Hynkova, *Talanta* 33 (1986) 825.
27. D. J. Legett, *Computational Methods for the Determination of Formation Constants*. Plenum Press, New York, London 1985.
28. H. Gampp, M. Maeder, C. J. Meyer, A. D. Zuberbühler, *Talanta* 32 (1985) 95.
29. H. Gampp, M. Maeder, C. J. Meyer, A. D. Zuberbühler, *Talanta* 32 (1985) 257.
30. I. Bertini, A. Dei, C. Luchinat, R. Monnanni, *Inorg. Chem.* 24 (1985) 301.
31. T. Kaden, A. D. Zuberbühler, *Talanta* 18 (1971) 61.
32. D. J. Legett, W. A. E. McBryde, *Anal. Chem.* 47 (1975) 1065.
33. M. Meloun, J. Cernak, *Talanta* 23 (1976) 15.
34. F. Gaizer, A. Puskas, *Talanta* 28 (1981) 925.
35. M. Meloun, M. Javurek, *Talanta* 31 (1984) 1083.
36. M. Meloun, M. Javurek, *Talanta* 32 (1985) 1011.
37. J. Havel, M. Meloun, *Talanta* 33 (1986) 435.
38. M. Meloun, M. Javurek, J. Havel, *Talanta* 33 (1986) 513.
39. H. Gampp, M. Maeder, C. J. Meyer, A. D. Zuberbühler, *Talanta* 32 (1985) 1133.
40. J. Polster, *GIT Fachz. Lab.* 27 (1983) 390.
41. M. M. Lipschutz, *Differential Geometry*. Schaum's Outline Series, McGraw-Hill Book Company, New York 1969.
42. B. Baule, *Die Mathematik des Naturforschers und Ingenieurs*, mehrbändig. S. Hirzel Verlag, Leipzig 1950.
43. J. Dreszer, *Mathematik Handbuch für Technik, Naturwissenschaft*. Verlag Harri Deutsch, Zürich-Frankfurt/Main-Thun 1975.

9 *s*-Step Overlapping Acid-Base Equilibria

9.1 Basic Equations

As has previously been noted, potentiometric titrations continue to serve as the principle source of data for establishing dissociation constants, and a wide variety of computer programs is available to support the associated data analysis.[1–24] Metal-complex equilibria, which are formally quite similar, are also amenable to potentiometric analysis,[25–47] and numerous useful computer programs have been developed for this application as well.[48–62,24]

Nevertheless, it is becoming increasingly common to investigate titration systems by spectrometric means. Spectrometric data can be as reproducible as the results from potentiometry, and spectrometry has actually been found to be superior with respect to discriminating between various types of model systems.[63] Not surprisingly, powerful computer methods have long been applied here as well, both for acid-base equilibria[63–75,24] and for metal-complex equilibria.[24,63,76–85] No attempt will be made in the present discussion to compare different computer programs; rather, we concern ourselves with demonstrating the general utility of analytical methods based on A-diagrams, analogous to methods introduced previously in the context of two- and three-step acid-base equilibria.

We begin simply by noting that A-diagrams do in principle lend themselves to the investigation of *s*-step overlapping protolysis equilibria.[86] Since the A-diagram approach has been examined in detail in Chapters 7 and 8 it will suffice to treat its generalization here only briefly.

In the course of a complete pH-titration of an *s*-step (unbranched) protolysis system the substrate BH_s loses a total of *s* protons:

$$\begin{aligned} BH_s &\xrightleftharpoons{K_1} BH_{s-1}^- \;(H^+) \\ &\vdots \\ BH^{(s-1)-} &\xrightleftharpoons{K_s} BH^{s-} \;(H^+) \end{aligned} \qquad (9\text{–}1)$$

The relevant (mixed) dissociation constants may be expressed as:

$$K_1 := a_{H_3O^+} \cdot \frac{bh_{s-1}}{bh_s} = 10^{-pH} \cdot \frac{bh_{s-1}}{bh_s} = h \cdot \frac{bh_{s-1}}{bh_s}$$

$$\vdots$$

(9–2)

$$K_s := a_{H_3O^+} \cdot \frac{b}{bh} = 10^{-pH} \cdot \frac{b}{bh} = h \cdot \frac{b}{bh}$$

Mass balance considerations assure that

$$b_0 = bh_s + bh_{s-1} + \ldots b \qquad (9\text{–}3)$$

According to equations 9–2 and 9–3, the concentrations of the various species involved are all functions of h:

$$bh_{s-i} = \frac{b_0 h^{s-i} \prod_{l=0}^{i} K_l}{H_s} \qquad (9\text{–}4a)$$

where $\displaystyle H_s = \sum_{k=0}^{s} h^k \prod_{l=0}^{s-k} K_l$

and i must fall within the bounds

$$0 \leq i \leq s$$

The following definitive equations apply when $i = s$ and $l = 0$:

$$bh_0 = b \quad \text{and} \quad K_0 = 1 \qquad (9\text{–}4b)$$

Taking as an example the case $s = 3$, equation 9–4a states that the appropriate expression for the concentration b is (cf. p. 134):

$$bh_0 = b = \frac{b_0 h^0 K_1 K_2 K_3}{h^0 K_1 K_2 K_3 + h^1 K_1 K_2 + h^2 K_1 + h^3} = \frac{b_0 K_1 K_2 K_3}{H_3}$$

The Lambert-Beer-Bouguer law in turn provides the following description of the absorbance A_λ:

$$A_\lambda = l(\varepsilon_{\lambda BH_s} \cdot bh_s + \varepsilon_{\lambda BH_{s-1}} \cdot bh_{s-1} + \ldots \varepsilon_{\lambda B} \cdot b) \qquad (9\text{–}5)$$

If, with the aid of equation 9–3, one eliminates the concentration bh_s, it becomes apparent that

$$\Delta A_\lambda := A_\lambda - l b_0 \varepsilon_{\lambda BH_s} = q_{\lambda 1} bh_{s-1} + q_{\lambda 2} bh_{s-2} + \ldots q_{\lambda s} \cdot b \qquad (9\text{–}6a)$$

where $q_{\lambda j} = l(\varepsilon_{\lambda BH_{s-j}} - \varepsilon_{\lambda BH_s})$, with $1 \leq j \leq s$

In the event that $j = s$, then by definition

$$\varepsilon_{\lambda BH_0} = \varepsilon_{\lambda B} \qquad (9\text{–}6b)$$

The following relationships thus apply with respect to the absorbance differences ΔA_{λ_m} at wavelengths λ_m [$\lambda_m = 2, 3 \ldots (s + 1)$]:

$$\begin{pmatrix} \Delta A_2 \\ \vdots \\ \Delta A_{s+1} \end{pmatrix} = \begin{pmatrix} q_{21} & \cdots & q_{2s} \\ \vdots & & \vdots \\ q_{(s+1)1} & \cdots & q_{(s+1)s} \end{pmatrix} \begin{pmatrix} bh_{s-1} \\ \vdots \\ b \end{pmatrix} = Q \begin{pmatrix} bh_{s-1} \\ \vdots \\ b \end{pmatrix} \quad (9\text{–}7)$$

If $|Q| \neq 0$ Cramer's rule makes it possible to develop from this system of equations the various concentrations bh_{s-j} as functions of ΔA_{λ_m} and $q_{\lambda_m j}$.[87] Introduction of the resulting concentrations into equation 9–6a then provides the following expression for ΔA_1:

$$\Delta A_1 = \alpha_1 \Delta A_2 + \ldots \alpha_s \Delta A_{(s+1)} = \sum_{j=1}^{s} \alpha_j \cdot \Delta A_{j+1} \quad (9\text{–}8)$$

where:

$$\alpha_j |Q| = (-1)^{(j+1)} \begin{vmatrix} q_{11} & \cdots & q_{1s} \\ \cdot & & \cdot \\ \cdot & & \cdot \\ q_{j1} & \cdots & q_{js} \\ q_{(j+2)1} & \cdots & q_{(j+2)s} \\ \cdot & & \cdot \\ \cdot & & \cdot \\ q_{(s+1)1} & \cdots & q_{(s+1)s} \end{vmatrix}$$

Thus, absorbance differences measured at s wavelengths in an s-step titration system may be expected to behave in a linearly independent fashion. Equation 9–8 is always applicable provided the system in question contains s linearly independent concentration variables.

9.2 Characteristic Concentration Diagrams and Absorbance Diagrams

According to equation 9–3, s of the concentrations associated with the $(s+1)$ components of a multi-step system behave in a linearly independent fashion. Therefore, the characteristic concentration diagram for such a system, like the characteristic A-diagram, is s-dimensional. If one were to specify the dependent concentration to be the quantity bh_{s-1} (chosen at random), then the vector

$$\mathbf{c} = \begin{pmatrix} bh_s \\ bh_{s-2} \\ \cdot \\ \cdot \\ b \end{pmatrix} \quad (9\text{–}9)$$

would serve generally as the position vector within the relevant characteristic concentration space. Thus, according to equation 9–4a, the position vector $\mathbf{c_p}$ of the concentration curve takes the following form:

$$\mathbf{c_p} = \frac{b_0}{H_s} \begin{pmatrix} h^s \\ h^{s-2} K_1 K_2 \\ \cdot \\ \cdot \\ \prod_{l=0}^{s} K_l \end{pmatrix} \quad (9\text{–}10)$$

In the limiting case of $h \to \infty$, $bh_s = b_0$. The position vector applicable to the point BH_s must be associated with the coordinates

$$\mathbf{c}_{BH_s} = \begin{pmatrix} b_0 \\ 0 \\ \cdot \\ \cdot \\ 0 \end{pmatrix} \quad (9\text{–}11)$$

Similarly, in the limiting case $h \to 0$, where $b = b_0$, the following specification must apply to the point B^{s-}:

$$\mathbf{c_B} = \begin{pmatrix} 0 \\ 0 \\ \cdot \\ \cdot \\ b_0 \end{pmatrix} \quad (9\text{–}12)$$

The orientation of the tangent ($\mathbf{t}'$) to the curve may be obtained by differentiating the coordinates of $\mathbf{c_p}$ by the parameter h:

$$\mathbf{t}' = \frac{d\mathbf{c_p}}{dh} \quad (9\text{–}13)$$

Thus, the tangent to BH_s may be shown to be:

9.2 Characteristic Concentration Diagrams and Absorbance Diagrams

$$t'\bigg|_{h\to\infty} = \frac{b_0 K_1}{h^2} \begin{pmatrix} 1 \\ 0 \\ \cdot \\ \cdot \\ \cdot \\ 0 \end{pmatrix} \qquad (9\text{--}14)$$

In other words, the tangent vector for the point BH_s has the same orientation as the coordinate axis bh_s, and it would include all the titration points associated with the isolated equilibrium

$$BH_s \leftrightarrows BH_{s-1}^- \; (H^+)$$

Analogously for $h \to 0$:

$$t'\bigg|_{h\to 0} = \frac{b_0}{K_s} \begin{pmatrix} 0 \\ 0 \\ \cdot \\ \cdot \\ \cdot \\ +1 \\ -1 \end{pmatrix} \qquad (9\text{--}15)$$

from which it follows that all the titration points for the isolated equilibrium

$$BH^{(s-1)-} \leftrightarrows B^{s-} \; (H^+)$$

lie along the tangent to B^{s-}.

As has been demonstrated in Sections 7.3.5 and 8.3.4, titration systems may be dissected into their individual equilibria with the aid of appropriate pencils of rays. This principle also applies to s-step systems. Thus, if an s-step system (or, more simply, s-system) were reduced by the last of its equilibrium steps, then the remaining system would consist only of

$$s' = (s-1) \qquad (9\text{--}16)$$

equilibrium steps (hereafter referred to as the s'-system). The points $P_{s'}$ comprising the s'-system occur along the pencil of rays in the concentration diagram originating in the point B^{s-} and passing through the points P of the s-system. This conclusion may be demonstrated by expressing the equation for the line $\overline{P\,B^{s-}}$ in parameter form (cf. equation 7–40):

$$\mathbf{c} = \mathbf{c_B} + k(\mathbf{c_p} - \mathbf{c_B}) \qquad (9\text{--}17)$$

where $-\infty < k < \infty$

In the case $b = 0$, it follows from equations 9–10 and 9–12 that

$$b = 0 = b_0 + k\left(\frac{b_0}{H_s}\prod_{l=0}^{s} K_l - b_0\right) \tag{9-18}$$

The locations of all the points $P_{s'}$ may be determined by solving this equation for k and introducing the result into equation 9–17. This leads to the following expression for the vector for $P_{s'}$:

$$c\mathbf{P}_{s'} = \begin{pmatrix} bh_s \\ bh_{s-2} \\ \cdot \\ \cdot \\ bh \\ b \end{pmatrix} = \frac{b_0}{H_{s-1}} \begin{pmatrix} h^{s-1} \\ h^{s-3}K_1 K_2 \\ \cdot \\ \cdot \\ \prod_{l=0}^{s-1} K_l \\ 0 \end{pmatrix} \tag{9-19}$$

where $\quad H_{s-1} = \sum_{k=0}^{s-1} h^k \prod_{l=0}^{s-k-1} K_l \quad$ and $\quad K_0 := 1$

The coordinates of this vector in turn provide the concentrations applicable to the s'-system, a conclusion that follows from equation 9–4a after introduction of the analytical expressions for the s'-system in place of the concentrations bh_s, bh_{s-2} ... b (s being replaced by $s-1$).

Thus, the points $P_{s'}$ fall along the lines $\overline{P\,B^{s-}}$, which constitute a *pencil of rays*.[88] It will be demonstrated in Sec. 9.4 how the actual locations of these points $P_{s'}$ may be unambiguously established.

According to equation 9–8, absorbance differences for an s-system measured at s wavelengths must be linearly independent. Just as is done with concentrations, absorbance differences may be utilized to construct an s-dimensional vector space. Since these two vector spaces may be interconverted by linear transformations (cf. Sections 7.6 and 8.5), all the relationships already developed for concentrations are equally valid with respect to an AD-diagram. For example, the titration points for the first and last equilibria may be sought along the tangents to the points BH_s and B^{s-}, respectively. Moreover, geometric relationships identical to those described above link the s- and s'-systems.[88]

The foregoing observations also apply with respect to A-diagrams, which are simply AD-diagrams that have been displaced by the vector **a** (cf. equation 7–64 in Sec. 7.6):

9.2 The Generalized Form of an ADQ-Diagram

$$\mathbf{a} = \mathbf{a}' + \mathbf{a}_0 \qquad (9\text{--}20)$$

where $\mathbf{a} = \begin{pmatrix} A_1 \\ \cdot \\ \cdot \\ A_s \end{pmatrix}$, $\mathbf{a}' = \begin{pmatrix} \Delta A_1 \\ \cdot \\ \cdot \\ \Delta A_s \end{pmatrix}$, and $\mathbf{a}_0 = l\, b_0 \begin{pmatrix} \varepsilon_{1\text{BH}_s} \\ \cdot \\ \cdot \\ \varepsilon_{s\,\text{BH}_s} \end{pmatrix}$

9.3 The Generalized Form of an ADQ-Diagram

Single-step protolysis equilibria lead to linear AD-diagrams, while two-step systems produce linear ADQ-diagrams (cf. Sections 6.2 and 7.2). Analogous two-dimensional diagrams may also be constructed for s-step systems, as is demonstrated below.

In systems consisting of s independent concentration variables, the absorbance differences derived from s wavelengths may generally be expected to behave independently:

$$\Delta A_1 = \alpha_1 \Delta A_2 + \ldots \alpha_s \Delta A_{(s+1)} \qquad (9\text{--}8)$$

In order to illustrate the corresponding relationships two-dimensionally it is necessary that all summation terms $(s - 2)$ be caused to disappear. This result may be achieved by employing absorbance differences measured at $(s - 2)$ different pH-values for the set of wavelengths $\lambda_u\ [u = 1 \ldots (s + 1)]$. It then becomes possible to eliminate all coefficients $\alpha_v\ [v = 1 \ldots (s - 2)]$. If the $(s + 1)(s - 2)$ absorbance differences are distinguished from one another by the use of double subscripts (i.e., ΔA_{uv}), where the first subscript refers to the wavelength and the second to the pH of the solution, then equation 9–8 translates into the following system of equations:

$$\Delta A_{11} - \alpha_{(s-1)} \Delta A_{s1} - \alpha_s \Delta A_{(s+1)1} = \alpha_1 \Delta A_{21} + \ldots \alpha_{(s-2)} \Delta A_{(s-1)1}$$
$$\vdots \qquad (9\text{--}21)$$
$$\Delta A_{1(s-2)} - \alpha_{(s-1)} \Delta A_{s(s-2)} - \alpha_s \Delta A_{(s+1)(s-2)} = \alpha_1 \Delta A_{2(s-2)} + \ldots \alpha_{(s-2)} \Delta A_{(s-1)(s-2)}$$

Such a formulation permits coefficients to be calculated with the aid of Cramer's rule[87] as a function of ΔA_{uv}, $\alpha_{(s-1)}$, and α_s. Replacing the coefficients α_v in equation 9–8 by the appropriate expressions and rearranging terms leads to the following relationship:

$$\frac{\begin{vmatrix} \Delta A_1 & \Delta A_{11} & \cdots & \Delta A_{1(s-2)} \\ \Delta A_2 & \Delta A_{21} & \cdots & \Delta A_{2(s-2)} \\ \vdots & \vdots & & \vdots \\ \Delta A_{(s-1)} & \Delta A_{(s-1)1} & \cdots & \Delta A_{(s-1)(s-2)} \end{vmatrix}}{|D|} = \alpha_{(s-1)} + \alpha_s \frac{\begin{vmatrix} \Delta A_{(s+1)} & \Delta A_{(s-1)1} & \cdots & \Delta A_{(s+1)(s-2)} \\ \Delta A_2 & \Delta A_{21} & \cdots & \Delta A_{s(s-2)} \\ \vdots & \vdots & & \vdots \\ \Delta A_{(s-1)} & \Delta A_{(s-1)1} & \cdots & \Delta A_{(s-1)(s-2)} \end{vmatrix}}{|D|}$$

(9–22)

$$\text{where } |D| = \begin{vmatrix} \Delta A_s & \Delta A_{s1} & \cdots & \Delta A_{s(s-2)} \\ \Delta A_2 & \Delta A_{21} & \cdots & \Delta A_{2(s-2)} \\ \vdots & \vdots & & \vdots \\ \Delta A_{(s-1)} & \Delta A_{(s-1)1} & \cdots & \Delta A_{(s-1)(s-2)} \end{vmatrix} \neq 0$$

If the two quotients constituting this equation are now plotted against each other, the result is a generalized ADQ-diagram (*ADQs-diagram*). A straight line in such a diagram ensures that the conditions of equation 9–8 have been fulfilled (*graphic matrix-rank analysis*).

Relatively small experimental errors may result in considerable scatter in an ADQs-diagram, so use of this method is actually somewhat problematic. However, the situation may be improved by first subjecting all data to a smoothing routine,[89,90] and by employing what is known as an *ADQz-diagram* to select the most appropriate ΔA_{uv} values. A diagram of this type is created by first changing the designation of s to z (to prevent confusion) and then treating the quantity z as a variable. What results, as shown by equation 9–22, is a set of ADQz-quotients containing determinants of the order $(z - 1)$. The corresponding diagrams generally consist of curves. Points lying on the tangents to the origins of the various ADQz-curves are presumably more suitable for use as ΔA_{uv} values than the experimental data points themselves.

The special properties associated with tangents to the origins of ADQz-curves, as well as the linear-geometric relationships linking an s-system with an s'-system (described in Sec. 9.2 for the case of an s-dimensional A-diagram) are applicable in an analogous way to two-dimensional AD- and ADQ2...ADQ(s–1)-diagrams. If, for example, one bases the absorbance differences required by equation 9–6a on the point BH$_s$, it is entirely possible—albeit complicated and time-consuming—to demonstrate in the general case that the curves constituting a set of ADQz-diagrams [$z = 1, 2, \ldots (s-1)$] possess the following characteristics:

(1) isolation of the first z equilibria of the s-step system produces titration points that fall along the tangent to $h \to \infty$;

(2) the tangent to $h \to 0$ contains the hypothetical titration points associated with the isolated last equilibrium $BH^{(s-1)-} \leftrightarrows B^{s-} (H^+)$; and

(3) points from the s- and s'-systems that share a common pH intersect at the limiting end of the curve ($h \to 0$).

An analogous set of observations applies to an ADQz-diagram based on the point B^{s-} ($\Delta A_\lambda = A_\lambda - lb_0 \varepsilon_{\lambda B}$).

As has been shown in Sec. 7.8, every point in an ADQ2-diagram may be used to construct a line in the corresponding (two-dimensional) A-diagram. Similarly, it may easily be verified that every point in an ADQz-diagram produces a straight line in an ADQ(z–1)-diagram. This means that various ADQz diagrams may be correlated in a manner similar to that encountered in the case of the A- and ADQ-diagrams of a three-step protolysis system (cf. Sec. 8.3.6). It is therefore possible in principle to use two-dimensional representations in a complete characterization and analysis of an s-step acid-base equilibrium system.

9.4 The Establishment of Absorbance Polyhedra

In Sec. 9–3 it has been shown that an s-step titration system may be completely analyzed using only two-dimensional diagrams. However, such a system may also be evaluated with the aid of pencils of rays within an s-dimensional A-diagram (cf. Sec. 9.5) provided the location of the appropriate *absorbance polyhedron* is known.

The vertices of an absorbance polyhedron are defined in terms of the absorbances of the pure components BH_s, BH_{s-1}^- ... B^{s-} (cf. the absorbance triangles and absorbance tetrahedra discussed in Sections 7.3.3 and 8.3.3, respectively), which may be ascertained solely on the basis of spectrometric data: for example, from tangents to the ends of ADQz-diagram curves. According to the results presented in Sec. 9.3 the tangent to $h \to \infty$ in such a diagram specifies the direction of the line containing titration points for the (isolated) first z equilibria of an s-step system (where absorbance differences are based on the point BH_s). This means that in an AD-diagram ($z = 1$) titration points for the isolated first equilibrium $BH_s \leftrightarrows BH_{s-1}^-$ (H^+) will fall along the tangent to BH_s ($h \to \infty$). The general equation for such a tangent is as follows:

$$z = 1: \quad \Delta A_1 = \alpha_{11} \Delta A_2 \tag{9–23a}$$

In an ADQ2-diagram ($z = 2$) the titration points for the (isolated) first two equilibria will similarly occur along the tangent to $h \to \infty$, the corresponding general equation for which is

$$\frac{\Delta A_1}{\Delta A_2} = \alpha_{21} + \alpha_{22} \frac{\Delta A_3}{\Delta A_2}$$

This leads to the relationship

$$z = 2: \quad \Delta A_1 = \alpha_{21} \Delta A_2 + \alpha_{22} \Delta A_3 \tag{9-23b}$$

Carrying this procedure forward to an ADQ3-diagram, one obtains the following equation for the tangent to $h \to \infty$ (cf. equation 9–22 with $s = z = 3$, or equation 8–10; the ordinate intercept and the slope are here designated as α'_{31} and α'_{32}, respectively):

$$\frac{\begin{vmatrix} \Delta A_1 & \Delta A_{11} \\ \Delta A_2 & \Delta A_{21} \end{vmatrix}}{\begin{vmatrix} \Delta A_3 & \Delta A_{31} \\ \Delta A_2 & \Delta A_{21} \end{vmatrix}} = \alpha'_{31} + \alpha'_{32} \frac{\begin{vmatrix} \Delta A_4 & \Delta A_{41} \\ \Delta A_2 & \Delta A_{21} \end{vmatrix}}{\begin{vmatrix} \Delta A_3 & \Delta A_{31} \\ \Delta A_2 & \Delta A_{21} \end{vmatrix}}$$

From this it follows that:

$z = 3$:

$$\Delta A_1 = \frac{1}{\Delta A_2}\left(\Delta A_{11} - \Delta A_{31}\alpha'_{31} - \Delta A_{41}\alpha'_{32}\right)\Delta A_2 + \Delta A_{21}\alpha'_{31}\Delta A_3 + \Delta A_{21}\alpha'_{32}\Delta A_4$$
$$= \alpha_{31}\Delta A_2 + \alpha_{32}\Delta A_3 + \alpha_{33}\Delta A_4 \tag{9-23c}$$

Finally, for the case $z = (s - 1)$ the corresponding expression is

$$z = (s-1): \quad \Delta A_1 = \alpha_{(s-1)1}\Delta A_2 + \ldots \alpha_{(s-1)(s-1)}\Delta A_s \tag{9-23d}$$

Equations 9–23 (a–d) show how ADQz-diagrams may be used to determine spectrometric characteristics of the individual equilibria by establishing the nature of the linear relationships between absorbance differences for the equilibria $z = 1$ through $z = (s - 1)$. These relationships then lead directly to the corresponding absorbance equations. Since the absorbance differences utilized above were based on the point BH_s, it follows in general for a set of z equilibria that:

$$A_1 = \alpha_{z1} A_2 + \alpha_{z2} \Delta A_3 + \ldots \alpha_{zz} A_{(z+1)} + \alpha_{z(z+1)} \tag{9-24}$$

where $\alpha_{z(z+1)} = l\, b_0\, (\varepsilon_{1BH_s} - \alpha_{z1}\varepsilon_{2BH_s} - \ldots \alpha_{zz}\varepsilon_{(z+1)BH_s})$

Thus, equations 9–23 (a–d) may be rewritten in terms of absorbances:

$z = 1: \quad A_1 = \alpha_{11} A_2 + \alpha_{12}$
$z = 2: \quad A_1 = \alpha_{21} A_2 + \alpha_{22} A_3 + \alpha_{23}$
$\vdots$
$$z = (s-1): \quad A_1 = \alpha_{(s-1)1} A_2 + \ldots \alpha_{(s-1)(s-1)} A_s + \alpha_{(s-1)s} \tag{9-25}$$

In many cases the coefficients a_{ij} may be determined directly from the absorbance matrix **A** (cf. Sec. 3.2.1), constructed on the basis of absorbances arranged according to pH-values and wavelengths. Frequently the various (more or less strongly overlapping) equilibria reveal themselves spectrometrically one after another. If this does, in fact, occur the coefficients α_{ij} may be determined directly from the ranks of the submatrices of which the absorbance matrix **A** is composed.[86] Strictly speaking,

9.4 The Establishment of Absorbance Polyhedra

this procedure cannot be regarded as general, but it may sometimes be more practical than resorting to the use of ADQz-diagrams.

A system of equations like 9–25, which may be based on a variety of wavelength combinations, makes it possible to characterize the desired absorbance polyhedron. The results presented in Sec. 9.2 ensure that absorbances due to the isolated last equilibrium $BH^{(s-1)-} \leftrightarrows B^{s-} (H^+)$ are to be found along the tangent ($\mathbf{t_B}$) to the point B^{s-} ($h \to 0$). The directional vector $\mathbf{t_B}$ and the vector $\mathbf{a_B}$ extending from the origin of the s-dimensional A-diagram to the point B^{s-} are known to have the following coordinates:

$$\mathbf{t_B} = \begin{pmatrix} t_1 \\ \vdots \\ t_s \end{pmatrix} \qquad (9\text{–}26a)$$

and

$$\mathbf{a_B} = \begin{pmatrix} A_{1B} \\ \vdots \\ A_{sB} \end{pmatrix} = l\,b_0 \begin{pmatrix} \varepsilon_{1B} \\ \vdots \\ \varepsilon_{sB} \end{pmatrix} \qquad (9\text{–}26b)$$

A line passing through B^{s-} in the direction of $\mathbf{t_B}$ must be subject to the condition:

$$\mathbf{a} = \mathbf{a_B} + k\,\mathbf{t_B} \qquad (9\text{–}27)$$

where $-\infty < k < \infty$

At the point $BH^{(s-1)-}$ the only species present is the ampholyte $BH^{(s-1)-}$. Therefore:

$$\mathbf{a} = \mathbf{a_{BH}} = \begin{pmatrix} A_{1BH} \\ \vdots \\ A_{s\,BH} \end{pmatrix} = l\,b_0 \begin{pmatrix} \varepsilon_{1BH} \\ \vdots \\ \varepsilon_{s\,BH} \end{pmatrix} \qquad (9\text{–}28)$$

If the s-step system is now reduced by the last equilibrium (such that $z = s - 1$), then the point $BH^{(s-1)-}$ becomes the limiting point along the curve constituting the A-diagram. The coordinates of $\mathbf{a_{BH}}$ thus fulfill the conditions of equation 9–25:

$$A_{1BH} = \alpha_{(s-1)1} A_{2BH} + \ldots \alpha_{(s-1)(s-1)} A_{s\,BH} + \alpha_{(s-1)s} \qquad (9\text{–}29)$$

Substituting this relationship into equation 9–28 permits equation 9–27 to be expanded as follows:

$$\begin{pmatrix} \alpha_{(s-1)1} A_{2BH} + \ldots \alpha_{(s-1)(s-1)} A_{s\,BH} + \alpha_{(s-1)s} \\ A_{2BH} \\ \vdots \\ A_{s\,BH} \end{pmatrix} = \begin{pmatrix} A_{1B} \\ A_{2B} \\ \vdots \\ A_{s\,B} \end{pmatrix} + k \begin{pmatrix} t_1 \\ t_2 \\ \vdots \\ t_s \end{pmatrix} \qquad (9\text{–}30)$$

This system of linear equations makes it possible to assign values to k, $A_{2BH} \ldots A_{sBH}$. The quantity A_{1BH} may then be calculated using either of the equations 9–27 or 9–29. It is thus proven that $A_{\lambda BH}$ is accessible in principle using only spectrometric data.

In Sec. 9.2 it was demonstrated that all points of the s' and s-systems are to be found along a pencil of rays whose members intersect at the point B^{s-}. Position vectors $\mathbf{a}_{P_{s'}}$ leading from the coordinate origin to the points constituting the s' system may be established by a procedure analogous to that used for the vector $\mathbf{a}_{BH}$. In this case the vector $\mathbf{a}$ in equation 9–27 is replaced by the position vector $\mathbf{a}_{P_{s'}}$ with the directional vector ($\mathbf{a}_{P_s} - \mathbf{a}_B$) taking the place of the vector $\mathbf{t}_B$ (where $\mathbf{a}_{P_s}$ is the position vector of the s-system). Since equation 9–25 with $z = (s - 1)$ is also applicable to the coordinates of $\mathbf{a}_{P_{s'}}$, expressing $\mathbf{a}_{P_{s'}}$ in this way permits a system of equations to be developed from which the absorbances of the points $P_{s'}$ may be calculated (the *generalized pencil of rays method*).

Thus, a procedure has been described for reducing an s-system by one equilibrium, converting it into the corresponding s'-system. This s'-system may now itself be subjected to the same treatment, and by continuing in this fashion it is possible in principle to ascertain the absorbances of all the ampholytes present—and thus to define the boundaries of an absorbance polyhedron—using only spectrometric data.[86] Moreover, one may then proceed to establish all the pK values associated with the system (cf. Sec. 9.5).

An absorbance polyhedron may also be defined starting from the opposite end of the A-diagram curve. This requires use of a set of ADQz-diagrams based on the point B^{s-}, where the system must again be successively reduced by the first, second, … $(s - 1)$th equilibria.

9.5 The Determination of Absolute pK-Values

Once an absorbance polyhedron has been established as described in Sec. 9.4 it becomes possible to investigate in isolation the individual equilibria comprising the s-step system. A demonstration of the validity of this assertion begins with the concentration diagram, affine distortion of which produces the appropriate A-diagrams (cf. Sec. 9.4).

In order to determine the concentrations associated with the isolated last equilibrium $BH^{(s-1)-} \leftrightarrows B^{s-}$ (H^+) it is necessary to establish position vectors for the vertices of the *concentration polyhedron*—just as was done previously with the concentration triangle (Sec. 7.5.3) and the concentration tetrahedron (Sec. 8.5.3) in the analogous cases of rank $s = 2$ and $s = 3$, respectively. The desired vectors must fulfill the conditions of equation 9–9:

9.5 The Determination of Absolute pK-Values

$$\mathbf{c}_{BH_s} = \begin{pmatrix} b_0 \\ 0 \\ \vdots \\ 0 \end{pmatrix}, \quad \mathbf{c}_{BH_{s-1}} = \begin{pmatrix} 0 \\ 0 \\ \vdots \\ 0 \end{pmatrix}, \quad \ldots \quad \mathbf{c}_B = \begin{pmatrix} 0 \\ 0 \\ \vdots \\ b_0 \end{pmatrix} \qquad (9\text{-}31)$$

The position vectors $\mathbf{c}_p$, which according to equation 9–10 lead from the coordinate origin to points on the concentration curve, and the vectors $\mathbf{c}_{BH_s}, \mathbf{c}_{BH_{s-1}} \ldots \mathbf{c}_{BH_2}$ define the following "subspace" within the corresponding s-dimensional domain (cf. the analogous situation described by equation 8–45 in Sec. 8.5.3):

$$\begin{aligned}\mathbf{c} = \ & \mathbf{c}_{BH_s} + k_1(\mathbf{c}_{BH_{s-1}} - \mathbf{c}_{BH_s}) + \ldots k_{(s-2)}(\mathbf{c}_{BH_2} - \mathbf{c}_{BH_s}) + \\ & + k_{(s-1)}(\mathbf{c}_p - \mathbf{c}_{BH_s})\end{aligned} \qquad (9\text{-}32)$$

where $-\infty < k_z < \infty$, $z = 1, 2 \ldots (s-1)$, and $s \geq 3$

Within this subspace are located all the concentrations that would obtain if h and K_s were held constant but $K_1, K_2 \ldots K_{(s-1)}$ were allowed to vary over the applicable range. If these equilibrium constants are now allowed to approach the limit $K_i \to \infty$, the only protolytes remaining in solution will be $BH^{(s-1)-}$ and B^{s-}, a conclusion that follows from equation 9–4a. Therefore, $\mathbf{c}$ terminates along the tangent $B^{s-} BH^{(s-1)-}$, which is also one of the edges of the desired polyhedron (cf. equation 9–15). Each point on this tangent fulfills the conditions specified by the following equation:

$$\mathbf{c} = \mathbf{c}_B + k(\mathbf{c}_{BH} - \mathbf{c}_B) \quad \text{with} \quad -\infty < k < \infty \qquad (9\text{-}33)$$

Setting equations 9–32 and 9–33 equal makes it possible to locate points corresponding to any particular pH-value in the s-step system for the isolated sth equilibrium $BH^{(s-1)-} \leftrightarrows B^{s-} (H^+)$:

$$\begin{pmatrix} b_0 \\ \vdots \\ 0 \\ 0 \\ 0 \end{pmatrix} + k_1 \begin{pmatrix} -b_0 \\ \vdots \\ 0 \\ 0 \\ 0 \end{pmatrix} + \ldots k_{(s-2)} \begin{pmatrix} -b_0 \\ \vdots \\ b_0 \\ 0 \\ 0 \end{pmatrix} +$$

$$+ k_{(s-1)} \frac{b_0}{H_s} \begin{pmatrix} h^s - H_s \\ \vdots \\ h^2 \prod_{l=0}^{s-2} K_l \\ h \prod_{l=0}^{s-1} K_l \\ \prod_{l=0}^{s} K_l \end{pmatrix} = \begin{pmatrix} 0 \\ \vdots \\ 0 \\ 0 \\ b_0 \end{pmatrix} + k \begin{pmatrix} 0 \\ \vdots \\ 0 \\ b_0 \\ -b_0 \end{pmatrix} \quad (9\text{--}34)$$

The last two equations in this system of equations provide the following expression for the parameter k:

$$k = \frac{h}{h + K_s} \quad (9\text{--}35)$$

It follows from equation 9–33 for the vector **c** that:

$$\mathbf{c} = \frac{b_0}{h + K_s} \begin{pmatrix} 0 \\ \vdots \\ 0 \\ h \\ K_s \end{pmatrix} \quad (9\text{--}36)$$

Thus, the coordinates of **c** have the same functional values as the coordinates of a single-step acid-base equilibrium (cf. Sec. 6.3). The vector **c** itself provides the titration points of the isolated sth equilibrium. This result shows in general how the last equilibrium in a titration system may be investigated separately, using a method that corresponds to the pencil of lines and pencil of planes methods already introduced for the cases $s = 2$ and $s = 3$, respectively, (cf. Sections 7.3.5 and 8.3.5).[86]

The relationships developed above also hold true within A-diagrams, which are simply affine distortions of concentration diagrams (cf. Sec. 9.2). Nevertheless, the task remains of locating the vertices of the appropriate absorbance polyhedron. After this is accomplished (cf. Sec. 9.4), absorbances associated with an isolated sth equilibrium may be evaluated like the titration points of a single-step acid-base equilibrium.

It has been demonstrated in Sec. 9.4 how A-diagram curves may be obtained for a protolysis system which is successively reduced by one equilibrium after another. Each of these curves may be evaluated by the method described, which makes it possible in principle to establish all pK-values for an s-step system.

Alternatively, one may examine s-step titration systems by replacing the linearly independent concentration variables by what are known as "barycentric coordinates", coordinates that take into consideration the complete set of relevant concentrations.[91] In the case of a two-step system these coordinates have been referred to as "Gibbs triangular coordinates" (cf. Sec. 7.5.4). We have chosen to forego any further discussion of this procedure, however, due to the convenience of the special relationships that exist between characteristic concentration- and A-diagrams.

9.6 Summary

The relationships described in earlier chapters with respect to two- and three-step acid-base systems may be generalized so that they apply to s-step systems. *ADQs-diagrams* make it possible to determine whether or not a system is comprised of s linearly independent equilibria (or concentration variables), a further application of the method of graphic matrix-rank analysis.

The experimental curves comprising an s-dimensional A-diagram may be dissected into z discrete curves, which in turn fall within a characteristic "subspace". These curves correspond to the data one would obtain if it were possible to titrate successively in the sequence $z = 1, 2 \ldots (s-1)$ the various equilibria comprising the overall system. The location of the corresponding subspace may be ascertained either from a set of *ADQz-diagrams* or by successive rank analyses of the *absorbance matrix* **A**. *Molar absorptivities* for all ampholytes are accessible once the appropriate subspace regions have been identified, providing a complete set of results based entirely on spectrometric data.

Finally, molar absorptivities permit construction of an *absorbance polyhedron.* An absorbance polyhedron may be used to dissect a system into s partial equilibria, each of which may be evaluated as if it were a single-step acid-base equilibrium. Thus, all the pK values associated with the system may be ascertained.

Literature

1. L. G. Sillen, *Acta Chem. Scand.* 18 (1964) 1085.
2. L. G. Sillen, *Acta Chem. Scand.* 18 (1964) 1085-1098.
3. N. Ingri, L.G. Sillen, *Ark. Kemi* 23 (1964) 97.
4. A. R. Emery, *J. Chem. Educ.* 42 (1965) 131.
5. I. G. Sayce, *Talanta* 15(1968) 1397-1411.
6. R. Arnek, L. G. Sillen, O. Wahlberg, *Ark. Kemi* 31 (1969) 353.
7. P. Branner, L. G. Sillen, R. Whiteker, *Ark. Kemi* 31 (1969) 365.
8. T. Meites, L. Meites, *Talanta* 19 (1972) 1131.
9. P. Gans, A. Vacca, A. Sabatini, *Inorg. Chim. Acta* 18 (1976) 237.
10. H. Stünzi, G. Anderegg, *Helv. Chim. Acta* 59 (1976) 1621.
11. G. Arena, E. Rizzarelli, S. Sammartan, C. Rigano, *Talanta* 26 (1979) 1.
12. A. Ivaska, I. Nagypal, *Talanta* 27 (1980) 721.
13. H. Gampp, M. Maeder, A. D. Zuberbühler T. Kaden, *Talanta* 27 (1980) 513.
14. F. Gaizer, A. Puskas, Talanta 28(1981) 565.
15. C. A. Chang, B. E. Douglas, *J. Coord. Chem.* 11 (1981) 91.
16. R. J. Motekaitis, AE. Martell, *Can. J. Chem.* 60 (1982) 168.
17. P. M. May, D. R. Williams, P. W. Linder, R. G. Torrington, *Talanta* 29 (1982) 249.
18. M. C. Garcia, G. Ramis, C. Mongay, *Talanta* 29 (1982) 435.
19. J. Kostrowicki, A. Liwo, *Computers & Chemistry* 8 (1984) 91.
20. J. Kostrowicki, A. Liwo, *Computers & Chemistry* 8 (1984) 101.
21. A. Laouenan, E. Suet, *Talanta* 32 (1985) 245.
22. J. Havel, M. Meloun, *Talanta* 33 (1986) 525.
23. T. Hofman, M. Krzyzanowska, *Talanta* 33 (1986) 851.
24. D. J. Legett, *Computational Methods for the Determination of Formation Constants.* Plenum Press, New York, London 1985.
25. S. Ahrland, P. Bläuenstein, B. Tagesson, D. Tuhtar, *Acta Chem. Scand.* A 34 (1980) 265.
26. S. Ahrland, I. Persson, R. Portanova, *Acta Chem. Scand.* A 35 (1981) 49.
27. S. Ahrland, N.-O. Bjork, I. Persson, *Acta Chem. Scand.* A 35 (1981) 67.
28. B. Holmberg, G. Thome, *Acta Chem. Scand.* A 34 (1980) 421.
29. I. Granberg, S. Sjöberg, *Acta Chem. Scand.* 35 (1981) 193.
30. G. Anderegg, *Helv. Chim. Acta* 64 (1981) 1790.
31. W. Forsling, I. Granberg, S. Sjöberg, *Acta Chem. Scand.* A 35 (1981) 473.
32. L.-O. Ohmann, S. Sjöberg, *Acta Chem. Scand.* A 35 (1981) 201.
33. L.-O. Ohmann, S. Sjöberg, *Acta Chem. Scand.* A 36 (1982) 47.
34. I. Granberg, W. Forsling, S. Sjöberg, *Acta Chem. Scand.* A 36 (1982) 819.
35. Z. Szeverenyi, U. Knopp, A. D. Zuberbühler, *Helv. Chim. Acta* 65 (1982) 2529.
36. Y. Wu, Th. Kaden, *Helv. Chim. Acta* 66 (1983) 1588.
37. M. Wilgocki, J. Bjerrum, *Acta Chem. Scand.* A 37(1983) 307.
38. H. Saarinen, M. Orama, T. Raikas, J. Korvenranta, *Acta Chem. Scand.* A 37 (1983) 631.
39. R. J. Motekaitis, A. E. Martell, J.-P. Lecomte, J.-M. Lehn, *Inorg. Chem.* 22 (1983) 609.
40. I. Yoshida, R. J. Motekaitis, A. E. Martell, *Inorg. Chem.* 22 (1983) 2795.
41. L.-O. Ohmann, S. Sjöberg, N. Ingri, *Acta Chem. Scand.* A 37 (1983) 561.
42. L.-O. Ohmann, S. Sjöberg, *Acta Chem. Scand.* A 37 (1983) 875.

43. J. Granberg, S. Sjöberg, *Acta Chem. Scand.* A 37 (1983) 415-.
44. A. W. Hamburg, M. T. Nemeth, D. W. Margerum, *Inorg. Chem.* 22 (1983) 3535.
45. S. Ishigura, Y. Oka, H. Ohtaki, *Bull. Chem. Soc. Jpn.* 57 (1984).
46. S. Sjöberg, L.-O. Ohmann, N. Ingri, *Acta Chem. Scand.* A 39 (1985) 93.
47. S. Ahrland, S.-I. Ishiguro, A. Marton, I. Persson, *Acta Chem. Scand.* A 39 (1985) 227.
48. T. Kaden, A. D. Zuberbühler, *Talanta* 18 (1971) 61.
49. A. Sabatini, A. Vacca, *J. Chem. Soc. Dalton* 1972, 1693.
50. G. A. Cumme, *Talanta* 20 (1973) 1009.
51. A. Sabatini, A. Vacca, P. Gans, *Talanta* 21 (1974) 53.
52. G. Ginzburg, *Talanta* 23 (1976) 149.
53. R. Karlsson, L. Kullberg, *Chem. Scr.* 9 (1976) 54.
54. A. M. Corrie, G. K. R. Makar, M. L. D. Toucho, D. R. Williams, *J. Chem. Soc. Dalton* 1975, 105.
55. D. J. Legett, *Talanta* 24 (1977) 535.
56. M. Wozniak, G. Nowogrocki, *Talanta* 25 (1978) 643.
57. H. Gampp, M. Maeder, A. D. Zuberbühler, T. Kaden, *Talanta* 27 (1980) 513.
58. E. Johansen, O.Jons, *Acta Chem. Scand.* A 35 (1981) 233.
59. R. J. Motekaitis, A. E. Martell, *Can. J. Chem.* 60 (1982) 2403.
60. M. Meloun, J. Havel, *Computation of Solution Equilibria.* 1. Spectrometry. Folia Fac. Sci. Nat., Univ. Purkynianae Brunensis, Brno, 1983.
61. J. Havel, M. Meloun, *Talanta* 32 (1985) 171.
62. H. Gampp, M. Maeder, C. J. Meyer, A. D. Zuberbühler, *Talanta* 32 (1985) 257.
63. H. Gampp, M. Maeder, C. J. Meyer, A. D. Zuberbühler, *Talanta* 32 (1985) 95.
64. K. Nagano, D. E. Metzler, *J. Amer. Chem. Soc.* 89 (1967) 2891.
65. F. Göbber, H. Lachmann, Hoppe-Seylers *Z. Physiol. Chem.* 359 (1978) 269.
66. R. M. Alcock, F. R. Hartley, D. E. Rogers, *J. Chem. Soc. Dalton Trans.* 1978, 115.
67. M. Meloun, J. Cernak, *Talanta* 26 (1979) 569.
68. H. Gampp, M. Maeder, A. D. Zuberbühler, *Talanta* 27 (1980) 1037.
69. M. Meloun, J. Cernak, *Talanta* 31 (1984) 947.
70. M. Meloun, M. Javurek, *Talanta* 31 (1984) 1083.
71. M. Meloun, M. Javurek, *Talanta* 32 (1985) 973.
72. M. Meloun, M. Javurek, A. Hynkova, *Talanta* 33 (1986) 825.
73. H. Gampp, M. Maeder, C. J. Meyer, A. D. Zuberbühler, *Talanta* 32 (1985) 95.
74. H. Gampp, M. Maeder, C. J. Meyer, A. D. Zuberbühler, *Talanta* 32 (1985) 257.
75. I. Bertini, A. Dei, C. Luchinat, R. Monnanni, *Inorg. Chem.* 24 (1985) 301.
76. T. Kaden, A. D. Zuberbühler, *Talanta* 18 (1971) 61.
77. D. J. Leggett, W. A. E. McBryde, *Anal. Chem.* 47 (1975) 1065.
78. M. Meloun, J. Cernak , *Talanta* 23 (1976) 15.
79. F. Gaizer, A. Puskas, *Talanta* 28 (1981) 925.
80. M. Meloun, M. Javurek, *Talanta* 31 (1984) 1083.
81. M. Meloun, M. Javurek, *Talanta* 32 (1985) 1011.
82. J. Havel, M. Meloun, *Talanta* 33 (1986) 453.
83. M. Meloun, M. Javurek, J. Havel, *Talanta* 33 (1986) 513.
84. H. Gampp, M. Maeder, C. J. Meyer, A. D. Zuberbühler, *Talanta* 32 (1985) 257.
85. H. Gampp, M. Maeder, C. J. Meyer, A. D. Zuberbühler, *Talanta* 32 (1985) 1133.
86. J. Polster, *Habilitationsschrift.* Tübingen 1980.
87. S. Lipschutz, *Lineare Algebra.* McGraw Hill Book Company, Düsseldorf, New York 1979.

88. J. Polster, *Z. Physik. Chem. N. F.* 97 (1975) 55.
89. D. Danil, G. Gauglit, H. Meier, *Photochem. Photobiol.* 26 (1977) 225.
90. G. Gauglitz, *GIT Fachz. Lab.* 3 (1982) 205.
91. K.-D. Willamowski, O. E. Rössler, *Z. Naturforsch.* 31 a (1976) 408.

10 Non-Linear Curve-Fitting of Spectrometric Data from s-Step Acid-Base Equilibrium Systems

10.1 Basic Equations

In an s-step sequential acid–base titration of the type

$$BH_s \overset{K_1}{\leftrightarrows} BH_{s-1}^- \overset{K_2}{\leftrightarrows} BH_{s-2}^{2-} \overset{K_3}{\leftrightarrows} \ldots \overset{K_{s-1}}{\leftrightarrows} BH^{(s-1)-} \overset{K_s}{\leftrightarrows} B_s \quad (10\text{–}1)$$

the following equation defines the relationship between the concentration of a given species and the pH or h of the system (cf. Chapter 9, equation 9–4a):

$$bh_{s-i} = \frac{b_0 \cdot h^{s-i} \cdot \prod_{l=0}^{l=i} K_l}{\sum_{k=0}^{k=s} h^k \cdot \prod_{l=0}^{l=s-k} K_l} \quad (10\text{–}2)$$

where $0 \leq i \leq s$, $bh_0 := b$, and $K_0 := 1$

For example, in the case $s = 2$:

if $i = 0$ $\quad bh_2 = \dfrac{b_0 \cdot h^2}{K_1 \cdot K_2 + hK_l + h^2}$

if $i = 1$ $\quad bh = \dfrac{b_0 \cdot h \cdot K_1}{K_1 \cdot K_2 + hK_l + h^2}$

if $i = 2$ $\quad b = \dfrac{b_0 \cdot K_1 \cdot K_2}{K_1 \cdot K_2 + hK_l + h^2}$ $\qquad$ (cf. Chapter 7, equation 7–5)

The generalized Lambert–Beer–Bouguer law for titration system 10–1 requires that

$$A_\lambda = l(\varepsilon_{\lambda BH_s} \cdot bh_s + \varepsilon_{\lambda BH_{s-1}} \cdot bh_{s-1} + \ldots \varepsilon_{\lambda BH} \cdot bh + \varepsilon_{\lambda B} \cdot b) \quad (10\text{–}3)$$

If each of the concentrations in this equation is replaced by the general expression 10–2, rearrangement of terms leads to a useful general relationship between the independent variable pH (or h) and the dependent variable absorbance (A_λ):

$$A_\lambda = l \cdot b_0 \cdot \frac{\varepsilon_{\lambda BH_s} \cdot h^s + \varepsilon_{\lambda BH_{s-1}} \cdot h^{s-1} \cdot K_1 + \ldots + \varepsilon_{\lambda BH} \cdot h \cdot \prod_{i=1}^{i=s-1} K_i + \varepsilon_{\lambda B} \cdot \prod_{i=1}^{i=s} K_i}{h^s + h^{s-1} \cdot K_1 + \ldots + h \cdot \prod_{i=1}^{i=s-1} K_i + \prod_{i=1}^{i=s} K_i}$$

$$= l \cdot \frac{A_{\lambda BH_s} \cdot h^s + A_{\lambda BH_{s-1}} \cdot h^{s-1} \cdot K_1 + \ldots + A_{\lambda BH} \cdot h \cdot \prod_{i=1}^{i=s-1} K_i + A_{\lambda B} \cdot \prod_{i=1}^{i=s} K_i}{h^s + h^{s-1} \cdot K_1 + \ldots + h \cdot \prod_{i=1}^{i=s-1} K_i + \prod_{i=1}^{i=s} K_i} \quad (10-4)$$

Thus, the functional relationship between the desired information (pK) and the experimental parameters A_λ and pH (or h) is nonlinear. By contrast, equation 10–3 reveals that there is a linear relationship between A_λ and the molar absorptivity ε_{λ_i}.

10.2 Iterative Curve-Fitting Procedures

For more than 20 years it has been common practice to conduct *pK determinations* by applying non-linear curve-fitting routines to spectrometric A_λ(pH) titration curves.[1-3] It is normally not possible to determine by direct experiment the molar absorptivities $\varepsilon_{\lambda i}$ of all the species present (or even all the ampholytes), so the pK values and all coefficients are initially taken to be unknowns.

What follows is a brief description of the logic employed by a typical non-linear curve-fitting program, TIFIT (cf. also Fig. 10–1).[4] Experimental data are supplied in the form of A_λ(pH) measurements acquired at one or more wavelengths. One must also supply estimates pK_{0i} for all the non-linear parameters (the unknown pK values).

In a first *iterative step* the pK_{0i} estimates are used to calculate with the aid of equation 10–2 a complete set of concentrations corresponding to each measured value of pH (or h). With the exception of the molar absorptivities $\varepsilon_{\lambda i}$, numerical values thereby become available for all the quantities in equation 10–3 (the pathlength l is set equal to 1 cm). Since the absorptivities $\varepsilon_{\lambda i}$ are, according to equation 10–3, linearly dependent upon the measured absorbances A_λ, these may in principle be calculated

10.2 Iterative Curve-Fitting Procedures

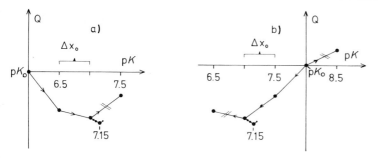

Fig. 10–1. Optimization strategy employed by the program TIFIT in analyzing data derived from *p*-nitrophenol; (a) search proceeding initially in the right direction; (b) search involving a false start.

from the experimental data $A_\lambda(\text{pH})$ using the standard techniques for solving systems of linear equations (see below).

Next, the calculated molar absorptivities and the estimated pK values are used to compute (by equation 10–4) absorbances at every available wavelength and for each pH value. The calculated results, $A_{\lambda_{\text{calc}}}$, are then compared with the experimental values so that differences $A_{\lambda_{\text{exp}}} - A_{\lambda_{\text{calc}}}$ may be taken as a measure of the "goodness" of the result from this iteration step. Such differences may be either positive or negative, so it is appropriate that they be squared, and that the sum of the squares be taken as the measure of an "error function":

$$Q = f\left[\sum \left(A_{\lambda_{\text{exp}}} - A_{\lambda_{\text{calc}}}\right)^2\right]$$

The complete set of available experimental data (e.g., data collected at l wavelengths λ_i with m separate titration steps pH$_j$) may be utilized in this operation as follows:

$$Q = \frac{1}{2} \cdot \sum_{i=1}^{i=l} \sum_{j=1}^{j=m} \left(A_{i\,j\text{exp}} - A_{i\,j\text{calc}}\right)^2 \tag{10–5}$$

This is an application of the "method of least squares" introduced by Gauss. The error function Q acquires a smaller value the more closely the estimated values for pK approach the "true" pK values. The goal of the iteration process is obviously to minimize the function Q:

$$Q(\text{p}K, \varepsilon) = \min \tag{10–6}$$

Although each of the pK values is a non-linear parameter that must be established through an iterative process, the molar absorptivities $\varepsilon_{\lambda i}$ (for any particular set of pK assumptions) are directly calculable, because the minimization conditions $(\partial Q/\partial \varepsilon_{\lambda i} = 0)$ result in a system of linear equations.

In a *second stage of iteration* one of the assumed values pK_{0i} (e.g., pK_{01}) is increased by a specific increment ΔX_0 (e.g., 0.5 pH units), all others being held constant. This revised set of pK values is then used to perform another set of calculations like those of the first iteration, and a corresponding error function Q is again evaluated. If the resulting Q is smaller than that from the first step it is assumed that progress has been made, so in a *subsequent iteration* pK_1 is increased again by the increment ΔX_0. This process is repeated until Q itself begins to show an increase (cf. Fig. 10–1a). In the event that Q increases in the very first step (as shown in Fig. 10–1b), the direction of incrementation is simply reversed.

Figure 10–1 illustrates the course of the iteration process for the *single-step titration* of *p*-nitrophenol, a substance reported to have a pK of 7.14–7.15. A crude search utilizing a pK increment (ΔX_0) of 0.5 pH units has in this example been pursued to the point of a minimum at pK = 7.0, after which the search was continued with the smaller increment $\Delta X_1 = 0.1 \cdot \Delta X_0 = 0.05$. Such a search for a precise minimum may be continued until some particular pre-determined level of accuracy has been reached (e.g., $\Delta X_{\text{final}} = 0.001$ pH unit). Application of the process to *p*-nitrophenol in the form of a coupled analysis of 9 sets of A_λ (pH) data, each consisting of measurements at 11 pH values, resulted in a pK estimate of 7.15_0 ($\Delta X_{\text{final}} = 0.0001$). Nineteen iteration steps were required when the initial value pK_0 was taken to be 6.0, while 29 steps were necessary starting from p$K_0 = 13.0$.

In the case of a *multi-step titration* a minimum is required for a function Q approached from several (i.e., s) different coordinate directions, so the search strategy must be modified somewhat. In the simplest approach the value of Q is optimized initially only in terms of the first pK, followed by the second, etc., but this leads to success only if the true pK values differ greatly from one another; that is, only in the case of non-overlapping titration systems.[3] The program TIFIT approaches the problem in a somewhat different way. A coarse increment ΔX_0 is used to optimize in a crude way the first pK value (leading to the estimate pK_{11}), after which pK_{11}, pK_{12}, pK_{13}, ... are used to roughly optimize the second pK (giving pK_{21}), and so forth. Only when each pK has been assigned an approximate value (pK_{11}, pK_{12}, pK_{13}, ...) is the increment reduced (e.g., $\Delta X_1 = 0.1 \cdot \Delta X_0$), and then a new optimization cycle is initiated.

When the *three-step titration* of pyridoxal (cf. Table 10–1 and Sec. 6.4) was made the subject of a coupled evaluation of 7 A_λ(pH) data series, each consisting of points taken at 130 pH values, the pK results that emerged (with $\Delta X_{\text{final}} = 0.0001$) were 4.06_9, 8.48_1, and 13.31_3. Using 4.5, 8.5, and 13.0 as the initial estimates for pK_{0i} the process required a total of 80 iteration steps. By contrast, the assumptions 1.0, 7.0, and 14.0 necessitated 120 iterations, although the final answers proved to be independent of the starting points (see below).

The procedures so far described may be characterized as employing a "trial and error" methodology; that is, a program like TIFIT compares any particular value of the error function Q only with that obtained from the most recent step in the process, which necessarily represents the current "best value" for Q. Such a strategy is inapplicable with respect to a great many optimization problems, including ones encountered in chemical and biochemical kinetics, as well as in industrial production and process control. For this reason the literature describes a wide variety of more complex approaches to non-linear regression, including the Gauss–Newton method,[5] the Newton–Raphson method,[5] the Powell method,[5,6] the Marquardt method,[5] and the "multiple shooting" method.[7] Powerful search algorithms are particularly important in the case of so-called "ill-conditioned" systems of equations,[8] or when several relative minima are present, in which case the particular minimum identified may depend on the values used as starting points (cf. Fig. 10–2).[9]

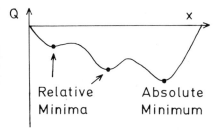

Fig. 10–2. Different types of minima that may be encountered during an optimization process.

The fact that a simple "trial and error" method (e.g., TIFIT) frequently provides excellent results in the analysis of multi-wavelength spectrometric titrations may be ascribed largely to the following causes:

(1) Spectrometric acid–base titration curves are typically associated with rather well-behaved ("well-conditioned") systems of equations.

(2) Plausible titration mechanisms may be established in advance by means of matrix-rank analysis; thus, the curve-fitting program need not be called to serve as a significant test of the titration mechanism.

(3) Correlations may be made at several wavelengths (preferably as many as possible), producing an error function Q that is also based on multi-wavelength information (cf. equation 10–5).

(4) In the case of multiple-step titrations the data set may be expanded to include measurements at an appropriately large number of pH values (cf. Table 10–1).

(5) Spectrometric titration data themselves are characterized by a very low noise level and a high degree of precision.

(6) An adequate degree of signal variation is normally present somewhere within a complete set of $A_\lambda(pH)$ data. Even if at some particular wavelength the absorbance is found to remain essentially invariant in the course of one titration step, compensation for the missing information is usually supplied by large absorbance changes at some other wavelength.

An example of this situation is provided by the spectrometric titration of pyridoxal (cf. Fig. 6–18), in which the first titration step produces virtually no change in absorbance at 380 nm, even though changes at other wavelengths are quite large. If only $A_\lambda(pH)$ data at 380 nm were to be evaluated, the results would inevitably be poor. Even highly sophisticated analyses (e.g, the Powell method) will in this case produce values for pK and $\varepsilon_{\lambda i}$ that are utterly incorrect, especially if an unfortunate choice is made for the starting value pK_{0i}.

As soon as a minimum for the function Q has been located with the aid of an iterative curve-fitting function it is wise to introduce the resulting pK- and ε-values into equation 10–4 and calculate a theoretical $A_\lambda(pH)$ curve, and then to compare this curve with the experimental data (i.e., to perform a *control simulation*). For this reason the program TIFIT was designed to operate in an interactive graphic mode, with results appearing first on a high-resolution computer screen prior to subsequent output (if desired) to a plotter. All the figures referred to in Table 10–1 were prepared in this way.

Table 10–1. Examples of experimental data analyses accomplished with the aid of the program TIFIT.

Titration system	Rank	No. of wavelengths examined	No. of pH measurements	Calc. pK	Calc. ΔpK	Ref.
p-Nitrophenol	$s = 1$	9	11	7.15± 0.01		Sec. 6.3, Fig. 6–8
4-Desoxypyridoxol	$2 \cdot (s = 1)$	6	46	5.26± 0.01		
					4.56	Sec. 6.4, Fig. 6–14
				9.82± 0.01		
Pyridoxal	$3 \cdot (s = 1)$	7	130	4.07± 0.01		
					4.41	
				8.48± 0.01		Sec. 6.4, Fig. 6–18
					4.83	
				13.31± 0.02		
Cyclization product of pyridoxal with histamine	$s = 4$ (with overlap)	6	116	1.46± 0.02		
					2.69	
				4.15± 0.01		
					3.46	Sec. 6.4, Fig. 6–22
				7.63± 0.01		
					2.49	
				10.10± 0.02		

TIFIT permits experimental and simulated data to be displayed either in the form of $A_\lambda(\mathrm{pH})$ diagrams or as A-diagrams (i.e., devoid of pH information). Since in the latter case the calculated molar absorptivities $\varepsilon_{\lambda i}$ (or absorbances $A_{\lambda \mathrm{BH}_s} = l \cdot b_0 \cdot \varepsilon_{\lambda \mathrm{BHs}}$ etc.) for the pure species may be regarded as the vertices of absorption triangles, the corresponding numerical values are also incorporated into the TIFIT output.

10.3 Statistical Analysis

It has been suggested above that the degree of similarity between a set of experimental results and a particular simulation may be assessed either by computing an error function Q or by preparing a graphical comparison. There is another possible approach, however: *statistical analysis*. For example, the program TIFIT permits one to take an entire set of both experimental and simulated $A(\lambda,\mathrm{pH})$ data and compute directly a *standard deviation of the absorbance measurements*.

Since pH data are introduced into the computation process as independent variables (no account being taken of errors in measurement), they are not associated initially with a standard deviation of their own. Instead, a separate program module must be used to perform a further iteration sequence in which the following expression is minimized for each of the m pH data points pH_j by variation of the

$$Q^* = \left(\mathrm{pH}_{(\exp)j} - \mathrm{pH}_{0j}\right)^2 + \sum_{i=1}^{i=l} \left(A_{ij(\exp)} - A_{ij(\mathrm{calc})}\right)^2 \tag{10-7}$$

This results in a set of "error-free" values pH_{0j} ($j = 1, 2, \ldots m$). The differences $\mathrm{pH}_{(\exp)j} - \mathrm{pH}_{0j}$ are then used to calculate a *standard deviation of the pH measurements*.

Finally, yet another portion of the program examines the quality of the experimental data and the statistical significance of the resulting interpretation by introducing random errors into the control simulation (a *Monte Carlo Simulation*; cf. references 4 and 10). Thus, a normally distributed set of random numbers is imposed upon the simulated data. The range for the artificial "scatter" is so chosen that the corresponding standard deviation is identical to that evidenced by the experimental data. The result is a set of perhaps 10 "pseudo-data" points, which is then subjected to a complete systematic evaluation like that applied to ordinary experimental data.

The purpose of the exercise is to ascertain whether anywhere in the course of the analysis real data behave differently from the artificial data, the "errors" in which are governed by a normal distribution pattern. In other words, the process is designed to point to the presence of non-random errors within a set of experimental data, as well as to signal the possibility of an erroneous mechanistic assumption. In none of the examples listed in Table 10–1 was any such problem encountered (apart from the anomalous effects associated with pyridoxal data collected at 380 nm above pH 12, as discussed in Sec. 6.4).

Compared with the very straightforward "trial and error" curve-fitting phase of the program, the Monte Carlo simulation (including subsequent repetition of the entire computation for 10 pseudo-data sets) requires an enormous number of computations; indeed, this aspect of the data workup consumes more than 90% of the total computer time required. It is thus common practice to forego exhaustive statistical analysis in routine cases, especially when the validity of a particular titration mechanism has already been verified through graphic matrix-rank analysis.

10.4 Summary

The principle advantages and disadvantages of the data analysis program TIFIT may be summarized as follows:

Advantages

(1) The program is applicable to any type of sequential acid–base titration, irrespective of the extent of overlap between individual titration steps ($\Delta pK \geq 0$).

(2) Straightforward program modifications permit evaluation of exceptional cases such as simultaneous titrations.

(3) The program was specifically designed to accomplish coupled multi-wavelength analysis, and it is applicable to any type of spectrometric titration satisfying the conditions of equation 10–4 (cf. Chapter 18).

(4) The procedure may be adapted to the evaluation of individual titration curves, such as those resulting from a potentiometric titration. In this case all coefficients in equation 10–4 assume the form $\varepsilon_\lambda = 1$ and $l = 1$. Such a simplification obviously eliminates the advantages associated with multi-wavelength information.

(5) Since even in the case of multi-wavelength investigations it is possible to evaluate various $A_\lambda(pH)$ curves individually or in combination, the information content of various spectral regions may be compared.

10.4 Summary

(6) The program is effective even when initial pK estimates are quite crude (e.g., estimates that are as much as ±1–2 pH units away from the true values, or sometimes farther; cf. the examples discussed above).

(7) Initial calculations employ rather large search increments (0.5 pH units), and thus lead to rapid convergence; thus, reducing the accuracy of the original pK estimates results in a relatively small increase in overall computation time.

(8) No prior knowledge is required of the values $A_{\lambda BH_s}$ and $A_{\lambda B}$ corresponding to the titration limits. Indeed, one of the results of the computation is the complete set of absorbances $A_{\lambda i}$ for all pure species relevant to the titration mechanism (or the molar absorptivities $\varepsilon_{\lambda i}$, provided the starting concentrations are known). It is thus not mandatory that the acidic or basic limits of the titration be quantitatively accessible (cf. the problems associated with pyridoxal at pH > 12, as described in Sec. 6.4)

(9) Evaluations are possible even in cases where the sample is unstable or virtually insoluble within specific pH regions (i.e., gaps are permitted within the titration curves[4]) provided all the relevant titration steps are reflected somewhere in the available $A_\lambda(pH)$ data.

(10) Provision exists for visual control over the results of a computation in the form of appropriate $A_\lambda(pH)$- or A-diagrams. In the latter case boundaries are also indicated for absorbance triangles.

(11) Various statistical tests may be performed, and randomly distributed errors may be introduced to permit a Monte Carlo simulation.

Disadvantages

(1) Access to a mainframe computer is required, preferably one equipped with a high resolution screen output and/or a plotter. Nevertheless, the actual iteration routine (excluding Monte Carlo simulations) entails so little computation time that it should also be suitable for implementation on a fast personal computer.

(2) As with all curve-fitting routines, there is relatively little assurance that TIFIT will recognize an erroneous mechanistic assumption. For this reason the presumed titration mechanism and the corresponding number s of spectroscopically observable titration steps should always be subjected to prior verification on the basis of a graphic matrix-rank analysis.

Literature

1. C. J. Sullivan, J. Rydberg, W. F. Miller, *Acta Chem. Scand.* 13 (1959) 2023.
2. L. G. Sillen, *Acta Chem. Scand.* 16 (1962) 159; N. Ingri, L. G. Sillen, *Acta Chem. Scand.* 16 (1962) 173; L. G. Sillen, *Acta Chem. Scand.* 18 (1964) 1085.
3. K. Nagano, D. E. Metzler, *J. Amer. Chem. Soc.* 89 (1967) 2891.
4. F. Göbber, H. Lachmenn, *Hoppe-Seylers Z. Physiol. Chem.* 359 (1978) 269.
5. U. Hoffmann, H. Hofmann, *Einführung in die Optimierung*. Verlag Chemie, Weinheim 1971.
6. M. J. D. Powell, *Comput. J.* 7 (1965) 303.
7. P. Deuflhard, G. Bader, in: *Numerical Treatment of Inverse Problems in Differential and Integral Equations*. Birkhäuser, Boston-Basel-Stuttgart 1983, p. 74.
8. G. E. Forsythe, C. B. Moler, *Computer-Verfahren für lineare algebraische Systeme*. R.Oldenbourg-Verlag, München 1971, Chap.8
9. M. Markus, T. Plesser, A. Boiteux, B. Hess, M. Malcovath, *Biochem. J.* 189 (1980) 225.
10. F. Göbber, J. Polster, *Anal. Chem.* 48 (1976) 1546.

11 Simultaneous Titrations

The attainable accuracy in a pK determination based on spectrometric titration data is normally limited not by the precision of the spectrometric measurements but rather by theoretical and practical problems associated with the conventional pH scale. For this reason a set of techniques has been developed for establishing internal, *relative* pK values in the course of investigating two-step and multiple-step overlapping titration systems (cf. Sec. 7.4 and the references cited therein).

However, this expedient is not available in the case of a single-step titration, nor can it be used with a non-overlapping multi-step titration system. Rather, one is here forced to add an *external* standard protolyte whose pK is known to a high degree of accuracy, and then to determine relative pK values with respect to this standard. The actual evaluation is conducted strictly on the basis of photometric data, with no further reliance on electrochemical measurements. Thus, the fundamental problem of assigning pK values to the standard substances may be left to others, as in the analogous case of standard DIN or NBS pH-calibration buffers.

11.1 Basic Equations

Simultaneous titration of the single-step protolyte BH in the presence of the single-step reference protolyte CH produces a titration curve each point of which represents the achievement of equilibrium conditions for two separate reactions:

$$BH + H_2O \leftrightarrows B^- + H_3O^+ \qquad (11\text{--}1a)$$

$$CH + H_2O \leftrightarrows C^- + H_3O^+ \qquad (11\text{--}1b)$$

Consideration of the mass balance requirements

$$b_0 = bh + b \tag{11-2a}$$

$$c_0 = ch + c \tag{11-2b}$$

leads to the following relationships for the respective (mixed) dissociation constants:

$$K_B = h \cdot \frac{b}{bh} \tag{11-3a}$$

$$K_C = h \cdot \frac{c}{ch} \tag{11-3b}$$

The quantity h appears in both equations, and if this is eliminated by combining the two, an expression results which may be rearranged as follows:[1]

$$b = \frac{K_B \cdot b_0 \cdot c}{K_C \cdot c_0 + (K_B - K_C) \cdot c} \tag{11-4}$$

This equation expresses a functional relationship that must exist between b and c at every point in a simultaneous titration (and constituting the basis for a *concentration diagram*). Subsequent introduction of the reduced variables $\beta := b/b_0$ and $\gamma := c/c_0$ representing relative extents of dissociation then leads to the following equation:

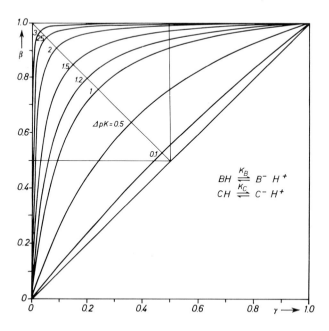

Fig. 11-1. Dissociation diagrams for a simultaneous titration, shown as a function of ΔpK.

11.1 Basic Equations

$$\beta = \frac{K_B \cdot \gamma}{K_C + (K_B - K_C) \cdot \gamma} = \frac{\gamma}{\frac{K_C}{K_B} + \left(1 - \frac{K_C}{K_B}\right) \cdot \gamma} \tag{11-5}$$

Fig. 11–1 is a *dissociation-degree diagram*, in which the function β vs. γ is displayed graphically for various values of the parameter $\Delta pK := pK_C - pK_B$. Since the two concentration variables b and c, as well as the reduced variables β and γ, are linearly independent, a simultaneous titration must have the rank $s = 2$. This means that a dissociation-degree diagram, like the corresponding concentration diagram (not included in Fig. 11–1), must follow a curved path. The only exception occurs in the limiting case $\Delta pK = 0$, in which simultaneous titration becomes degenerate with a homogeneous system ($s = 1$). The slope at any point along a dissociation-degree curve is given by the derivative $d\beta/d\gamma$:

$$\frac{d\beta}{d\gamma} = \frac{K_B \cdot K_C}{[K_C + (K_B - K_C) \cdot \gamma]^2} \tag{11-6}$$

For $\Delta pK = 0$, $K_B = K_C$, so in this case $d\beta/d\gamma = 1$.

If the pK value of the reference substance C were smaller than that of B, the curves in Fig. 11–1 would need to be reflected through the diagonal (0/0; 1/1), thus appearing in the lower right triangle. Provided all definitions remained unchanged the value corresponding to ΔpK would then be negative.

Assuming only the species BH, B⁻, CH, and C⁻ are subject to absorption, and assuming further that the Lambert–Beer–Bouguer law is obeyed, the following relationship (based on equations 11–2a and 11–2b) must be valid for any selected wavelength:

$$A_\lambda = l \cdot (\varepsilon_{\lambda BH} \cdot bh + \varepsilon_{\lambda B} \cdot b + \varepsilon_{\lambda CH} \cdot ch + \varepsilon_{\lambda C} \cdot c)$$

$$= l \cdot [\varepsilon_{\lambda BH} \cdot b_0 + (\varepsilon_{\lambda B} - \varepsilon_{\lambda BH}) \cdot b + \varepsilon_{\lambda CH} \cdot c_0 + (\varepsilon_{\lambda C} - \varepsilon_{\lambda CH}) \cdot c] \tag{11-7}$$

Thus, absorbance in this system is a function of $s = 2$ linearly independent concentration variables (here represented by b and c). If equations 11–3a and 11–3b are now used to eliminate the quantities b and c it may be shown that

$$A_\lambda = l \cdot \left(\varepsilon_{\lambda BH} \cdot b_0 + (\varepsilon_{\lambda B} - \varepsilon_{\lambda BH}) \cdot \frac{K_B \cdot b_0}{h + K_B} + \varepsilon_{\lambda CH} \cdot b_0 + (\varepsilon_{\lambda C} - \varepsilon_{\lambda CH}) \cdot \frac{K_C \cdot c_0}{h + K_C} \right) \tag{11-8}$$

Equation 11–8 is the analytical expression for the set of spectrometric titration curves $A_\lambda(pH)$ obtained in a simultaneous titration.

11.2 Graphic Matrix-Rank Analysis

Since the simultaneous titration described above has been shown to have the rank $s = 2$, the corresponding titration spectra would be expected to contain no isosbestic points, and all A- and AD-diagrams should be curved.

In order to subject the method of relative pK determination with an external reference protolyte to the severest possible test, a sample solution was prepared containing a *mixture of o- and p-nitrophenol*, two compounds differing in pK by less than 0.1 unit. According to the literature,[2] the pK of *p*-nitrophenol lies between 7.14 and 7.15 (cf. Sec. 6.3), and that of *o*-nitrophenol between 7.21 and 7.23.

The *titration spectra* for pure *o*-nitrophenol are shown in Fig. 11–2. They differ significantly throughout their course from those for *p*-nitrophenol, depicted in Fig. 6–1 (Sec. 6.2). Titration of a mixture $2.5 \cdot 10^{-5}$ molar in *p*-nitrophenol and $5 \cdot 10^{-5}$ molar in *o*-nitrophenol produces the titration spectra illustrated in Fig. 11–3, which will be seen to intersect four times between 200 and 600 nm. At first glance these intersections appear to represent isosbestic points, but a more careful examination reveals that the points of intersection shift slightly during the titration, an effect made somewhat more apparent in Fig. 11–3 by the fact that on several occasions the plotter pen was lifted momentarily from the paper.

The corresponding set of *AD-diagrams* (Fig. 11–4) shows much more clearly that the titration does not follow a uniform course. Most of the lines are at least slightly curved. Given that the pK difference here is < 0.1, and that the limiting case for a simultaneous titration (with Δp$K = 0$) displays the rank $s = 1$, Fig. 11–4 may be taken as strong evidence for the contention that linearity within an A- or AD-diagram is an extremely sensitive test of the uniformity of a titration.

11.3 The Determination of Relative pK Values

While a two-step *sequential* titration involves three concentration variables, two of which are linearly independent, in a *simultaneous* titration of two single-step systems there are four concentration variables, though again only two are independent. A simultaneous titration of rank $s = 2$ would thus appear to present a more complex picture than a sequential titration of the same rank, although the potential problem of branching is here excluded.

11.3 The Determination of Relative pK Values

11.3.1 Creation of a Dissociation-Degree Diagram

Determining a relative pK value by simultaneous titration of two single-step subsystems is a relatively straightforward process provided one has access to the appropriate dissociation-degree diagram β vs. γ. Such a diagram may be easily generated from spectrometric titration data by taking advantage of isosbestic points within the two subsystems. Wavelengths corresponding to isosbestic points for the pure components o- and p-nitrophenol may be identified by reference to Figures 6–1 and 11–2, respectively; they are also indicated in Fig. 11–3 by the abscissa markings "IP".

The change in absorbance ΔA_λ during a simultaneous titration may be calculated from equation 11–7; for example, if change is defined relative to the acid titration limit (where $bh = b_0, ch = c_0$):

$$\Delta A_\lambda := A_\lambda - l \cdot (\varepsilon_{\lambda BH} \cdot b_0 + \varepsilon_{\lambda CH} \cdot c_0)$$

$$= l \cdot [(\varepsilon_{\lambda B} - \varepsilon_{\lambda BH}) \cdot b + (\varepsilon_{\lambda C} - \varepsilon_{\lambda CH}) \cdot c]$$

$$= l \cdot [(\varepsilon_{\lambda B} - \varepsilon_{\lambda BH}) \cdot b_0 \cdot \beta + (\varepsilon_{\lambda C} - \varepsilon_{\lambda CH}) \cdot c_0 \cdot \gamma] \qquad (11\text{–}9)$$

At an isosbestic point of one of the subsystems the only changes in absorbance during a simultaneous titration must be due to the other subsystem:

$$\Delta A_{IP(CH \rightleftarrows C^-)} = l \cdot b_0 \cdot (\varepsilon_{\lambda B} - \varepsilon_{\lambda BH}) \cdot \beta, \text{ since } \varepsilon_{\lambda C} - \varepsilon_{\lambda CH} = 0 \qquad (11\text{–}10a)$$

$$\Delta A_{IP(BH \rightleftarrows B^-)} = l \cdot b_0 \cdot (\varepsilon_{\lambda C} - \varepsilon_{\lambda CH}) \cdot \gamma, \text{ since } \varepsilon_{\lambda B} - \varepsilon_{\lambda BH} = 0 \qquad (11\text{–}10b)$$

Thus, experimental data derived from the mixture but recorded at wavelengths corresponding to isosbestic points provide measures of β and γ, the degrees of dissociation of the separate constituents. Once again A- and AD-diagrams suffice for the purpose; actual pH data are not required.

Figure 11–5 shows an AD-diagram in which absorbance changes at 370 nm (an isosbestic point for the o-nitrophenol system) are plotted against those at 321.8 nm (an isosbestic point for the p-nitrophenol system). The ordinate (ΔA_{370}) records only changes due to p-nitrophenol ($\cong$ BH $\leftrightarrows$ B$^-$). According to equation 11–9a these absorbance changes are directly proportional to the corresponding degree of dissociation β. Similarly (cf. equation 11–9b), abscissa ($\Delta A_{321.8}$) values are directly proportional to γ, the degree of dissociation of o-nitrophenol. According to equation 11–9, the proportionality constants $l \cdot b_0 \cdot (\varepsilon_{\lambda B} - \varepsilon_{\lambda BH})$ and $l \cdot c_0 \cdot (\varepsilon_{\lambda C} - \varepsilon_{\lambda CH})$ must equal the total change in absorbance in the course of the titration ($\beta = \gamma = 1$). Thus, if absorbance values corresponding to the titration limits at the two wavelengths in question are known with sufficient precision, measured absorbances may be converted directly into data points (β, γ) for a dissociation-degree diagram.

> An alternative possibility would be to perform a (statistical) multi-component analysis of the system on the basis of equation 11–7 for each titration increment. However, this approach requires precise knowledge of the molar absorptivities of all four components, together with the initial weights of the samples.

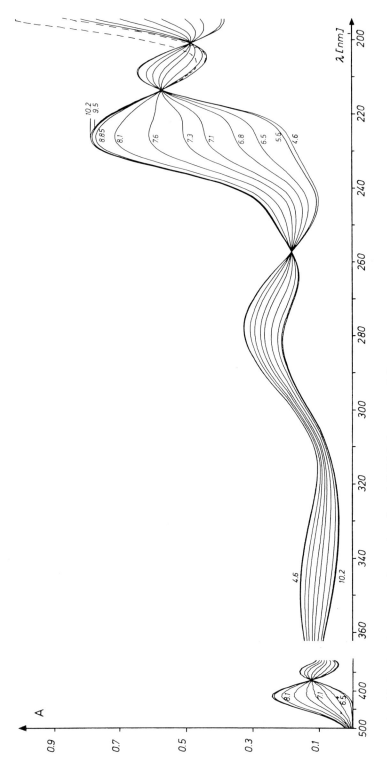

Fig. 11-2. Titration spectra for *o*-nitrophenol (25 °C, $b_0 = 5 \cdot 10^{-5}$ M).

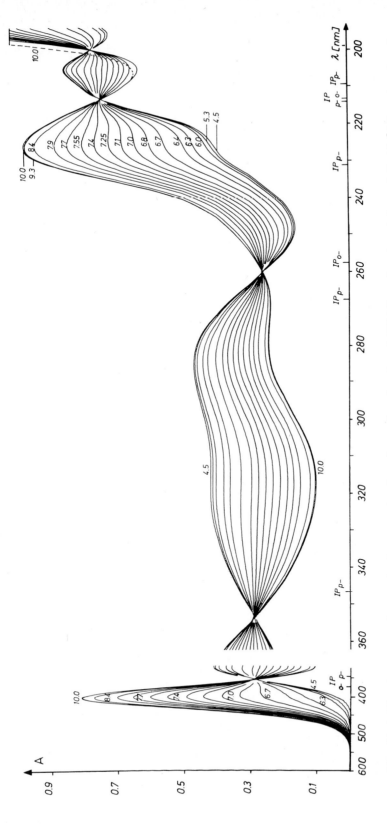

Fig. 11-3. Titration spectra from a simultaneous titration of *p*- and *o*-nitrophenol (25 °C; *p*-nitrophenol: $b_0 = 2.5 \cdot 10^{-5}$ M; *o*-nitrophenol: $c_0 = 5 \cdot 10^{-5}$ M).

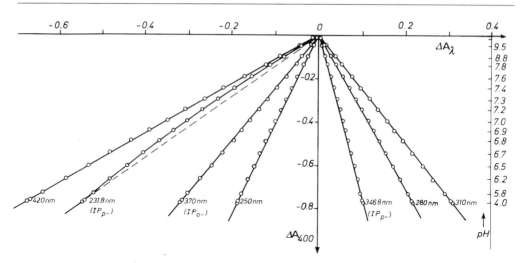

Fig. 11–4. AD-Diagrams for the system of Fig. 11–3.

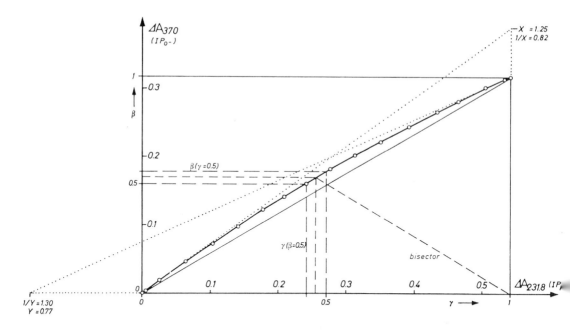

Fig. 11–5. AD-Diagram for the system of Fig. 11–3. The two wavelengths employed correspond respectively to isosbestic points for the two compounds.

Several methods have been reported for extracting from a dissociation diagram either a pK difference $\Delta pK = pK_C - pK_B$ or else the quotient K_B/K_C of the two dissociation constants, as described below.[1]

11.3 The Determination of Relative pK Values

11.3.2 Points of Bisection in a Dissociation-Degree Diagram

It follows from equation 11–5 that

$$\text{for } \beta = 0.5: \quad \gamma_{(\beta=0.5)} = \frac{K_C}{K_B + K_C} = \frac{1}{1 + \dfrac{K_B}{K_C}} \qquad (11\text{–}11a)$$

$$\text{for } \gamma = 0.5: \quad \beta_{(\gamma=0.5)} = \frac{K_B}{K_B + K_C} = \frac{1}{1 + \dfrac{K_C}{K_B}} \qquad (11\text{–}11b)$$

Since $\beta = 0.5$ and $\gamma = 0.5$ are functions only of the K values of the two subsystems, the corresponding points in a dissociation-degree diagram (or in appropriate A- or AD-diagrams) may be used together with either equation 11–11a or 11–11b to establish the quotient of the two dissociation constants.

From Fig. 11–5 for the simultaneous titration of o- and p-nitrophenol the following results may be extracted:

using $\beta = 0.5$: $K_{o\text{-}}/K_{p\text{-}} = 0.81$

using $\gamma = 0.5$: $K_{o\text{-}}/K_{p\text{-}} = 0.80$

11.3.3 Terminal Slope in a Dissociation-Degree Diagram

Equation 11–6 yields the following expressions for the slopes at the limits ($\gamma = 0$ and $\gamma = 1$) of the diagram β vs. γ:

$$\left.\frac{d\beta}{d\gamma}\right|_{\gamma=0} = \frac{K_B}{K_C} =: X$$

$$\left.\frac{d\beta}{d\gamma}\right|_{\gamma=1} = \frac{K_C}{K_B} =: Y$$

These slopes may be evaluated in either an AD-diagram (cf. Fig. 11.5) or a dissociation-degree diagram. In the present case:

for $\beta = \gamma = 0$, $K_{o\text{-}}/K_{p\text{-}} = 0.82$

for $\beta = \gamma = 1$, $K_{o\text{-}}/K_{p\text{-}} = 0.77$

The latter value is subject to a systematic error, because at 231.8 nm the titrant NaOH has already begun to absorb, a factor that is particularly troublesome at the alkaline titration limit.

11.3.4 Complete Evaluation of a Dissociation-Degree Diagram

Dividing equation 11–3a by equation 11–3b leads to the expression

$$\frac{K_B}{K_C} = \frac{b}{bh} \cdot \frac{ch}{c} \tag{11-12}$$

A formally analogous relationship has been presented in Sec. 7.4.1 for the case of a sequential two-step titration (equation 7–22).

Replacing the concentrations in equation 11–12 by the degree of dissociation and rearranging terms leads to

$$\frac{\beta}{1-\beta} = \frac{K_B}{K_C} \cdot \frac{\gamma}{1-\gamma} \tag{11-13}$$

This means that every data point in Fig. 11–5 is capable of yielding a value for the quotient K_B/K_C. Evaluation of ten such points between pH 6.2 and 7.6 and taking the arithmetic mean results in the following ratio:

$K_{o\text{-}}/K_{p\text{-}} = 0.80 \pm 0.02$

Larger deviations are found near the limits of the titration since the absorbance differences there are quite small and the relative errors correspondingly large.

11.3.5 The Diagonal of a Dissociation-Degree Diagram

Points along the diagonal line passing through the upper-left corner of the diagram in Fig. 11–1 are subject to the condition $\beta + \gamma = 1$. Introduction of this relationship into equation 11–13 produces the following expressions:

$$\left.\frac{\beta}{1-\beta}\right|_{\beta=1-\gamma} = \pm\sqrt{\frac{K_B}{K_C}} \tag{11-14a}$$

$$\left.\frac{\gamma}{1-\gamma}\right|_{\gamma=1-\beta} = \pm\sqrt{\frac{K_C}{K_B}} \tag{11-14b}$$

Thus, the point of intersection of this diagonal with the curve β vs. γ provides two additional values for the quotient of the dissociation constants. For reasons of clarity the diagonal in the AD-diagram for the nitrophenol system (Fig. 11–5) is shown only in the lower right quadrant.

Evaluation by this method provides the following results:

for $\beta = 1 - \gamma$, $K_{o\text{-}}/K_{p\text{-}} = 0.80$

for $\gamma = 1 - \beta$, $K_{o\text{-}}/K_{p\text{-}} = 0.80$

11.3.6 A Comparison of the Various Methods

In order to perform a control simulation, a dissociation-degree diagram β vs. γ for K_o/K_p was constructed on the basis of equation 11–5 and then adjusted in scale so that it matched the AD-diagram. The result is the solid line in Fig. 11–5, which essentially passes through all the experimental points.

A comparison of the various methods for extracting K_B/K_C leads to the following evaluative conclusion:

> The bisection point method (Sec. 11.3.2), the diagonal intercept method (Sec. 11.3.5), and the method utilizing experimental data points from within the central portion of a dissociation-degree diagram (Sec. 11.3.4) all provide equally satisfactory results.

That this should be so is not surprising; equations 11–11 and 11–14 are, after all, simply special cases of equation 11–13. One distinction should be noted, however: the method of Sec. 11.3.4 leads directly from experimental data points to the desired information, while the methods of Sections 11.3.2 and 11.3.5 require some type of interpolation of the curve β vs. γ (cf. Fig. 11–5).

The terminal slope method of Sec. 11.3.3 usually produces poorer results. However, in the event that one of the limiting values is subject to a systematic error, this method is the only one that makes that fact readily apparent. Since the method based on individual data points is fundamentally inapplicable in the limiting regions it is always advisable to examine the terminal slopes, both for diagnostic reasons and as a test of consistency.

Thus, combined application of all four evaluation methods serves as a valuable test of the entire dissociation-degree diagram.

Figure 11–1 makes it possible to identify the most suitable application range for each of the relative evaluation methods, and it also suggests limitations that may be associated with each. In the event that the difference between the two pK values of interest is greater than ca. 1.2–1.5 the tangent extrapolation method of Sec. 11.3.3 must be regarded as inapplicable. For $\Delta pK > 1.5$ the bisection method of Sec. 11.3.2 will also lead to results that are imprecise, since corresponding values drawn from the other axis lie too near the value 1, and are thus associated with large relative errors. Data taken in the vicinity of the diagonal may be utilized up to a ΔpK of perhaps 2.5, permitting continued use within this broader region of the methods of Sections 11.3.4 and 11.3.5.

The region of applicability of relative evaluation methods has been shown to extend over the entire interval $0.1 \leq \Delta pK \leq 2$. However, this assertion relates only to data derived from a dissociation-degree diagram β vs. γ. In the case of AD- and A-diagrams—representations like Fig. 11–5 and based directly on experimental data—the scale factors for the two axes may differ significantly due to differences between the two proportionality constants $b_0 \cdot (\varepsilon_{\lambda B} - \varepsilon_{\lambda BH})$ and $c_0 \cdot (\varepsilon_{\lambda C} - \varepsilon_{\lambda CH})$. Thus, it may be that a very different degree of relative error will be encountered in the course of calculating β as opposed to γ.

In contrast to the situation with a two-step sequential titration, the affine distortion associated with an A- or an AD-diagram in a simultaneous titration is a consequence not only of a difference between corresponding molar absorptivities; it results also from the relationship between the starting concentrations (b_0/c_0). This complicates the task of assessing probable errors, but at the same time means that the ratio b_0/c_0 may be so chosen as to provide for maximum absorbance changes along both axes, thereby assuring optimal A- and AD-diagrams.

Another method for determining K_B/K_C has also recently been described.[3] This method does not rely on isosbestic points, but it does require knowledge of the directions of the lines in the A-diagram referring to the two individual equilibria.

The computer program TIFIT discussed in Chapter 10 could easily be modified to accommodate the demands of a simultaneous titration. This would have the advantage that one would no longer be required to select a special set of wavelengths (i.e., those corresponding to isosbestic points in the subsystems) prior to conducting an analysis. Multi-step simultaneous titrations might also be analyzed in this way. Furthermore, such modifications would open the way to convenient investigation of more complex problems, including analysis of the pH-dependence of spectra derived from isomeric mixtures possessing multiple dissociable groups, or the pH-dependence of intermolecular interactions, as between protic dyes and macromolecules (e.g., DNA, proteins).

Literature

1. R. Blume, H. Lachmann, H. Mauser, F. Schneider, *Z. Naturforsch.* 29 b (1974) 500.
2. G. Kortüm, W. Vogel, K. Andrussow, *Pure Appl. Chem.* 1 (1960) 190.
3. J. Polster, *Talanta* 31 (1984) 113.

12 Branched Acid-Base Equilibrium Systems

12.1 Macroscopic and Microscopic pK Values: The Problem of Assignment

Protolysis equilibria involving two or more dissociable groups introduce into our discussion the problem of *branching*: for such a system there is no *a priori* way to establish in what order the groups will dissociate.

Consider the example of *p-aminobenzoic acid*,[1,2] whose behavior might be described in terms of the following five equilibria (protons or hydronium ions have been omitted in the interest of clarity):

$$H_3\overset{+}{N}-\underset{}{\bigcirc}-COO^- \quad \rightleftarrows \quad H_3\overset{+}{N}-\underset{}{\bigcirc}-COOH \quad \rightleftarrows \quad H_2N-\underset{}{\bigcirc}-COO^- \quad \rightleftarrows \quad H_2N-\underset{}{\bigcirc}-COOH \tag{12-1}$$

This particular scheme may easily be generalized so that it applies to any dibasic acid in which the two dissociating groups differ:

$$BH_2 \underset{K_{II}}{\overset{K_I}{\rightleftarrows}} \begin{matrix} B_1H^- \\ \updownarrow K_V \\ B_2H^- \end{matrix} \underset{K_{IV}}{\overset{K_{III}}{\rightleftarrows}} B^{2-} \tag{12-2}$$

where

$$K_\mathrm{I} = \frac{b_1 h \cdot h}{bh_2}; \quad K_\mathrm{II} = \frac{b_2 h \cdot h}{bh_2}; \quad K_\mathrm{III} = \frac{b \cdot h}{b_1 h}; \quad K_\mathrm{IV} = \frac{b \cdot h}{b_2 h};$$

$$K_\mathrm{V} = \frac{b_1 h}{b_2 h} = \frac{K_\mathrm{I}}{K_\mathrm{II}} = \frac{K_\mathrm{IV}}{K_\mathrm{III}} \tag{12-3}$$

The constants K_I through K_IV are referred to as *microscopic dissociation constants*. The remaining constant, K_V, is not a dissociation constant at all since it characterizes a *tautomeric* equilibrium. If the two ampholytes B_1H^- and B_2H^- are treated collectively as being two representations of the single species "BH^-" then the system becomes equivalent to an unbranched, sequential titration:

$$BH_2 \overset{K_1}{\leftrightarrows} BH^- \overset{K_2}{\leftrightarrows} B^{2-} \tag{12-4}$$

In this formulation $bh = b_1 h + b_2 h$, so the corresponding *macroscopic dissociation constants* K_1 and K_2 may be expressed as

$$K_1 = \frac{bh \cdot h}{bh_2} = \frac{(b_1 h + b_2 h) \cdot h}{bh_2} \quad \text{and}$$

$$K_2 = \frac{b \cdot h}{bh} = \frac{b \cdot h}{(b_1 h + b_2 h)} \tag{12-5}$$

The relationship between these macroscopic constants and the previously defined microscopic dissociation constants follows directly from equations 12–3 and 12–5:

$$K_1 = K_\mathrm{I} + K_\mathrm{II}; \quad \frac{1}{K_2} = \frac{1}{K_\mathrm{III}} + \frac{1}{K_\mathrm{IV}} \tag{12-6}$$

The number of discrete species involved in a branched titration system is always 2^s, marking a sharp contrast to the $s + 1$ species comprising an unbranched titration. Moreover, the number of microscopic dissociation constants increases astronomically with only a modest increase in the rank s, as may be seen from Table 12–1. Only if the dissociating groups in question are indistinguishable will problems due to branching be eliminated, as in unsubstituted phthalic acid or 1,3,5-benzenetricarboxylic acid:

Phthalic acid

1,3,5-Benzenetricarboxylic acid

12.1 Macroscopic and Microscopic pK Values

Table 12–1. Information regarding the number of macroscopic and microscopic dissociation constants required to characterize various branched titration systems (adapted from ref. 1).

Number of dissociable groups	2	3	4	s
Potential species in the absence of branching	3	4	5	$s+1$
Potential species in a branched system	4	8	16	2^s
Macroscopic dissociation constants	2	3	4	s
Microscopic dissociation constants (ignoring tautomeric equilibria)	4	12	32	$s \cdot 2^{s-1}$
Pieces of supplemental information required for a complete analysis	1	4	11	$2^s - s - 1$

It has already been established in the preceding chapters that *in the absence of supplementary information* a spectrometric titration is only capable of furnishing macroscopic dissociation constants, a limitation that also applies to potentiometric titrations.[1,3] Unfortunately, a complete molecular interpretation of dissociation phenomena, including any attempt to assign pK values to particular dissociable groups, or to interpret the pH-dependence of specific absorption bands, requires a knowledge of microscopic dissociation constants. In fact, any concrete assertion about the mechanism of a reaction involving two or more dissociable groups is suspect so long as the corresponding microscopic pK values remain uncertain. This limitation is particularly severe with respect to interpreting the pH dependence of a chemical reaction. Even in a kinetic investigation of a dissociation phenomenon, where equilibrium constants K are to be expressed in terms of rate constants k (e.g., with the aid of relaxation methods), prior knowledge of microscopic dissociation constants is normally regarded as an absolute prerequisite.

For reasons such as these, countless attempts have been made over the years to develop analytical tools that would provide the extra bits of information necessary (cf. Table 12–1) to extract microscopic constants from the readily available macroscopic constants. For example, in the case of a two-step titration all that is required is the single constant K_V, which specifies the concentration ratio of the two species B_1H^-

and B_2H^-. Unfortunately, even today there is no generally applicable procedure for determining microscopic pK values, or for solving the assignment problem in a way that is both precise and free of assumptions.[3]

In the sections that follow we provide a series of examples to illustrate various approaches so far reported and discussed in the literature.

12.2 pK Assignments Based on Extrapolations from Unbranched Systems

The first procedure—and the oldest—starts from the assumption that chemical blocking of one dissociable group (e.g., by esterification) will leave the pK of another group essentially unchanged. For example, consider the case of glycine:

$$
\begin{array}{c}
\overset{+}{H_3N}-CH_2-COO^- \\
B_1H^-
\end{array}
$$

$$
\overset{+}{H_3N}-CH_2-COOH \quad\quad H_2N-CH_2-COO^-
$$
$$BH_2 \quad\quad\quad\quad B^{2-}$$

$$H_2N-CH_2-COOH$$
$$B_2H^-$$

with rate constants K_I, K_{II}, K_{III}, K_{IV}, K_V. (12-7)

If one assumes K_{II} for glycine to be of the same magnitude as the dissociation constant for glycine methyl ester (with p$K = 7.73$[4,5]), then it is a straightforward matter to calculate all the relevant microscopic dissociation constants. Two estimates have been made on this basis for K_V, the ratio of dipolar to neutral forms: ca. $2.2 \cdot 10^2$ (ref. 6) and ca. $2.6 \cdot 10^2$ (ref. 1). A similar evaluation of the dissociation equilibria applicable to p-aminobenzoic acid (cf. Sec. 12.1) leads to a K_V of 0.16, suggesting that here the neutral form predominates.[1,2]

These results may appear promising, but the fundamental assumption upon which they rest is only an approximation, so the corresponding numbers can only be regarded as rough estimates at best. Indeed, similar treatments of certain aromatic amino acids lead to utterly meaningless conclusions, such as negative values for K_V.[1,7]

12.3 Assignment of Spectral Features to Specific Dissociable Groups

The spectroscopic techniques upon which the following arguments depend are themselves a subject of discussion in Part III of this book; nevertheless, a few methodological details will of necessity surface here as well.

12.3.1 UV–VIS Spectroscopy

The compound *cysteine* provides an assignment problem that has been the subject of discussion for decades, and even today the debate has not been fully resolved.[8-10]

$$
\begin{array}{c}
^{-}\text{OOC—CH—}^{+}\text{NH}_3 \\
| \\
\text{CH}_2\text{—S}^{-} \\
\text{B}_1\text{H}^{-}
\end{array}
\qquad (12\text{–}8)
$$

with equilibria K_I, K_{II}, K_{III}, K_{IV}, K_V connecting species BH_2, B_1H^-, B_2H^-, and B^{2-}:

- BH_2: $^-\text{OOC—CH(}^+\text{NH}_3\text{)—CH}_2\text{—SH}$
- B_1H^-: $^-\text{OOC—CH(}^+\text{NH}_3\text{)—CH}_2\text{—S}^-$
- B_2H^-: $^-\text{OOC—CH(NH}_2\text{)—CH}_2\text{—SH}$
- B^{2-}: $^-\text{OOC—CH(NH}_2\text{)—CH}_2\text{—S}^-$

($pK_1 = 1.71$; $pK_2 = 8.33$; $pK_3 = 10.78$; cf. ref. 4)

One begins by assuming that the first stage of dissociation (i.e., from the cysteine cation) can be ascribed exclusively to the carboxyl group; that is, the first titration step is regarded as unbranched. Some authors then proceed to interpret the macroscopic dissociation constants K_2 and K_3 in a similar way, identifying them directly with the ammonium and sulfhydryl groups—but not always in the same sequence.[11,12] However, since 1955 a number of attempts has also been made to evaluate true microscopic dissociation constants for this system.[13-15]

Figure 12–1 is an illustrative set of titration spectra for L-cysteine in 0.1 M NaOH over the pH range ca. 5–11. No isosbestic points are present. The corresponding A-diagrams (Fig. 12–2) are distinctly curved, an observation consistent with a rank $s = 2$ for the subsystem encompassed by K_2 and K_3.

If one now assumes (as is done in ref. 13) that absorption in the region between 230 nm and 240 nm is attributable *solely* to ionic –S⁻ groups in the two species B_1H^-

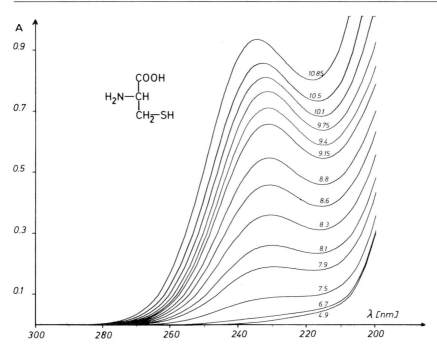

Fig. 12–1. Titration spectra obtained during a titration of L-cysteine with 0.1 M NaOH under nitrogen at 25 °C ($c_0 = 2 \cdot 10^{-4}$ M).

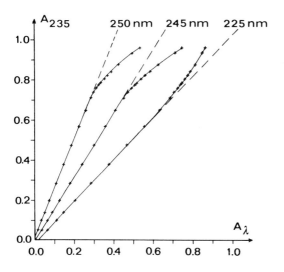

Fig. 12–2. A-Diagrams for the L-cysteine titration described in Fig. 12–1.

12.3 Assignment of Spectral Features to Specific Dissociable Groups

and B^{2-}, and also that the two species share a common molar absorptivity, then spectrometric titration data may be used to evaluate not only K_2 and K_3, but also the complete set of microscopic dissociation constants K_I–K_V. Values for K_V established in this way range from about 2.1 to 2.25.[13,15] Using the method described in Sec. 12.2, and with the aid of the analogs S-methylcysteine and cysteine betaine, a rather different value of 1.3 has been suggested for K_V.[16,17]

Unfortunately, the premises related to the absorption band at 240 nm have so far not yielded to verification, and they remain a subject of controversy.[14,16,18] Even laser-Raman and ^{13}C-NMR studies have failed to provide a conclusive solution to the branching problem for cysteine (see below).

Cysteine is not the only molecule that has been subjected to intensive study of this sort with respect to branching. Other examples include tyrosine[19,20] and DOPA (or dopamine).[21,22]

12.3.2 IR and Raman Spectroscopy

Electronic excitation spectra recorded in aqueous solution are almost invariably rather featureless, so assignment of absorption bands to specific dissociable groups is a problematic exercise at best. By contrast, vibrational spectra are rich in detail, and they frequently make important contributions to structural studies. It is therefore natural to anticipate that infrared spectrometry would be ideally suited to providing insight into problems of branching.

Unfortunately, *quantitative infrared spectrometry*, especially in dilute aqueous solution, continues to suffer from a lack of precision, and it is subject to numerous experimental complications (cf. Sec. 18.4). For this reason there are only a few reports of IR-spectrometric titration studies, and nearly all are semi-quantitative in nature, including work at very high concentration (ca. 0.5 M)—for example with the vitamin-B_6 aldehydes (cf. Sec. 12.3.4).[23,24]

A more promising technique is *Raman spectroscopy*, which was called upon as early as 1958 in the context of a semi-quantitative investigation of the dissociation of cysteine.[25] The subsequent development of laser-Raman instruments (cf. Sec. 18.5) has opened the way to quantitative experiments in more dilute aqueous solution. Figure 12–3 shows a Raman spectrum of L-cysteine at pH 7. A prominent feature such as the SH-stretching band at 2575 cm^{-1} should lend itself nicely to selective—perhaps even quantitative—analysis.[10]

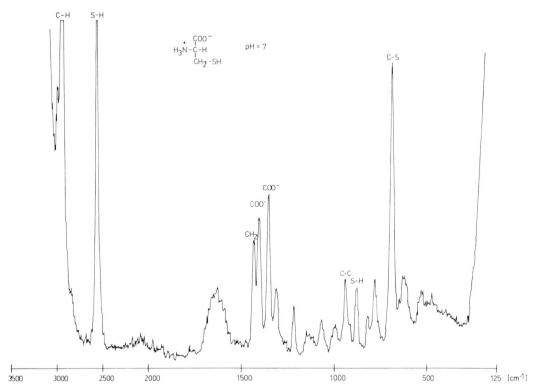

Fig. 12–3. Laser-Raman spectrum of a solution of L-cysteine in phosphate buffer, pH 7 ($c_0 = 400$ mg/mL).

12.3.3 NMR Spectroscopy

Aside from vibrational spectroscopy, nuclear magnetic resonance spectroscopy distinguishes itself as perhaps the most powerful tool available for use in chemical structure determinations, and thus for providing quantifiable correlations between spectral features and specific atomic groupings within a molecule.

In Sec. 18.6 we discuss one of the unique aspects of NMR-spectrometric titration, namely the fact that in the interpretation of a rapid equilibrium reaction the important spectral variable as a function of pH is not signal intensity (plotted on the y-axis), but *chemical shift* (in Hz or ppm), which consititutes the x-axis of the spectrum. Chemical shift (frequency) changes in NMR are ascertainable to a very high degree of precision, so the potential clearly exists for obtaining a very precise measure of pK.

Once again, however, concentration poses a problem. NMR spectrometry continues to be best suited to the investigation of relatively concentrated solutions, particularly if recording times are to be held within defensible limits. Most NMR studies involve substrate at concentrations of 0.1–1 M. Moreover, it is common

12.3 Assignment of Spectral Features to Specific Dissociable Groups

practice to introduce an extra substance such as NaCl at concentrations ≥ 1 M as a way of holding ionic strength relatively constant during titration. For these reasons, most NMR titrations (like their IR and Raman counterparts) provide results that are not subject to direct comparison with UV–VIS spectrometric titration data; indeed, the concentrations employed in the two cases typically differ by as much as 2–4 orders of magnitude. It is extremely difficult to predict the effect of high concentrations on activity coefficients for a complicated branched titration system containing both polar and non-polar species. One can only assume that microscopic pK values applicable at a concentration $c_0 = 10^{-4}$–10^{-5} are likely to bear little resemblance to the same quantities established using a different method in which $c_0 = 10^0$–10^{-1}. The severity of the problem becomes most apparent when one attempts to convert a set of mixed-mode constants derived from NMR spectra into a set of true thermodynamic dissociation constants.

NMR titration is sometimes used as a source of *supplementary information* for converting macroscopic dissociation constants into their microscopic counterparts, in which case it is convenient to be able to invoke an important premise:[26-30] that somewhere within the substrate it is possible to locate an NMR-active nucleus subject to dissociation-dependent chemical shift effects from only *one* dissociable group. That is, a search is conducted for a "nuclear probe" which is insulated from dissociation involving other parts of the molecule, and thus fulfills the condition of *unique resonance*.[28]

The significance of this assumption can be illustrated with the aid of a generalized two-step titration system involving two different dissociable groups, H–B~ and H–C~:

$$
\begin{array}{c}
   ^-\text{B}\sim\text{C–H} \phantom{K_{III}} \\
 \nearrow \text{B}_1\text{H}^- \searrow \\
\text{H–B}\sim\text{C–H} \underset{K_I}{\nearrow} \Big\Updownarrow K_V \underset{K_{III}}{\searrow} ^-\text{B}\sim\text{C}^- \\
\text{BH}_2 \searrow \nearrow \text{B}^{2-} \\
 \underset{K_{II}}{\searrow} \text{H–B}\sim\text{C}^- \underset{K_{IV}}{\nearrow} \\
 \text{B}_2\text{H}^-
\end{array}
\tag{12-9}
$$

Unique resonance implies that there is some nucleus in the substrate whose chemical shift is affected significantly, and to the *same extent*, by dissociation of the group H–B in *both* BH_2 and B_2H^-, even though it is completely insensitive to dissociation of the group H–C. Alternatively, one must seek some resonance line sensitive only to the dissociation state of the group H–C, this time irrespective of whether the species in question is BH_2 or B_1H^-. One signal of either type suffices for the establishment of a complete set of microscopic dissociation constants provided values are available for K_1 and K_2.

The high structural specificity associated with NMR spectroscopy, coupled with the fact that it is possible to examine any of several NMR-active isotopes (e.g., ^{1}H, ^{13}C, ^{15}N, etc.), greatly increases the probability that a line displaying unique resonance may be found. Moreover, the validity of the basic premise is subject to verification by examination of a series of analogous compounds, or by careful spin-spin coupling analysis.

Perhaps the best *example* to date of a study such as that described above comes from work with amino acids, their esters, and various small peptides ($c_0 \approx 0.15$–0.3 M).[26-30] Nevertheless, even NMR has so far failed to resolve the issues raised in Sections 12.3.1 and 12.3.2 with respect to cysteine. Indeed, several authors have deliberately refrained from suggesting a quantitative interpretation of the pH-dependent NMR spectra of cysteine ($c_0 \approx 0.3$ M), and no attempt appears to have been made to use these spectra to establish a set of microscopic pK values.[10,18,27]

A great many ^{1}H-, ^{13}C-, and ^{31}P-NMR studies of vitamin-B$_6$ aldehydes and related compounds (with $c_0 \approx 0.3$–1 M; cf. Sections 6.4 and 12.3.2) were reported more than a decade ago,[31-35] but again the authors were unwilling to propose quantitative interpretations of the pH-dependency of the data, or to use the data as a source for microscopic constants (cf. also Sec. 12.3.4). There is every reason to hope that further technical advances, perhaps at the level of instrumentation, will someday facilitate work at substantially lower substrate concentrations c_0 so that the great potential of NMR can be further exploited, ultimately resolving some of the open questions surrounding branched equilibria.

12.3.4 The Dependence of Spectra on Solvent and Temperature; Band Shape Analysis

None of the methods discussed in Sections 12.3.2 and 12.3.3 has produced a satisfactory solution to the assignment problems posed by compounds in the important vitamin-B$_6$ series. However, certain special techniques adapted to use with these particular substances have been the subject of considerable attention since the mid-1950s, and it is to these techniques that we devote the remainder of the chapter.

Relevant titration spectra, A-diagrams, and A(pH)-diagrams have already been presented in Sec. 6.4. The (macroscopic) dissociation equilibria for most members of the vitamin-B$_6$ series appear to be relatively free of overlap (Δp$K > 3.0$–3.5), so a complete titration can usually be reduced to a set of single-step sub-systems. It should be noted, however, that the vitamin-B$_6$ aldehydes are subject to a further complication in the form of a pH-dependent equilibrium between free aldehyde and aldehyde hydrate. Moreover, in the case of pyridoxal there is the additional possibility of internal condensation to a hemiacetal (cf. equation 12–10).

12.3 Assignment of Spectral Features to Specific Dissociable Groups

$$\text{Hydrate} \rightleftharpoons \text{Aldehyde} \rightleftharpoons \text{Hemiacetal} \quad (12\text{–}10)$$

Our discussion here will be centered on 3-hydroxypyridine (cf. equation 12–11), the simplest representative of the family, in order to avoid unnecessary complications.

$$(pK_1 = 4.8\text{–}5.1; \quad pK_2 = 8.6\text{–}8.7)^{36,37} \quad (12\text{–}11)$$

(a) The Effects of Solvent on the Spectra

The analytical method to be described first was introduced as early as 1955 in the course of an attempt to establish a value for K_V, which in conjunction with K_1 and K_2 provides the key to all the desired microscopic constants.[38] Because the macroscopic pK difference is relatively large ($\Delta pK \approx 3.6$) it is reasonable to assume that at the isoelectric point $(pK_1 + pK_2)/2 \approx 6.8\text{–}7.0$ only two species, B_1H^- and B_2H^-, will be present in solution. If one assumes further that in an organic solvent such as dioxane or alcohol only the nonpolar form B_2H^- will be stable, then it should be possible to determine the missing constant K_V by examining differences in spectra as a function of solvent (e.g., using spectra recorded in dioxane–water mixtures covering a dioxane range of 0–100%). However, it should be recognized that any such procedure is subject to certain conditions:[39,40]

(1) Pure dioxane must in fact contain only the species B_2H^- (see above).

(2) The pH of the solution (i.e., pH ≈ 7) must remain constant despite changes in solvent.

(3) The molar absorptivity of B_2H^- must be the same in both H_2O and dioxane, and all solutions should obey the Lambert–Beer–Bouguer law. This implies, for example, that titration spectra obtained in the course of a change in solvent should correspond to the rank $s = 1$, and that sharp isosbestic points should be observed. Unfortunately, these prerequisites are not strictly fulfilled.

(4) Expectations with respect to the dielectric constants of the solvent mixtures must be met. Once again, the true situation is somewhat disappointing,[38] and partly for this reason the final result, $K_V = 1.2$, has been acknowledged to be only a rough estimate.

(b) *Temperature Effects on the Spectra*

In an analogous fashion, attempts have been made to deduce the position of the tautomeric equilibrium linking B_1H^- and B_2H^- on the basis of changes in spectra as a function of temperature (see below).[36, 39–41]

(c) *Band-Shape Analysis*

More recently, the research group that has been most active in this area has tried to combine the study of solvent and temperature dependencies with what they describe as a "band-shape analysis".[36,39–42] The key assumption in this case is that it is possible to dissect certain strongly overlapping UV–VIS absorption bands into "isolated" segments that reflect the concentrations of individual species. Several types of mathematical functions have been examined in this context (e.g., Gauss, log normal, Poisson, and spline functions).[36,39,40,42] Indeed, appropriate "isolated" segments corresponding to individual components of the complicated, branched titration system were ultimately identified on the basis of a search involving analogous compounds, spectroscopic precedents, and even quantum-mechanical calculations.

In order to ascertain microscopic pK values starting from the available macroscopic information, changes in the area of appropriate "isolated" absorption bands were then examined as a function of solvent and temperature. The analysis entailed the following assumptions:

(1) It is possible to subdivide spectra accurately into segments that reflect the concentrations of the individual species constituting the titration system (see above).

(2) An *integrated* form of the Lambert–Beer–Bouguer law is applicable to each of the segments so identified. That is, instead of the usual formulation of this law as it relates to a given species B, i.e.:

$$A_\lambda = l \cdot \varepsilon_{\lambda B} \cdot b \tag{12–12}$$

it is here necessary that one work with an *integral absorption*, defined as the

12.3 Assignment of Spectral Features to Specific Dissociable Groups

surface area S_B of an appropriate "isolated" segment of an absorption spectrum (where area is usually specified in terms of the unit *wavenumber* $\tilde{v}$, cf. Fig. 12–4):

$$S_B = \int_{\tilde{v}_1}^{\tilde{v}_2} A \, d\tilde{v} = l \cdot b \cdot \int_{\tilde{v}_1}^{\tilde{v}_2} \varepsilon_B \, d\tilde{v} = l \cdot s_B \cdot b \tag{12–13}$$

The *integral molar absorptivity* s_B, sometimes referred to as the "molar area" then replaces the more usual molar absorptivity term ε_B.[37,41,42]

(3) The Lambert–Beer–Bouguer law (in its integral form) is unaffected by changes in pH, temperature, and solvent; i.e., integral molar absorptivities s_i maintain constant values with respect to all the solutions examined.

(4) The pH of the solution is independent of temperature and solvent composition.

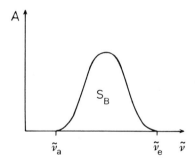

Fig. 12–4. Integral absorption ("molar area") S_B corresponding to an "isolated" segment of an absorption spectrum.

Provided all four of these assumptions are adequately met, one can establish a value for the tautomerization constant K_V (cf. the set of equations 12–11) on the basis of how a spectrum changes as a function, for example, of solvent. Such a change is depicted schematically in Fig. 12–5.

The procedure is relatively straightforward. One begins with the relationship governing the concentrations $b_1 h$ and $b_2 h$ in neutral aqueous solution:

$$K_V := \frac{b_1 h}{b_2 h} = \frac{S_{B_1 H}}{s_{B_1 H}} \cdot \frac{s_{B_2 H}}{S_{B_2 H}} \tag{12–14}$$

where $S_{B_1 H} = l \cdot s_{B_1 H} \cdot b_1 h$ and $S_{B_2 H} = l \cdot s_{B_2 H} \cdot b_2 h$

In a methanol–water mixture containing, say, 90% methanol, equilibrium should be shifted significantly toward the non-polar form $B_2 H^-$. The new ratio of concentrations may also be expressed in terms of molar areas (cf. Fig. 12–5):

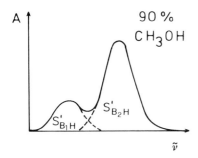

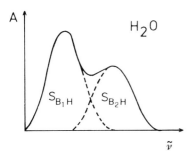

Fig. 12–5. Solvent effect upon the tautomeric equilibrium of a 3-hydroxypyridine derivative (schematic representation based on ref. 36).

$$K_V := \frac{b_1 h'}{b_2 h'} = \frac{S'_{B_1H}}{S_{B_1H}} \cdot \frac{S_{B_2H}}{S'_{B_2H}} \tag{12–15}$$

Changes in area within the sensitive segment of the spectrum on going from pure water to the methanol–water mixture may now be computed:

$$\Delta S_{B_1H} := S_{B_1H} - S'_{B_1H} \quad \text{and} \quad \Delta S_{B_2H} := S_{B_2H} - S'_{B_2H}$$

Taking into account the mass-balance relationship $b_1 h + b_2 h = $ constant (which translates into $\Delta b_1 h = -\Delta b_2 h$), it follows from equations 12–14 and 12–15 that:[37]

$$-\frac{\Delta S_{B_1H}}{S_{B_1H}} = \frac{\Delta S_{B_2H}}{S_{B_2H}} \tag{12–16}$$

hence

$$K_V = \frac{b_1 h}{b_2 h} = -\frac{\Delta S_{B_2H}}{S_{B_2H}} \cdot \frac{S_{B_1H}}{\Delta S_{B_1H}} \tag{12–17}$$

In other words, by comparing the areas S_{B_1H} and S_{B_2H} obtained in neutral aqueous solution with those from any desired water–methanol mixture it should be possible to establish a value for K_V.

In principle, this technique is applicable to any type of spectroscopy (e.g, IR, Raman, etc.) so long as appropriate isolated segments can be assigned to the key

12.3 Assignment of Spectral Features to Specific Dissociable Groups

molecular species, or equivalent information can be obtained through a band-shape analysis. The accuracy of the resulting constant K_V and of the microscopic pK values to which it leads will be dependent upon how well assumptions (1) – (4) are met. In the case of 3-hydroxypyridine, tautomeric equilibrium constants K_V elucidated by the research group cited vary between 0.88 and 1.27.

This pioneering work has been followed by reports describing attempts to establish orders of dissociation and sets of microscopic pK values for most of the compounds in the vitamin-B$_6$ series. In the case of aldehydes, data have also been made available for hydration and hemiacetal-formation equilibria.[40,43] Based on these results, kinetic aspects of the protolysis and tautomerization equilibria have been further investigated with the aid of relaxation methods.[31,44]

12.4 Summary

As has been demonstrated in the preceding sections, spectrometric or potentiometric data *alone* can provide values only for *macroscopic* dissociation constants. However, any molecular interpretation of dissociation-dependent phenomena requires a knowledge of *microscopic* dissociation constants, a factor that is particularly relevant to the assignment of pK values to *specific* dissociable groups, or to a quantitative understanding of changes in absorption spectra as a function of pH. The problem of *branching* in a titration system, together with inevitable complications with respect to pK assignment, may in principle arise even with systems in which ΔpK is greater than three.

Although macroscopic pK values are typically known to an accuracy of ca. 0.01 pH units, values for the tautomeric equilibrium constant K_V often represent only order-of-magnitude approximations. None of the methods for establishing microscopic dissociation constants described in Sections 12.2 and 12.3 is free of assumptions, and it is not unusual to find that results obtained in different ways are in serious disagreement.

The most powerful tools for assigning discrete spectral bands to individual functional groups (as a prelude to determining microscopic dissociation constants) would appear to be laser-Raman and NMR spectroscopy. The latter shows particularly great promise, but the important assumption of "unique resonance" of an appropriate "atomic probe" (^{1}H, ^{13}C, ^{31}P, etc.) must always be subjected to critical testing.

Literature

1. E. J. King, *Acid-Base Equilibria*. Pergamon Press, London 1965, p. 219.
2. R. A. Robinson, A. J. Biggs, *Austral. J. Chem.* 10 (1957) 128.
3. R. Winkler-Oswatitsch, M. Eigen, *Angew. Chem.* 91 (1979) 20.
4. *Handbook of Biochemistry*, 2nd ed. The Chemical Rubber Co., Cleveland 1970, Section J.
5. D. D. Perrin, *Dissociation Constants of Organic Bases in Aqueous Solutions*. Butterworths, London 1965.
6. H. R. Mahler, E. H. Cordes, *Biological Chemistry*. Harper and Row, New York 1969.
7. A. Bryson, N. R. Davies, E. P. Serjant, *J. Amer. Chem. Soc.* 85 (1963) 1933.
8. E. T. Edsall, J. Wyman, *Biophysical Chemistry*. Vol.1. Academic Press, New York 1966.
9. G. Maaß, F. Peters, *Angew. Chem.* 84 (1972) 430.
10. G. Jung, E. Breitmaier, W. A. Günzler, M. Ottnad, W. Voelter, L. Flohé, *Glutathione*. Georg Thieme Verlag, Stuttgart 1974, Vol. 1.
11. M. Calvin, *Glutathione*. Academic Press, New York 1954, Vol. 3.
12. E. J. Cohn, J. T. Edsall, *Proteins, Amino Acids and Peptides*. Academic Press, New York 1943, p. 84.
13. R. E. Benesch, R. Benesch, *J. Amer. Chem. Soc.* 77 (1955) 5877.
14. K. Wallenfels, Ch. Streffer, *Biochem. Z.* 346 (1966) 119.
15. E. Coates, C. Marsden, B. Rigg, *Trans. Farad. Soc.* 65 (1969) 863.
16. F. Peters, *Dissertation*. Braunschweig 1971.
17. M.A. Grafins, J.B. Neilands, *J. Amer. Chem. Soc.* 77 (1955) 3389.
18. L. Flohé, E. Breitmaier, W. A. Günzler, W. Voelter, G. Jung, *Hoppe Seylers Z. Physiol. Chem.* 353 (1972) 1159.
19. T. Ishimitsu, S. Hirose, H. Sakurai, *Chem. Pharm. Bull.* 24 (1976) 3195.
20. T. Kiss, B. Tóth, *Talanta* 29 (1982) 529.
21. T. Ishimitsu, S. Hirose, *Talanta* 24 (1977) 555.
22. T. Ishimitsu, S. Hirose, H. Sakurai, *Chem. Pharm. Bull.* 26 (1978) 74.
23. A. E. Martell, *Chemical and Biological Aspects of Pyridoxal-Catalysis*. Pergamon Press, London 1963, p. 13.
24. F. J. Anderson, A. E. Martell, *J. Amer. Chem. Soc.* 86 (1964) 715.
25. D. Garfinkel, J.T. Edsall, *J. Amer. Chem. Soc.* 80 (1958) 3823.
26. R. I. Shrager, J. S. Cohen, S. R. Heller, D. H. Sachs, A. N. Schechter, *Biochemistry* 11 (1972) 541.
27. D. L. Rabenstein, *J. Amer. Chem. Soc.* 95 (1973) 2797.
28. D. L. Rabenstein, T. L. Sayer, *Anal. Chem.* 48 (1976) 1141.
29. T. L. Sayer, D. L. Rabenstein, *Can. J. Chem.* 54 (1976) 3392.
30. D. L. Rabenstein, M. S. Greenberg, C. A. Evans, *Biochemistry* 16 (1977) 977.
31. M. L. Ahrens, G. Maaß, P. Schuster, H. Windler, *J. Amer. Chem. Soc.* 92 (1970) 6134.
32. E. H. Abbott, A. E. Martell, *J. Amer. Chem. Soc.* 93 (1971) 5852.
33. R. Katz, H. J. C. Yeh, D. F. Johnson, *Biochem. Biophys. Res. Commun.* 58 (1974) 316.
34. T. H. Witherup, E. H. Abbott, *J. Org. Chem.* 40 (1975) 2229.
35. R. C. Harruff, W. T. Jenkins, *Org. Magn. Reson.* 8 (1976) 548.
36. D. E. Metzler, C. M. Harris, R. J. Johnson, D. B. Siano, J. S. Thomson, *Biochemistry* 12 (1973) 5377.

37. O. Schuster, K. Torshanoff, H. Winkler, *Z. Naturforsch.* 31 c (1976) 219.
38. D. E. Metzler, E. E. Snell, *J. Amer. Chem. Soc.* 77 (1955) 2431.
39. R. J. Johnson, D. E. Metzler, *Meth. Enzymol.* 18 A (1970) 433.
40. D. E. Metzler, *BBA* 421 (1976) 181.
41. D. E. Metzler, C. M. Harris, R. L. Reeves, W. H. Lawton, M. S. Maggio, *Anal. Chem.* 49 (1977) 864 A.
42. D. B. Siano, D. E. Metzler, *J. Chem. Phys.* 51 (1969) 1856.
43. Y. V. Morozov, N. P. Bazhulina, L. P. Cherkashina, M. Y. Karpeiskii, *Biophysics* 12 (1968) 454.
44. M. L. Ahrens, G. Maaß, P. Schuster, H. Winkler, *FEBS Lett.* 5 (1969) 327.

13 Metal-Complex Equilibria

13.1 The System BH $\rightleftarrows$ B (H), M + B $\rightleftarrows$ MB

13.1.1 Concentration Diagrams

Metal complexes play an important role not only in inorganic qualitative and quantitative analysis, but also in the area of metal-ion catalysis with its wide-ranging chemical and biochemical implications. One dramatic example of the increasing interest in metal complexes is found in the category of substances known as "cage compounds".[1-3] These materials, including coronates, catapinates, and cryptates,[4] have been shown to facilitate (through phase transfer or phase transfer catalysis) numerous chemical transformations that once would have been regarded as extremely difficult if not inconceivable.[5-12] Metal complexes and their associated equilibria also proved the key to the highly selective liquid-membrane electrodes now enjoying broad application throughout analytical and biochemistry.[13-21]

The investigation of metal-complex solutions generally involves either electrometric or spectrometric measurements, with the electrometric approach normally being preferred. Most of the ligands found in metal complexes are protolytes, so it is not surprising that pH titration is often called upon to supply useful information. Thus, a knowledge of the pH of a typical metal-complex system at each step in a titration, together with a record of the amounts of titrant introduced, permits one to assign a numerical value to pK, and this in combination with initial concentration data leads to a stability constant. The literature describes several techniques for evaluating such data, including use of the "complex formation function" introduced by J. Bjerrum,[22-25] as well as a number of powerful computer programs that facilitate the data-analysis task both in general[26-42] and with respect to specific titration systems.[42-65]

Effective data-analysis procedures are also available for the evaluation of spectrometric data, and some have been the subject of extensive treatment in standard works.[23,24,66] Recently, reports have appeared describing new and exceptionally comprehensive computer programs, ones that promise to have far-reaching consequences.[40,67-80] For example, attempts have now been made to perform spectrometric investigations on systems containing as many as ten different absorbing species,[77] and systems under the influence of up to six metal-complex equilibria.[81] The procedures themselves appear applicable to the analysis of nearly any metal-complex system, ranging from the simplest complexes (MB, MB$_2$ ...) to extensive polynuclear species (M$_2$L$_2$H, M$_2$L$_3$, M$_3$L$_4$(OH)$_2$ ...).

In the present discussion we limit our attention to questions of the following types:

(1) How can geometric relationships like those established in the context of acid-base equilibria be applied to metal-complex systems?

(2) What types of information may be derived from a metal-complex system using only spectrometric titration data?

(3) To what extent is it possible to correlate such information with electrometric data (e.g., pH measurements)?

Once again, A-diagrams and related representations (such as the so-called "excess absorbance diagram") play a central part in the development.

As will be demonstrated in Sec. 13.4, it is possible to perform a complete evaluation of a metal-complex equilibrium system using techniques analogous to those applied to protolysis systems. Thus, both the molar absorptivities of intermediates and the ratios of stability constants are accessible by strictly spectrometric means. This finding is of considerable interest, particularly for situations in which pH measurements are problematic (e.g., with non-aqueous systems).

Before turning to complicated problems that are best resolved with the aid of excess absorbance diagrams it will be useful to examine a few simpler cases—ones in which water is the solvent—in order to illustrate the types of information that can be obtained directly from an A-diagram. Consider the following system:

$$\mathrm{BH} \overset{K_1}{\leftrightarrows} \mathrm{B\,(H)}$$

$$\mathrm{B + M} \overset{k_1}{\leftrightarrows} \mathrm{MB} \tag{13-1}$$

(charges are here ignored)

Let us assume with respect to the equilibria 13–1 that only the deprotonated protolyte B complexes with the free metal ion M, and that only a 1:1 complex (MB) is formed. K_1 is the dissociation constant for the single-step protolyte, and k_1 is the desired (individual) stability constant. Therefore:

$$K_1 = \frac{b \cdot h}{bh} \quad \text{and} \tag{13-2a}$$

$$k_1 = \frac{mb}{b \cdot m} \tag{13-2b}$$

Many authors prefer to work not with individual stability constants (k_i) but rather with what are known as *overall stability constants* (β_i). The latter are functions of k_i of the form

13.1 The System BH ⇌ B (H), M + B ⇌ MB

$$\beta_i = \prod_{j=1}^{i} k_j$$

In the present case only one metal complex is involved, so $k_1 = \beta_1$; hence:

$$\beta_1 = \frac{mb}{b \cdot m} \tag{13-2c}$$

The two equilibria comprising system 13–1 may be incrementally displaced in any of the following ways:

(1) by successive changes in the pH (pH titration);

(2) by varying the initial concentrations (b_0 and m_0) at constant pH (cf. Sec. 13.4);

(3) by the method of continuous variation (the "Job method"),[23,24,66,82] in which the sum of the initial concentrations ($b_0 + m_0$) is held constant but the ratio of the two is varied; or

(4) by dynamic method,[83] whereby equilibrium is disturbed through sudden changes in temperature, pressure, or electric field strength, all of which affect the equilibrium constant. (If the ensuing reactions are followed as a function of time it is often possible to establish a complete set of rate constants, and these may in turn be used to calculate k_1 and β_1.)

Each of the above techniques has been successfully invoked in the study of equilibrium parameters for metal-complex systems. In many cases a knowledge of the molar absorptivities is presupposed, but their determination is often not trivial, as will presently be demonstrated.

The discussion that follows centers on elucidation of two types of information relevant to metal complexes:

- molar absorptivities, and

- individual (or overall) stability constants.

When a *pH titration* is used to displace the equilibria comprising equations 13–1 the following mass-balance relationships may be applied at each of the titration points:

$$b_0 = bh + b + mb \tag{13-3a}$$

and

$$m_0 = m + mb \tag{13-3b}$$

Equations 13–2a through 13–3b make it possible to express the concentrations bh and mb as functions of h, and the resulting formulations may be used to construct a *characteristic concentration diagram bh* vs. *mb* (cf. Fig. 13–1). If no complex whatsoever results during such a pH titration (i.e., $k_1 = \beta_1 = 0$), then the only consequence is increased dissociation of the protolyte BH. Thus, the titration points simply involve differing concentrations of bh, and all the points in Fig. 13–1 will fall along the line $\overline{BHB}$ (cf. the point P_1). However, if the situation is complicated by the formation of a metal complex, then the characteristic concentration diagram will consist of a curve, the precise nature of which depends upon the value of k_1. An analogous situation was encountered in the case of acid–base equilibria in Sec. 7.5.3, where it was further shown that all the concentrations in a diagram like Fig. 13–1 may be induced to fall along a straight line by holding h and K_1 constant while permitting k_1 to change.[84,85] This is easily demonstrated here by using equation 13–2a to eliminate the quantity b from equation 13–3a. What results is the relationship

$$bh = \frac{h}{K_1 + h}(b_0 - mb) \qquad (13-4)$$

which is the equation for a straight line so long as h and K_1 remain constant. If k_1 is now allowed to change (producing corresponding changes in bh and mb), equation 13–4 ensures that a linear dependency will govern the behavior of bh relative to mb. If not only k_1, but also h, is permitted to change, the result will be a pencil of rays emanating from a point P_0 along the abscissa (where $bh = 0$, and $mb = b_0$). This pencil of rays (cf. Fig. 13–1) intersects the ordinate at points P_1, each of which must represent a titration point for the equivalent "isolated" protolyte (cf. Chapter 6).

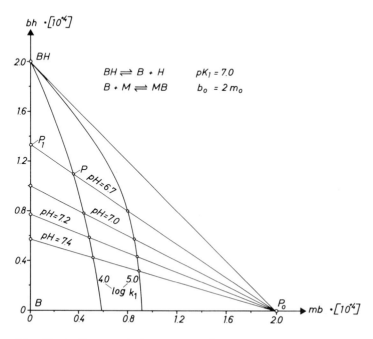

Fig. 13–1. Characteristic concentration diagrams bh vs. mb as a function of the parameter k_1 (log $k_1 = 4.0$ and 5.0; $b_0 = 2\,m_0 = 2 \cdot 10^{-4}$ M; $pK_1 = 7.0$).

13.1 The System BH $\rightleftarrows$ B (H), M + B $\rightleftarrows$ MB

Thus, a true set of titration points for the system 13–1, irrespective of the value of k_1, must fall along some line $P_1 P_0$ (cf. Fig. 13–1). Introducing the relationship $mb = b_0$ into equations 13–3a and 13–3b provides the complete set of concentrations associated with P_0:[84,85]

$$bh(P_0) = b(P_0) = 0, \quad mb(P_0) = b_0, \quad \text{and}$$

$$m(P_0) = m_0 - b_0 \tag{13-5}$$

It should be noted that the point P_0 itself is only attainable in the event that $m_0 > b_0$, because if $b_0 > m_0$ then $m(P_0)$ assumes a negative value. Even so, there is always merit in establishing the locations of points BH and P_0, because both may be used to obtain values for the concentration b_0 (cf. Fig. 13–1).

The points P_0, B, and BH in Fig. 13–1 constitute a triangle from which concentrations b, bh, and mb may be determined on the basis of simple distance relationships. The relationships that apply here are analogous to those established in Sec. 7.5.4 (cf. equation 7–59 and Fig. 7–21). The remaining concentration, m, may be calculated with the aid of equation 13–3b. Relationships such as these play an important part in the discussion of A-diagrams that follows.

13.1.2 Absorbance Diagrams

Provided there is no measurable absorption due to either the solvent or the titrant for a metal-complex system, then within the range of validity of the Lambert–Beer–Bouguer law the following expression characterizes the measured absorbance:

$$A_\lambda = l(\varepsilon_{\lambda BH} bh + \varepsilon_{\lambda B} b + \varepsilon_{\lambda MB} mb + \varepsilon_{\lambda M} m) \tag{13-6}$$

The curves in Fig. 13–2 indicate how the stability constant k_1 affects the appearance of an A-diagram. Such diagrams may be regarded as affine distortions of characteristic concentration diagrams bh vs. mb, an assertion that may be verified by first rearranging equations 13–3a and 13–3b in terms of b and m and then introducing the appropriate expressions into equation 13–6 for λ_i ($i = 1, 2$). The result is most conveniently written in matrix form:

$$\begin{pmatrix} A_1 \\ A_2 \end{pmatrix} = \begin{pmatrix} q_{11} & q_{12} \\ q_{21} & q_{22} \end{pmatrix} \begin{pmatrix} bh \\ mb \end{pmatrix} + \begin{pmatrix} l\left[\varepsilon_{1B} b_0 + \varepsilon_{1M} m_0\right] \\ l\left[\varepsilon_{2B} b_0 + \varepsilon_{2M} m_0\right] \end{pmatrix} \tag{13-7}$$

where $q_{\lambda_i 1} = l(\varepsilon_{\lambda_i BH} - \varepsilon_{\lambda_i B})$ and $q_{\lambda_i 2} = l(\varepsilon_{\lambda_i MB} - \varepsilon_{\lambda_i B} - \varepsilon_{\lambda_i M})$

Thus, the absorbances A_1 and A_2 are linearly dependent upon bh and mb. The arguments of Sec. 7.6 lead one to conclude further that diagrams bh vs. mb and A_1 vs. A_2 may be interconverted through affine transformations. Therefore, concentration and absorbance diagrams bear a special relationship to each other not only for acid–base equilibria, but also for metal-complex equilibria.

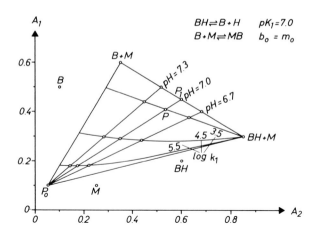

Fig. 13–2. The pencil of rays method: simulated A-diagrams as a function of k_1 [$\log k_1$ = 3.5, 4.5, and 5.5; $b_0 = m_0 = 10^{-4}$ M; molar absorptivities λ_i (i = 1, 2): $\varepsilon_{\lambda_i \text{BH}}$ = 2000 and 6000; $\varepsilon_{\lambda_i \text{B}}$ = 5000 and 1000; $\varepsilon_{\lambda_i \text{M}}$ = 1000 and 2500; $\varepsilon_{\lambda_i \text{MB}}$ = 1000 and 500, all in the units cm^{-1}M^{-1}].

As may be seen from Fig. 13–2, a system containing only the components BH and B would produce titration points along the line $\overline{\text{BH B}}$ (in the interest of clarity the only such points shown are BH and B themselves). If into this acid–base system a metal salt were to be introduced—one that absorbs, but is incapable of forming complexes—the titration points would continue to be restricted to a straight line, but this time a line parallel to the original line $\overline{\text{BH B}}$ and displaced from it by a distance determined by the magnitude of the absorbance components $l\varepsilon_{1\text{M}}m_0$ and $l\varepsilon_{2\text{M}}m_0$ (cf. the line (BH + M) (B + M) in Fig. 13–2).

Finally, let us suppose that pH titration is accompanied by metal-complex formation. This time the titration points describe a curve, the nature of which is a function of k_1. According to Sec. 13.1.1, points associated with a given pH but different values of k_1 must fall along a straight line, and the complete set of such lines must intersect at a single point (P_0), as shown in Fig. 13–2. The coordinates of this point in the A-diagram may be established by combining the set of expressions 13–5 with equation 13–6:[84,85]

$$A_\lambda (P_0) = l\,[\varepsilon_{\lambda\text{MB}}b_0 + \varepsilon_{\lambda\text{M}}(m_0 - b_0)] \tag{13-8}$$

Therefore, provided values are available for b_0, m_0, and $\varepsilon_{\lambda\text{M}}$, the molar absorptivity $\varepsilon_{\lambda\text{MB}}$ may be established directly from the point P_0.

The three points P_0, (BH + M), and (B + M) define a triangle analogous to the absorbance triangle for a two-step acid–base equilibrium system (cf. Fig. 7–13 in Sec. 7.3.3). Thus, the present triangle may serve as the basis for distance relationships that in turn provide values for the concentrations bh, b, and mb, as shown by the following equations (cf. Fig. 13–3):

13.1 The System $BH \rightleftarrows B(H)$, $M + B \rightleftarrows MB$

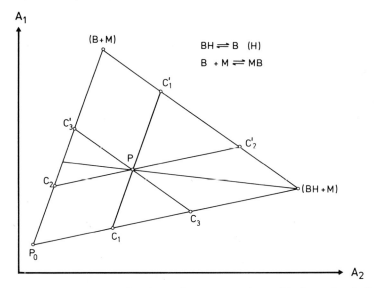

Fig. 13–3. Determination of concentrations bh, b, and mb for the system 13–1 using distance relationships within an absorbance triangle.

$$bh = \frac{\overline{C_1 P_0}}{\overline{(BH+M) P_0}} \cdot b_0 = \frac{\overline{C_1'(B+M)}}{\overline{(B+M)(BH+M)}} \cdot b_0$$

$$b = \frac{\overline{C_2 P_0}}{\overline{(B+M) P_0}} \cdot b_0 = \frac{\overline{C_2'(BH+M)}}{\overline{(BH+M)(B+M)}} \cdot b_0 \quad \text{and} \quad (13\text{–}9)$$

$$mb = \frac{\overline{C_3(BH+M)}}{\overline{(BH+M) P_0}} \cdot b_0 = \frac{\overline{C_3'(B+M)}}{\overline{(B+M) P_0}} \cdot b_0$$

The missing concentration, m, may be established from the further relationship $m = m_0 - mb$. Once the complete set of concentrations is known it is a straightforward matter to evaluate k_1 or β_1.[84,85]

We have therefore established that the entire *pencil of rays* methodology, which was derived in terms of acid–base equilibria (cf. Sec. 7.3.5), is equally applicable (after suitable modification) to systems involving metal complexes.

It is often assumed that the formation of a metal complex MB may be driven essentially to completion by adding a large excess of ligand to a solution of the appropriate metal salt (or *vice versa*). However, this assumption is valid only if k_1 is sufficiently large. Consider, for example, Fig. 13–4, which consists of a set of absorbance curves reflecting various initial concentrations m_0 for an arbitrary set of molar absorptivities and constant values of k_1 and b_0. The quantity $\varepsilon_{\lambda M}$ has been

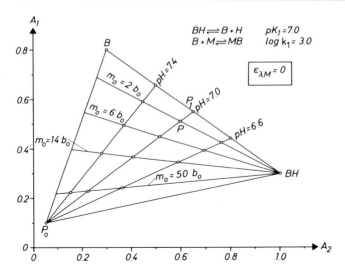

Fig. 13–4. Simulated A-diagrams as a function of m_0 [$m_0 = 2b_0, 6b_0, 14b_0, 50b_0$; $b_0 = 10^{-4}$ M; $\log k_1 = 3.0$; molar absorptivities λ_i ($i = 1, 2$): $\varepsilon_{\lambda_i\text{BH}} = 3\,000$ and $10\,000$; $\varepsilon_{\lambda_i\text{B}} = 8\,000$ and $3\,000$; $\varepsilon_{\lambda_i\text{M}} = 0$ and 0; $\varepsilon_{\lambda_i\text{MB}} = 1000$ and 500, all in the units $\text{cm}^{-1}\text{M}^{-1}$].

assigned the value zero in order to maintain a common point of intersection $A_\lambda(\text{P}_0)$ despite variations in m_0. It is apparent from the figure that even under alkaline conditions, and with a 50-fold excess of metal ion, complex formation remains incomplete (with $\log k_1 = 3.0$). Note that the assumption has also been made that the metal ion does not separate as an insoluble hydroxide under alkaline conditions, a favorable circumstance that cannot be taken for granted. Fortunately, the potential problems raised by these considerations may be circumvented by application of the pencil of rays method.[84,85]

Spectrometric investigations are directed primarily toward establishing concentrations, and thus to evaluating "classical" overall stability constants (β_i; cf. Sec. 6.1). In fact, these need not remain truly constant throughout a titration, at least not in the rigorous sense expected of a "thermodynamic" overall stability constant (k_1):

$$k_1 = \frac{mb}{b \cdot m} \cdot \frac{f_{\text{MB}}}{f_\text{B} \cdot f_\text{M}} = k_1 \cdot \frac{f_{\text{MB}}}{f_\text{B} \cdot f_\text{M}} \qquad (13\text{–}10)$$

where f_{MB} = activity coefficient of the metal complex MB (and analogously for f_B and f_M)

If the activity coefficients change in the course of a titration, and if a corresponding change occurs in the quotient $f_{\text{MB}}/(f_\text{B} \cdot f_\text{M})$, then k_1 cannot remain constant for a fixed value of k_1.

13.1 The System BH ⇌ B (H), M + B ⇌ MB

As it happens, the pencil of rays method retains its validity even if k_1 does change in the course of a titration, provided the additional restriction of constant molar absorptivity is met. Curve I in Fig. 13–5 illustrates the expected pathway for a process with constant k_1. By contrast, Curve II in the same figure reflects a situation in which log k_1 changes by 0.05 units over the course of each pH interval ΔpH = 0.1. If all the points in this diagram associated with a given pH are joined by a straight line it will again be observed that the lines intersect at the point P_0. Thus, despite changes in the quantity k_1 an absorbance triangle with the vertices (BH + M), (B + M), and P_0 may be constructed, and it may be used to determine values of k_1 applicable at various stages in the titration.[84,85]

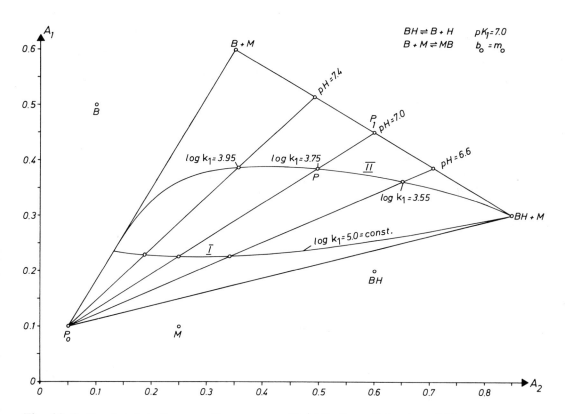

Fig. 13–5. Simulated A-diagrams ($b_0 = m_0 = 10^{-4}$ M). Curve I: log $k_1 = 5.00$ = constant. Curve II: log k_1 undergoes a consistent change of 0.05 units for each 0.1 unit change in pH (molar absorptivities as in Fig. 13–2).

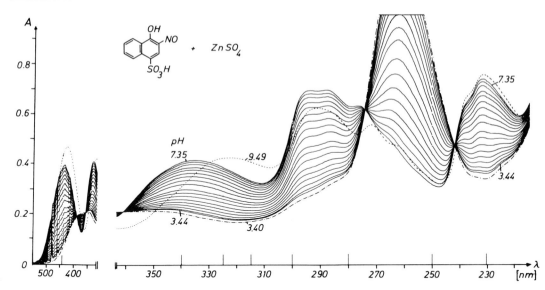

Fig. 13–6. Titration spectra for the system 2-nitroso-1-naphthol-4-sulfonic acid/ZnSO$_4$ (b_0 = 5 · 10^{-5} M; m_0 = 10^{-3} M; 0.05 M KCl; 20 °C). Broken lines indicate the spectra of BH and B. Marks along the abscissa signify wavelengths that were selected to serve as the basis for A-diagrams. Artifacts in the visible region are due to the D$_2$/H$_2$ lamp.

Figure 13–6 shows a set of titration spectra for the system 2-nitroso-1-naphthol-4-sulfonic acid/zinc sulfate (k_1 = constant), and it provides the basis for a concrete example of the methods so far discussed. Since it is known that this system may produce both mono- and bis-complexes (MB, MB$_2$),[86,87] zinc ion has been introduced in a 20-fold excess relative to ligand (m_0 = 20 b_0) to minimize the formation of MB$_2$. The points P in the corresponding A-diagram (Fig. 13–7) represent measured absorbances for the metal-complex solution, while the points P$_1$ refer to the single-step protolyte alone (i.e., in the absence of zinc salt). Points P$_1$ were computed with the aid of equation 6–11 from the known values

$$A_{\lambda BH} = l\,\varepsilon_{\lambda BH}b_0 = 0.181,$$

$$A_{\lambda B} = l\,\varepsilon_{\lambda B}b_0 = 0.385, \text{ and}$$

$$pK_1 = 6.38$$

and using the appropriate values of pH. There is no need here to worry about a potential displacement of the line $\overline{BH\,B}$ since $\varepsilon_{\lambda M}$ = 0 at the wavelengths investigated. The corresponding absorbance triangle provides the following mean result:[84,85]

$$\log k_1 = 3.72 \pm 0.06 \quad (20.0\,°C;\ 0.05\,M\,KCl)$$

13.1 The System BH $\rightleftarrows$ B (H), M + B $\rightleftarrows$ MB

O. Mäkitie and H. Saarinen[86,87] have reported a useful reference value:

$\log k_1 = 3.06$ (25 °C; 0.1 M KCl)

Geometric considerations of the type discussed above may also be applied to an A-diagram for a system in which the metal ion simultaneously binds with n ligand groups:

$$\text{BH} \overset{K_1}{\rightleftarrows} \text{B (H)}$$
$$n\text{B} + \text{M} \overset{k_1}{\rightleftarrows} \text{MB}_n \tag{13-11}$$

However, the difference in stoichiometry requires that equation 13–8 be replaced by the more complicated relationship[84]

$$A_\lambda (P_0) = l \left[\varepsilon_{\lambda MB_n} \frac{b_0}{n} + \varepsilon_{\lambda M} \left(m_0 - \frac{b_0}{n} \right) \right] \tag{13-12}$$

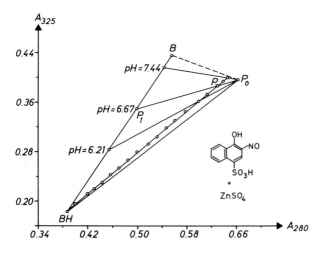

Fig. 13–7. The diagram A_{325} vs. A_{280} for the titration system of Fig. 13–6, with $\varepsilon_{\lambda M} = 0$. Calculated absorbances for the protolysis system alone (i.e., in the absence of ZnSO$_4$) are shown along the line $\overline{\text{BH B}}$, with $pK_1 = 6.38$ (0.05 M KCl; 20 °C). $\overline{\text{B P}_0}$ is shown as a dashed line because it does not come in contact with the experimental curve passing through the points BH and P (due to prior precipitation of the hydroxide). Evaluated pH range: $5.91 \leq \text{pH} \leq 7.44$.

13.2 Determination of Overall Stability Constants for Metal Complexes Related to Multi-Step Acid-Base Equilibria

We begin by considering a titration system of the form

$$BH_2 \underset{}{\overset{K_1}{\leftrightarrows}} BH \ (H)$$

$$BH \underset{}{\overset{K_2}{\leftrightarrows}} B \ (H) \tag{13–13}$$

$$B + M \underset{}{\overset{k_1}{\leftrightarrows}} MB$$

where it will again be assumed that only the deprotonated protolyte is capable of complexing with the metal ion.

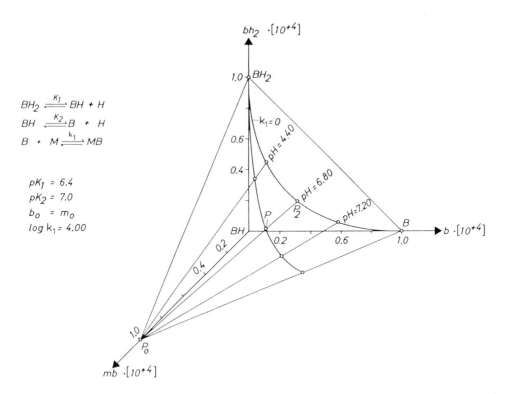

Fig. 13–8. Characteristic concentration diagram for the system 13–13 ($b_0 = m_0 = 10^{-4}$ M; $pK_1 = 6.4$; $pK_2 = 7.0$; $\log k_1 = 4.00$; the point P represents one of the data points). At points P_2 only the two-step protolysis system is involved ($k_1 = 0$).

13.2 Determination of Overall Stability Constants for Metal Complexes

The mass balance conditions in this case may be expressed as:
$$b_0 = bh_2 + bh + b + mb \quad \text{and} \quad m_0 = m + mb \tag{13-14}$$

The relevant equilibrium constants are defined as follows:
$$K_1 = \frac{bh \cdot h}{bh_2}, \quad K_2 = \frac{b \cdot h}{bh}, \quad \text{and} \quad k_1 = \frac{mb}{b \cdot m} \tag{13-15}$$

Equations 13–14 and 13–15 permit the concentrations of all components to be expressed as functions of h (or pH). Figure 13–8 is a characteristic concentration diagram constructed in terms of the concentrations bh_2, b, and mb. If no metal complexes were formed (i.e., $k_1 = 0$) then all titration points (P_2) would fall within the plane defined by the coordinate axes bh_2 and b. For $k_1 \neq 0$, titration produces a curve located somewhere within a concentration tetrahedron defined by the vertices BH_2, BH, B, and P_0 (cf. the curve described by the points P in Fig. 13–8).

Just as with system 13–1, an important pencil of rays may be shown to emerge from the point P_0 (cf. Fig. 13–8). We begin by establishing the vectors leading from the coordinate origin to the points P_2 and P (cf. equations 7–5, 13–14, and 13–15, although in this case it is not essential that mb be developed as a function of h):

$$\mathbf{c}_{P_2} = \begin{pmatrix} bh_2 \\ b \\ mb \end{pmatrix} = \frac{b_0}{H_2} \begin{pmatrix} h^2 \\ K_1 K_2 \\ 0 \end{pmatrix} \quad \text{and}$$

$$\mathbf{c}_P = \frac{1}{H_2} \begin{pmatrix} (b_0 - mb)\, h^2 \\ (b_0 - mb)\, K_1 K_2 \\ mb\, H_2 \end{pmatrix}, \text{ and } H_2 = h^2 + K_1 h + K_1 K_2 \tag{13-16}$$

The following relationship must apply to any line passing through a particular pair of points P and P_2 (cf. equation 8–32):

$$\mathbf{c} = \begin{pmatrix} bh_2 \\ b \\ mb \end{pmatrix} = \mathbf{c}_{P_2} + k(\mathbf{c}_P - \mathbf{c}_{P_2}) \quad \text{with} \quad -\infty < k < \infty \tag{13-17}$$

If this expression is used to establish the value for k that corresponds to $bh_2 = 0$, and the result is introduced into equation 13–17, one obtains the following:

$$\mathbf{c}_{P_0} = \begin{pmatrix} 0 \\ 0 \\ b_0 \end{pmatrix} \tag{13-18}$$

Thus, the location of the point P_0 is independent of pH. From this one may conclude that lines $\overline{P\,P_2}$ constructed for various values of pH must intersect in the point P_0, which means that such lines constitute a pencil of rays. The concentrations bh and m corresponding to the point P_0 may now be determined using equations 13–18 and 13–14. The point P_0 is thus defined as follows:[84]

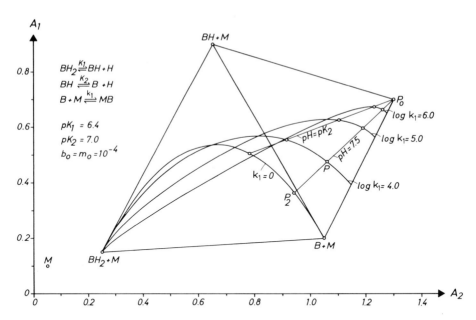

Fig. 13–9. Simulated A-diagrams for the system 13–13 as a function of k_1 (log k_1 = 4.0, 5.0, and 6.0; $b_0 = m_0 = 10^{-4}$ M; $pK_1 = 6.4$; $pK_2 = 7.0$; the point P represents a data point). At points P_2 only the two-step protolysis system is involved ($k_1 = 0$). Molar absorptivities λ_i ($i = 1,2$): $\varepsilon_{\lambda_i BH_2}$ = 500 and 2 000; $\varepsilon_{\lambda_i BH}$ = 8 000 and 6 000; $\varepsilon_{\lambda_i B}$ = 1 000 and 10 000; $\varepsilon_{\lambda_i M}$ = 1 000 and 500; $\varepsilon_{\lambda_i MB}$ = 7 000 and 13 000, all in the units cm^{-1}M^{-1}.

$$bh_2(P_0) = bh(P_0) = b(P_0) = 0, \quad mb(P_0) = b_0, \text{ and } m(P_0) = (m_0 - b_0) \quad (13-19)$$

Analogous results apply to two- and three-dimensional A-diagrams. As may be seen in the case of the two-dimensional A-diagram of Fig. 13–9, all titration points (P_2) for the simplified case $k_1 = 0$ fall within the triangle defined by the vertices (BH$_2$ + M), (BH + M), and (B + M). If metal complexes are formed (i.e., $k_1 \neq 0$), the curves penetrate beyond the bounds of this triangle (cf. the points P), but they remain confined within the absorbance tetrahedron of a corresponding three-dimensional A-diagram, and thus within the four-sided figure obtained by projecting this tetrahedron into a two-dimensional A-diagram. The point P_0 serves as the origin for a pencil of rays through the points P and P_2, where P_0 has the coordinates:[84]

$$A_\lambda(P_0) = l\,[\varepsilon_{\lambda MB} b_0 + \varepsilon_{\lambda M}(m_0 - b_0)] \quad (13-20)$$

This relationship is identical to that obtained with system 13–1 (cf. equation 13–8).

The procedure described in Sec. 8.3.3 may be used to establish distance relationships applicable to the concentrations of all components in the system, thereby permitting calculation of k_1, provided the points (BH$_2$ + M), (BH + M), (B + M), and P_0 have been identified in the appropriate three-dimensional A-diagram (cf. Sec. 13.1.2).

13.2 Determination of Overall Stability Constants for Metal Complexes

The titration system

$$BH_2 \overset{K_1}{\leftrightarrows} BH \; (H)$$

$$BH \overset{K_2}{\leftrightarrows} B \; (H) \tag{13-21}$$

$$nB + M \overset{k_1}{\leftrightarrows} MB_n$$

may be treated similarly; all that is required is replacement of equation 13–20 with the relationship:[84]

$$A_\lambda(P_0) = l\left[\varepsilon_{\lambda MB_n}\frac{b_0}{n} + \varepsilon_{\lambda M}\left(m_0 - \frac{b_0}{n}\right)\right] \tag{13-22}$$

The validity of the pencil of rays method is independent of which protolyte forms complexes with the metal ion. If, for example, only the protolyte BH_2 participates in such an interaction, then the relevant equilibria are:

$$BH_2 + M \overset{k_1}{\leftrightarrows} MBH_2$$

$$BH_2 \overset{K_1}{\leftrightarrows} BH \; (H) \tag{13-23}$$

$$BH \overset{K_2}{\leftrightarrows} B \; (H)$$

The key relationships are the same as those for system 13–13, except that references to the component MB (*mb*) must be rewritten in terms of MBH_2 (*mbh*$_2$). Similarly, if the only complex present is MBH, then the place of MB (*mb*) is taken by MBH (*mbh*).

The situation also remains unchanged for an *s*-step acid–base system. In other words, it is possible in principle to use the methods described here to analyze the generalized system[84]

$$BH_s \overset{K_1}{\leftrightarrows} BH_{s-1} \; (H)$$

$$BH_{s-1} \overset{K_2}{\leftrightarrows} BH_{s-2} \; (H)$$

$$\vdots \tag{13-24}$$

$$BH \overset{K_s}{\leftrightarrows} B \; (H)$$

$$B + M \overset{k_1}{\leftrightarrows} MB$$

13.3 The Method of Continuous Variation ("Job's Method")

The "method of continuous variation" is derived from a suggestion by P. Job,[82] and it proceeds from the assumption that a titration system contains but a *single* metal complex formed as a consequence of the equilibrium:

$$M + nB \leftrightarrows MB_n \qquad (13\text{-}25)$$

Variation of the initial concentration of the constituents M and B in such a way that the sum of their concentrations remains constant (i.e., $m_0 + b_0 = c_0$ = constant) provides a means of establishing the ratio of the components within the complex. Proof of this assertion is facilitated by defining a quantity known as the *excess absorbance* A_λ^E, equivalent to the absorbance of the sample solution minus the absorbance of a hypothetical solution of M and B which is identical in every way except that it contains no complex.[24,66] The titration itself is conducted such that, at each step, equimolar solutions (with concentrations c_0) of the components B and M are mixed so that the sum of their units of volume (x and $1 - x$, respectively) remains constant, after which the mixture is submitted to spectrometric measurement. This means that in every mixture the concentrations of M and B may be expressed as

$$m_0 = (1-x)c_0 \quad \text{and} \quad b_0 = x \cdot c_0 \qquad (13\text{-}26)$$

where $m_0 + b_0 = c_0$ = constant

The Lambert–Beer–Bouguer law specifies for the absorbance A_λ of such a solution:

$$A_\lambda = l(\varepsilon_{\lambda M} \cdot m + \varepsilon_{\lambda B} \cdot b + \varepsilon_{\lambda MB_n} \cdot mb_n) \qquad (13\text{-}27)$$

Assuming there were no complex formed the corresponding absorbance would be simply

$$A'_\lambda = l(\varepsilon_{\lambda M} \cdot m + \varepsilon_{\lambda B} \cdot b) = l(\varepsilon_{\lambda M} \cdot (1-x)c_0 + \varepsilon_{\lambda B} \cdot x \cdot c_0) \qquad (13\text{-}28)$$

Thus, with the restrictions imposed by equations 13–26, the excess absorbance A_λ^E must be:[24]

$$A_\lambda^E := A_\lambda - A'_\lambda = l[\varepsilon_{\lambda M} \cdot m + \varepsilon_{\lambda B} \cdot b + \varepsilon_{\lambda MB_n} \cdot mb_n$$
$$- \varepsilon_{\lambda M} \cdot (1-x)c_0 - \varepsilon_{\lambda B} \cdot x \cdot c_0]$$

$$= l[\varepsilon_{\lambda M} \cdot m + \varepsilon_{\lambda B} \cdot b + \varepsilon_{\lambda MB_n} \cdot mb_n - \varepsilon_{\lambda M}(c_0 - b_0) - \varepsilon_{\lambda B} \cdot b_0] \qquad (13\text{-}29)$$

Application of the method of continuous variation entails constructing the diagram A_λ^E vs. b_0 (or A_λ^E vs. x). The curve constituting such a diagram is always found to possess either a maximum or a minimum, and the magnitude of b_0 at this extreme may be used to ascertain the desired quantity n. To illustrate the principle we begin by developing the differential dA_λ^E / db_0:

13.3 The Method of Continuous Variation ("Job's Method")

$$\frac{dA_\lambda^E}{db_0} = l\left(\varepsilon_{\lambda M}\frac{dm}{db_0} + \varepsilon_{\lambda B}\frac{db}{db_0} + \varepsilon_{\lambda MB_n}\frac{dmb_n}{db_0} + \varepsilon_{\lambda M} - \varepsilon_{\lambda B}\right) \quad (13\text{--}30)$$

From the mass balance constraints

$$m_0 = m + mb_n = c_0 - b_0 \quad (13\text{--}31\text{a})$$

and

$$b_0 = b + n \cdot mb_n \quad (13\text{--}31\text{b})$$

it is clear that the differentials dm/db_0 and db/db_0 must take the forms:

$$\frac{dm}{db_0} = -1 - \frac{dmb_n}{db_0} \quad \text{and} \quad \frac{db}{db_0} = 1 - n \cdot \frac{dmb_n}{db_0} \quad (13\text{--}32)$$

If dmb_n/db_0 is now set equal to zero, it follows from equation 13–30 that:

$$\frac{dA_\lambda^E}{db_0} = l\left(-\varepsilon_{\lambda M} + \varepsilon_{\lambda B} + \varepsilon_{\lambda M} - \varepsilon_{\lambda B}\right) = 0 \quad (13\text{--}33)$$

Thus, if $dmb_n/db_0 = 0$, then dA_λ^E/db_0 must also equal zero. In order to demonstrate how n may be established from the maximum or minimum in the diagram A_λ^E vs. b_0 it is only necessary to prepare an implicit representation of the curve constituting the diagram mb_n vs. b_0. Starting with the overall stability constant β_n for the system described by equation 13–25:

$$\beta_n = \frac{mb_n}{m \cdot b^n} \quad (13\text{--}34)$$

it follows from equations 13–31a and 13–31b that

$$mb_n = \beta_n \cdot m \cdot b^n = \beta_n(c_0 - b_0 - mb_n)(b_0 - n\,mb_n)^n$$

or

$$F := \beta_n(c_0 - b_0 - mb_n)(b_0 - n\,mb_n)^n - mb_n = 0$$

Partial differentiation of the implicit function F leads to the desired expression for dmb_n/db_0:[88]

$$\frac{dmb_n}{db_0} = -\frac{\dfrac{\partial F}{\partial b_0}}{\dfrac{\partial F}{\partial A_\lambda^E}} = -\left[\frac{-\beta_n(b_0 - n\,mb_n)^n + \beta_n(c_0 - b_0 - mb_n)\,n\,(b_0 - n\,mb_n)^{n-1}}{-\beta_n(b_0 - n\,mb_n)^n + \beta_n(c_0 - b_0 - mb_n)\,n\,(b_0 - n\,mb_n)^{n-1}\cdot(-n) - 1}\right]$$

It follows for the case $dmb_n/db_0 = 0$ that

$$b_0 - n\,mb_n = n\,(c_0 - b_0 - mb_n)$$

or

$$b_0 = \frac{c_0 n}{1+n} \qquad (13\text{--}35)$$

and, from equation 13–26,

$$n = \frac{b_0}{c_0 - b_0} = \frac{x}{1-x} \qquad (13\text{--}36)$$

Since a maximum or minimum in a graph of A_λ^E vs. b_0 has been shown to fulfill the condition $dmb_n/db_0 = 0$ (cf. equation 13–33), equation 13–36 must be applicable to this point, permitting straightforward spectrometric evaluation of n.[23,24,66,82]

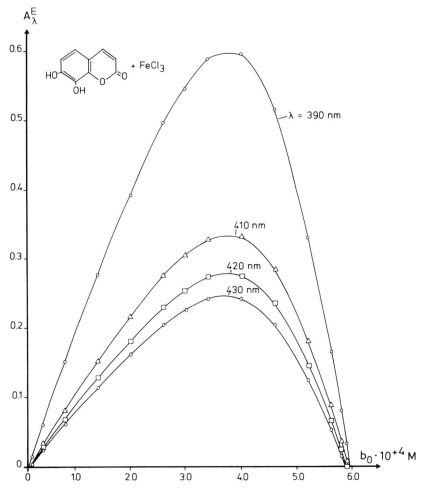

Fig. 13–10. The method of continuous variation (P. Job) applied to the system daphnetin/FeCl$_3$ (0.1 M acetate buffer; pH = 4.00; 25 °C).

Figure 13–10 illustrates a set of "Job curves" corresponding to measurements at various wavelengths for the system daphnetin (7,8-dihydroxycoumarin)/FeCl$_3$ in a pH 4 acetate buffer.[89,90] Application of equation 13–36 to the data shows that this system is consistent with the assignment $n = 2.0$.

13.4 The System M + B $\rightleftarrows$ MB, MB + B $\rightleftarrows$ MB$_2$

13.4.1 Concentration Diagrams

Stepwise association of multiple ligands with a metal ion may be formulated in such a way that the process becomes analogous to a set of protolysis equilibria. Thus, if a two-step protolysis system is written in the form

$$B^{2-} + H^+ \overset{K_2}{\leftrightarrows} BH^-$$

$$BH^- + H^+ \overset{K_1}{\leftrightarrows} BH_2 \tag{7–1}$$

then the protolyte B^{2-} reveals itself to be a species capable of acquiring two protons. Similarly, the pair of equilibria

$$M + B \overset{k_1}{\leftrightarrows} MB$$

$$MB + B \overset{k_2}{\leftrightarrows} MB_2 \tag{13–37}$$

reflects a metal ion M capable of establishing a relationship with two molecules of the ligand B (the role played by charges has once again been ignored as irrelevant). We retain here the standard definitions for the dissociation constants K_1 and K_2:

$$(K_1)^{-1} = \frac{bh_2}{bh \cdot h} \quad \text{and} \quad (K_2)^{-1} = \frac{bh}{b \cdot h} \tag{7–2}$$

Metal-complex systems are usually treated in terms of the corresponding stability constants k_1 and k_2:

$$k_1 = \frac{mb}{m \cdot b} \quad \text{and} \quad k_2 = \frac{mb_2}{mb \cdot b} \tag{13–38}$$

Note that these definitions result in indices that are used in different ways: while "K_1" is the dissociation constant for the species BH$_2$, the stability constant "k_1" refers to the species MB (not MB$_2$).

Throughout the course of a pH titration of system 7–1, mass balance considerations (assuming no dilution) dictate that

$$b_0 = b + bh + bh_2 \qquad (7\text{--}4)$$

If a solution of M were to be treated successively with portions of a solution of B (i.e., a b_0 *titration* were conducted), then a similar mass balance equation would be applicable:

$$m_0 = m + mb + mb_2 \qquad (13\text{--}39)$$

where m_0 = initial (molar) concentration of the metal salt

> In principle, one could investigate the metal-complex system equally well by starting with a solution of B and successively adding aliquots of M (i.e., an m_0 titration could be performed), but this approach would complicate comparisons with a set of acid–base equilibria, since studies of the latter essentially always entail incremental changes in the concentration of H$^+$ (induced by the addition, for example, of HCl or NaOH).

Thus, throughout the course of a b_0 titration the initial concentration of metal (m_0) may be regarded as the total concentration of metal, provided dilution effects are negligible. Moreover, at any given stage in the titration the following relationship applies to the ligand B:

$$b_0 = b + mb + 2\,mb_2 \qquad (13\text{--}40)$$

Since B is being introduced successively into the solution, the quantity b_0 does not remain constant; rather, it is a variable that changes from step to step (hence the term "b_0 titration"). For this reason, equation 13–40 must be regarded as containing three linearly independent concentration variables, in contrast to equation 13–39 which includes only two.

As will be demonstrated in Sec. 13.4.2, the experimentally accessible "excess absorbance diagram" for this system is an affine distortion of the characteristic concentration diagram associated with equation 13–39 (e.g., a diagram of m vs. mb_2). Various curves of the latter type are shown in Fig. 13–11 as a function of κ, where in analogy to equation 7–3 κ is defined as:

$$\kappa = \frac{k_1}{k_2} \qquad (13\text{--}41)$$

All the curves originate in the point M ($b_0 = 0, m = m_0$) and terminate at high b_0 values in the point MB$_2$ ($m = mb = 0$, $mb_2 = m_0$). As in the case of the concentration diagram bh_2 vs. b of Fig. 7–18 (Sec. 7.5.1), the coordinate axes are tangents to these key points. To put it somewhat differently, the line $\overline{\text{M MB}_2}$ in Fig. 13–11 is a polar, the pole for which is the point MB ($m = mb_2 = 0$, $mb = m_0$). This conclusion follows directly from the analogy that exists with acid–base equilibria, and it may be independently verified by the methods introduced in Sec. 7.5.2. We may thus regard it as established that, at the level of concentration diagrams, equivalent relationships exist within the two systems 7–1 and 13–37.

13.4 The System $M + B \rightleftarrows MB$, $MB + B \rightleftarrows MB_2$

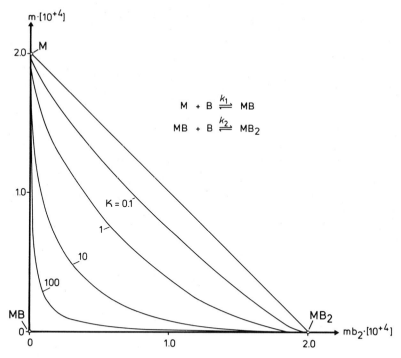

Fig. 13–11. Characteristic concentration diagrams m vs. mb_2 for the system 13–37 as a function of $\kappa = k_1/k_2$ ($\kappa = 0.1, 1, 10,$ and 100).

13.4.2 Excess-Absorbance Diagrams

The Lambert–Beer–Bouguer law provides the following absorbance equation for the reaction system 13–37:

$$A_\lambda = l(\varepsilon_{\lambda B} \cdot b + \varepsilon_{\lambda M} \cdot m + \varepsilon_{\lambda MB} \cdot mb + \varepsilon_{\lambda MB_2} \cdot mb_2) \tag{13–42}$$

Equations 13–39 and 13–40 permit this equation to be rewritten as:

$$A_\lambda = l\,\varepsilon_{\lambda B} \cdot b_0 + l\,[(\varepsilon_{\lambda M} + \varepsilon_{\lambda B} - \varepsilon_{\lambda MB})\,m + \\ + (\varepsilon_{\lambda MB_2} - \varepsilon_{\lambda MB} - \varepsilon_{\lambda B})\,mb_2 + (\varepsilon_{\lambda MB} - \varepsilon_{\lambda B})\,m_0] \tag{13–43}$$

If system 13–37 were to be investigated by varying b_0 (i.e., by conducting a b_0 titration, cf. Sec. 13.4.1), this equation might be regarded as supplying the functional relationship linking A_λ and the three linearly independent quantities b_0, m, and mb_2. The *excess absorbance* A_λ^E as defined in Sec. 13.3 shows itself here to be a useful and experimentally accessible quantity, because it is based upon only two linearly independent concentrations (m and mb_2):

$$A_\lambda^E := A_\lambda - l\,\varepsilon_{\lambda B} \cdot b_0 = l\,[(\varepsilon_{\lambda M} + \varepsilon_{\lambda B} - \varepsilon_{\lambda MB})\,m + \\ + (\varepsilon_{\lambda MB_2} - \varepsilon_{\lambda MB} - \varepsilon_{\lambda B})\,mb_2 + (\varepsilon_{\lambda MB} - \varepsilon_{\lambda B})\,m_0] \tag{13–44}$$

In what is known as an *excess absorbance diagram* (or "EA-diagram") two excess absorbances $A^E_{\lambda_1}$ and $A^E_{\lambda_2}$ are plotted against each other, as in Fig. 13–12. The resulting diagram may be regarded as an affine distortion of the concentration diagram m vs. mb_2 (cf. Fig. 13–11). To demonstrate the validity of this assertion one need only modify slightly the development presented in Sec. 7.6.

As is apparent from Fig. 13–12, the quantity A^E_λ may assume either positive or negative values, depending upon which of the terms A_λ or $(l \cdot \varepsilon_{\lambda B} \cdot b_0)$ dominates in equation 13–44. The curves shown in Fig. 13–12 correspond to various values of κ (cf. equation 13–41), but all lie within the triangle defined by the points M, MB, and MB$_2$, which is known as the *excess absorbance triangle*. The point MB is also seen to fall at the intersection of tangents to the curves at M and MB$_2$. Equation 13–44 reveals that the following relationships must apply to the points M, MB, and MB$_2$:

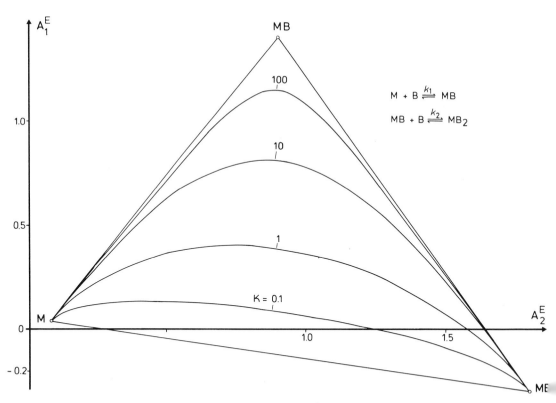

Fig. 13–12. Excess absorbance (EA-) diagrams A^E_1 vs. A^E_2 for the system 13–37 as a function of $\kappa = k_1/k_2$ (κ = 0.1, 1, 10, and 100). The system may be evaluated by a method analogous to that used with a two-step acid–base equilibrium system [cf. Fig. 7–1; $m_0 = 2 \cdot 10^{-4}$ M; molar absorptivities λ_i (i = 1, 2): $\varepsilon_{\lambda_i B}$ = 1 000 and 1 500; $\varepsilon_{\lambda_i M}$ = 200 and 400; $\varepsilon_{\lambda_i MB}$ = 8 000 and 6 000; $\varepsilon_{\lambda_i MB_2}$ = 500 and 12 000, all in units cm^{-1}M^{-1}].

13.4 The System $M + B \rightleftarrows MB$, $MB + B \rightleftarrows MB_2$

$$A_\lambda^E (M) := A_{\lambda M}^E = l\, \varepsilon_{\lambda M} \cdot m_0,$$
$$A_\lambda^E (MB) := A_{\lambda MB}^E = l\, (\varepsilon_{\lambda MB} - \varepsilon_{\lambda B})\, m_0, \quad \text{and} \quad (13\text{–}45)$$
$$A_\lambda^E (MB_2) := A_{\lambda MB_2}^E = l\, (\varepsilon_{\lambda MB_2} - 2\varepsilon_{\lambda B})\, m_0$$

Consequently, once the limiting points M and MB_2 have been established for a given titration, construction of an appropriate set of tangents provides a value for $A_{\lambda MB}^E$, and this in turn permits a value to be assigned to $\varepsilon_{\lambda MB}$. In other words, metal-complex equilibria may be evaluated by a straightforward adaptation of a procedure applied previously to protolysis systems. Excess absorbance diagrams play the same role with respect to metal complexes as absorbance diagrams do with acid–base equilibria, pH titration being replaced here by b_0 titration.

The preceding result confers increased significance upon the methods developed in Chapters 6–9. In principle, all the geometric relationships associated with (multi-dimensional) A-diagrams apply in an analogous fashion to EA-diagrams, and are thus applicable in evaluating metal-complex equilibria. For example, in a system consisting of the two metal-complex equilibria $M + B \rightleftarrows MB$ and $MB + B \rightleftarrows MB_2$, the complete set of concentrations m, mb, and mb_2 may be established through use of distance relationships like those described in Sec. 7.3.3 (cf. Fig. 7–13). The following relationships (analogous to those from equation 7–19) are illustrated in Fig. 13–13:

$$mb_2 = \frac{\overline{C_1\, MB}}{\overline{MB_2\, MB}} \cdot m_0 = \frac{\overline{C_1'\, M}}{\overline{MB_2\, M}} \cdot m_0$$

$$mb = \frac{\overline{C_2\, MB_2}}{\overline{MB_2\, MB}} \cdot m_0 = \frac{\overline{C_2'\, M}}{\overline{MB\, M}} \cdot m_0, \quad \text{and} \qquad (13\text{–}46)$$

$$m = \frac{\overline{C_3\, MB}}{\overline{MB\, M}} \cdot m_0 = \frac{\overline{C_3'\, MB_2}}{\overline{MB_2\, M}} \cdot m_0$$

The only concentration still missing is b, and this is easily computed by recalling the relationship

$$b = b_0 - mb - 2\, mb_2$$

(cf. equation 13–40), which in turn permits values to be assigned to k_1 and k_2 (with the aid of the equations 13–38).

As has been demonstrated in Sec. 7.6 for the case of a two-step protolysis equilibrium system, parallel displacement of the coordinate axes in an A-diagram leads to an AD-diagram. An analogous transformation in the context of metal complexes converts an excess absorbance (EA-) diagram into what is known as an *excess absorbance difference (EAD-) diagram*. For example, an EAD-diagram is produced when the coordinate axes in Fig. 13–12 are moved in a parallel fashion until they intersect at the point M. It then becomes true with respect to the *excess absorbance difference* ΔA_λ^E that:

$$\Delta A_\lambda^E = AA_\lambda^E - l\,\varepsilon_{\lambda M} \cdot m_0 \tag{13–47}$$

Absorbance differences ΔA_λ play the important general role in acid–base equilibria of facilitating the construction of ADQ- and ADQz-diagrams (cf. Sections 7.8, 8.3.6, and 9.3). In an analogous way, *excess absorbance difference quotient (EADQ- and EADQz-) diagrams* may be invoked in the analysis of systems involving metal complexes.

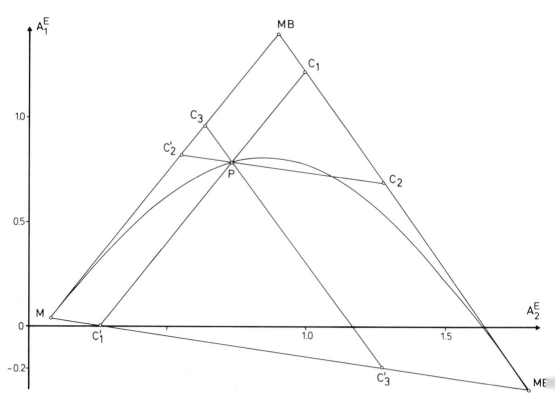

Fig. 13–13. Determination of the concentrations m, mb, and mb_2 from distance relationships analogous to those of equation 7–19 in Sec. 7.3.3.

13.4.3 The System Daphnetin/FeCl₃

A solution of the compound daphnetin (7,8-dihydroxycoumarin) in an acetate buffer at pH 4.00 forms metal complexes with $FeCl_3$, causing the solution to acquire a distinctive blue-green color (cf. ref. 91). It has already been established from Fig. 13–10 in Sec. 13.3 that under these conditions two daphnetin ligands ultimately become associated with each Fe^{3+} ion (i.e., $n = 2$). Thus, the corresponding titration system may evidently be taken as an example of reaction sequence 13–37.

In principle, an excess absorbance triangle like that shown in Fig. 13–12 should allow one to establish the two overall stability constants (β_1 and β_2) for this system. Unfortunately, the solubility of daphnetin is such that it is impossible to prepare a concentrated solution of the substance in acetate buffer at pH 4 like that required for a complete b_0 titration, so any titration contemplated must involve the addition of both daphnetin and $FeCl_3$. The equilibrium

$$M + B \overset{\beta_1}{\rightleftharpoons} MB \qquad (13\text{–}48a)$$

with

$$\beta_1 = \frac{mb}{m \cdot b} \qquad (13\text{–}48b)$$

may be analyzed separately over a relatively wide concentration range by introducing increasing amounts of daphnetin into a relatively concentrated ($3 \cdot 10^{-4}$ M) solution of $FeCl_3$ (i.e., by conducting a b_0 titration). Similarly, the equilibrium

$$M + 2B \overset{\beta_2}{\rightleftharpoons} MB_2 \qquad (13\text{–}49a)$$

with

$$\beta_2 = \frac{mb_2}{m \cdot b^2} \qquad (13\text{–}49b)$$

may be titrated separately by adding increasing amounts of $FeCl_3$ to a relatively concentrated ($1 \cdot 10^{-3}$ M) daphnetin solution (in an m_0 titration).

It is advantageous to employ EA-diagrams in trying to establish the desired constants β_1 and β_2. The procedure begins with an evaluation of the b_0 titration on the basis of equation 13–44, where excess absorbance is defined as the difference

$$A_\lambda^E = A_\lambda - l\,\varepsilon_{\lambda B} \cdot b_0 \qquad (13\text{–}44)$$

The quantity A_λ represents the absorbance of the complex solution at any stage in the proposed b_0 titration, while $l\,\varepsilon_{\lambda B} \cdot b_0$ is the absorbance that would be observed for the same solution in the absence of metal salt. If EA-diagrams are now constructed for various wavelength combinations they should be found to consist of straight lines—

provided 13–48a is the only equilibrium involved. Any deviations from linearity would imply the presence of other spectrometrically observable equilibria.

EA-diagrams also make it possible to determine the concentrations of the species M and MB once the titration limits have been characterized. The line shown in Fig. 13–14 corresponds to line $\overline{\text{M MB}}$ in Fig. 13–12. The following expressions specify the limiting points M and MB:

$$A^E_{\lambda M} = l\, \varepsilon_{\lambda M} \cdot m_0 \quad \text{and} \quad A^E_{\lambda MB} = l\,(\varepsilon_{\lambda MB} - \varepsilon_{\lambda B})\, m_0 \qquad (13\text{--}45)$$

Concentrations m and mb for any given titration point P may be established using the distance relationships (cf. Fig. 13–14)

$$m = \frac{\overline{\text{P MB}}}{\overline{\text{M MB}}} \cdot m_0 \quad \text{and} \quad mb = \frac{\overline{\text{M P}}}{\overline{\text{M MB}}} \cdot m_0 \qquad (13\text{--}50)$$

Once m and mb are known, the corresponding concentration of B may be calculated from the relationship $b = b_0 - mb$, thus providing all the information required for a determination of β_1 on the basis of equation 13–48b.

EA-Diagrams for the system daphnetin/FeCl$_3$ are shown in Fig. 13–15. The limiting points MB may be established from equation 13–45 assuming the appropriate values $\varepsilon_{\lambda MB}$ are known. These may be approximated by examining a solution that contains a large excess of FeCl$_3$ relative to daphnetin. Experimental data points deviate from the line $\overline{\text{M MB}}$ at high b_0 values, apparently because the second equilibrium MB + B $\leftrightarrows$ MB$_2$ begins to exert an influence under these conditions. Within

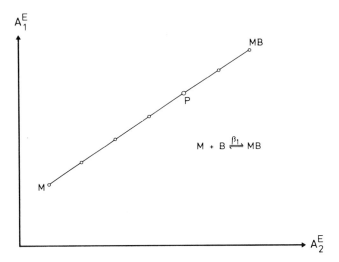

Fig. 13–14. Excess absorbance diagram for determining the concentrations m and mb resulting from the equilibrium M + B $\leftrightarrows$ MB (schematic representation). Excess absorbance is here defined as $A^E_\lambda = A_\lambda - l\, \varepsilon_{\lambda B} b_0$ (i.e., in terms of a b_0 titration).

13.4 The System M + B ⇌ MB, MB + B ⇌ MB$_2$

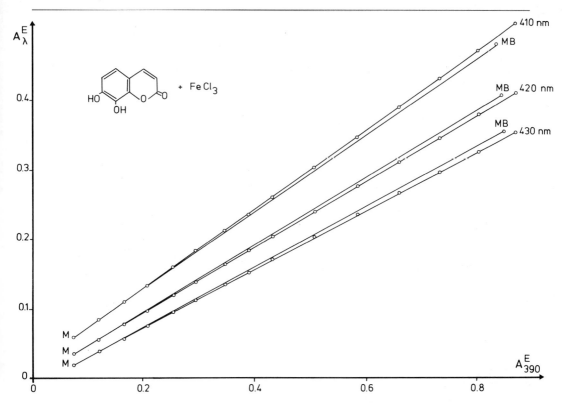

Fig. 13–15. Excess absorbance diagrams for the system daphnetin/FeCl$_3$ (b_0 titration). The concentrations m and mb at the titration points (which fall along the lines $\overline{\text{M MB}}$) may be determined using equation 13–50 and the appropriate limiting points M and MB.

the limits of experimental accuracy, points MB in Fig. 13–5 fall along the tangents to M, an observation consistent with the discussion in Sec. 13.4.2. Evaluation of the initial segments of Fig. 13–15 utilizing equation 13–50 provides the following result:[89,90]

$$\log \beta_1 \approx 4.3$$

EA-diagrams may also be used to evaluate the m_0 titration and thereby characterize the equilibrium M + 2 B ⇌ MB$_2$. In contrast to the preceding analysis involving equation 13–44, excess absorbance A_λ^E must this time be defined in terms of the difference

$$A_\lambda^E = A_\lambda - l \cdot \varepsilon_{\lambda M} \cdot m_0 \qquad (13\text{–}51)$$

The concentrations of B and MB$_2$ in an EA-diagram defined in this way are again associated with simple distance relationships. An EA-diagram for the general case

$$M + n B \underset{}{\overset{\beta_n}{\leftrightarrows}} MB_n \qquad (13\text{–}25)$$

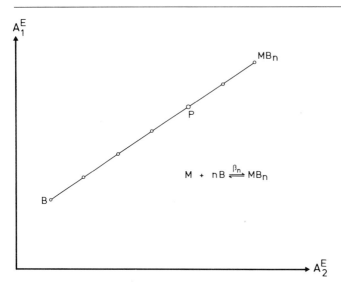

Fig. 13–16. Excess absorbance diagram for determining the concentrations b and mb_n associated with the equilibrium $M + nB \rightleftharpoons MB_n$ (schematic representation). Excess absorbance is here defined as $A_\lambda^E = A_\lambda - l\,\varepsilon_{\lambda M} m_0$ (i.e., in terms of an m_0 titration).

is depicted schematically in Fig. 13–16. The corresponding titration points P lie along the straight line joining the points B and MB_n, which in analogy with the conclusions presented as equation 13–45 are subject to the relationships

$$A_\lambda^E(B) = l\,\varepsilon_{\lambda B} \cdot b_0 \quad \text{and} \quad A_\lambda^E(MB_n) = l\left(\varepsilon_{\lambda MB_n} - \varepsilon_{\lambda M}\right)\frac{b_0}{n} \tag{13–52}$$

The concentrations of B and MB_n may be established geometrically with the aid of the further relationships

$$b = \frac{\overline{P\,MB_n}}{\overline{B\,MB_n}} \cdot b_0 \quad \text{and} \quad mb_n = \frac{\overline{B\,P}}{\overline{B\,MB_n}} \cdot \frac{b_0}{n} \tag{13–53}$$

Finally, m may be computed from the expression $m = m_0 - mb_n$, producing the last of the values required to determine β_n from equation 13–34 (or β_2 if $n = 2$, cf. equation 13–49b).

A set of EA-diagrams derived from an m_0 titration of daphnetin and $FeCl_3$ is depicted in Fig. 13–17. All titration points from the early stages of the titration are located along straight lines joining points B and MB_2 (MB_n, where $n = 2$). The points MB_2 may be established from equation 13–52 provided $\varepsilon_{\lambda MB_2}$ is known. This molar absorptivity may be approximated through measurements taken on solutions that contain a large excess of daphnetin relative to $FeCl_3$. Evaluation with the aid of equation 13–53 of the first stages of the curves in Fig. 13–17 leads to the result[89,90]

$$\log \beta_2 \approx 7.5$$

13.4 The System $M + B \rightleftarrows MB$, $MB + B \rightleftarrows MB_2$

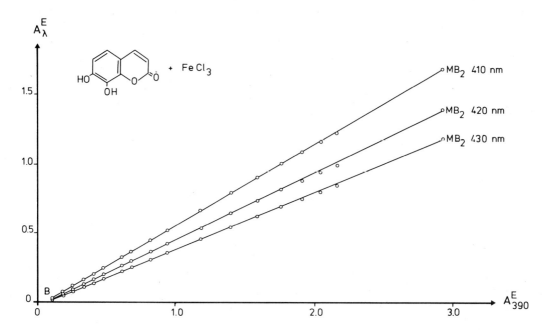

Fig. 13–17. Excess absorbance diagrams for the system daphnetin/FeCl$_3$ (m_0 titration). The concentrations b and mb_2 corresponding to the titration points (located along the lines $\overline{B\ MB_2}$) may be calculated from equation 13–53 with the aid of the appropriate limiting points B and MB$_2$.

In order to ascertain the degree of ionization of daphnetin at pH 4 the corresponding pK value must be known. It has been shown in this case that pK_1 = 7.37 (0.1 M KCl),[89,90] so daphnetin must remain essentially undissociated at pH 4. This means that the concentration b established on the basis of distance relationships must refer exclusively to the undissociated form of the substance.

13.5 Ion-Selective Electrodes in the Determination of Stability Constants

In all the titration systems so far described the only electrometric data utilized have been pH values. Such a limitation is by no means essential, however. Indeed, electrodes sensitive to the activities of individual ions can play an important role in the investigation of certain systems.

As an example, consider the generalized equilibrium

$$B + M \leftrightarrows MB \qquad (13\text{–}48a)$$

in which B represents a ligand and M is a metal ion. This system is to be examined with the aid of an electrode with specific sensitivity for the free, uncomplexed metal ion M. Unfortunately, such an ideal situation is not always presented in practice. Often it is found that other ions in the system cause interference. For example, the following expression is generally found to describe most accurately the potential E of a glass-membrane electrode selective for one of the monovalent cations (Li^+, Na^+, K^+, Rb^+, Cs^+, NH_4^+, NR_4^+, Ag^+, Tl^+):[14]

$$E = E' + \frac{RT}{F} \ln\left(a_M + \sum_d K_{M,d}\, a_d^{\frac{1}{z_d}} \right) \qquad (13\text{–}54)$$

where E' = electromotive force of the measuring circuit, determined in a standard solution
a_M = activity of the target ion
a_d = activity of a particular interfering ion
z_d = valence of the interfering ion
$K_{M,d}$ = selectivity coefficient for the target ion relative to the interfering ion
R = ideal gas constant (8.3144 J K^{-1} mol^{-1})
T = temperature expressed in Kelvin
F = Faraday's constant (96 496 A s/g-equivalent)

Such an electrode could be regarded as responding only to the target ion if the interference term

$$\sum_d K_{M,d}\, a_d^{\frac{1}{z_d}}$$

were negligibly small, an assumption we shall henceforth make.

If, in analogy to pH, one now defines a quantity pM:

$$pM := -\log a_M \qquad (13\text{–}55)$$

then equation 13–54 may be rewritten as follows:

13.5 Ion-Selective Electrodes

$$E = E' + \frac{RT}{F} \ln a_M = E' - \frac{2.303 \cdot RT}{F} pM \qquad (13\text{--}56)$$

Thus, the electrode potential derived from a sample solution should be directly proportional to pM, which means that an ion-selective electrode might provide access to the same type of information as a glass pH electrode. Therefore, it should be possible to examine and characterize the titration system

$$B + M \overset{k_1}{\rightleftharpoons} MB$$

in precisely the same way as a single-step acid–base equilibrium. The only difference is that equation 6–2 must be replaced by the expression

$$k_1 = \frac{mb}{b \cdot a_M} = \frac{mb}{b} \cdot 10^{+pM} \qquad (13\text{--}57)$$

permitting k_1 to be evaluated by the procedure used with equation 6–14b. In this case titration entails changing the amount of metal salt rather than altering the pH of the solution. Thus, given the conditions $\varepsilon_{\lambda M} = 0$, $A_{\lambda B} = lb_0 \varepsilon_{\lambda B}$, and $A_{\lambda MB} = lb_0 \varepsilon_{\lambda MB}$:

$$(A_\lambda - A_{\lambda B}) 10^{+pM} = -k_1 A_\lambda + k_1 A_{\lambda MB} \qquad (13\text{--}58)$$

An even more recent development is the "enzyme electrode", which is based on an appropriate immobilized enzyme or coenzyme that responds selectively to a specific type of substrate (ligand).[92-97] As a result of this latest advance a whole new dimension has been added to the field of chemical and especially biochemical analysis, one that is certain to open new avenues for spectrometric investigations. Electrodes in this class should make it possible to perform routine measurements of ligand concentrations just as pH electrodes are today used for determining hydronium ion activities. In the case of a b_0 titration, for example, an enzyme electrode might permit correlations based on EA-diagrams to be carried out just as easily as the analogous correlation of an A-diagram with pH following an acid–base titration.

13.6 Summary

After appropriate modification, the *pencil of rays method* that was developed in the context of acid–base equilibria may be applied to studies of metal-complex equilibria. The only restriction is that the system in question must involve a single well-defined complex (e.g., MB) of the metal ion with a specific protolyte from a single-step acid–base system. Titration points associated with a given pH for the protolyte system and the metal-complex solution produce a pencil of rays in the corresponding (two-dimensional) A-diagram, with the lines intersecting at the point P_0. From such lines it is possible to determine directly the molar absorptivity $\varepsilon_{\lambda MB}$. Concentrations of the various species, ascertained on the basis of an *absorbance triangle*, then provide the information necessary for establishing the desired *stability constant*, as has been demonstrated with the system 2-nitroso-1-naphthol-4-sulfonic acid/zinc sulfate.

This procedure has also been shown to be valid with respect to a multi-step protolysis system and a metal ion capable of simultaneously binding n ligands to produce the complex MB_n.

Alternatively, metal complex equilibria may be investigated spectrometrically, either by varying the concentration of metal ion (in an m_0 *titration*), or by varying the ligand concentration (in a b_0 *titration*). Using the *method of continuous variation* introduced by P. Job the sum of these two concentrations is held constant (i.e., $c_0 = m_0 + b_0$) throughout the analysis. It has also been demonstrated that the number n in the equilibrium $M + nB \leftrightarrows MB_n$ may be determined with the aid of the concept *excess absorbance*.

The relationships found within multi-dimensional A-diagrams for acid–base equilibria may be extended to include metal-complex equilibria, provided the latter are represented by *excess absorbance (EA-) diagrams*. The case ($M + B \leftrightarrows MB$; $MB + B \leftrightarrows MB_2$) has been used to illustrate that the EA-diagram obtained from a b_0 *titration* may be evaluated in the same way as the analogous A-diagram from the acid–base system ($B^{2-} + H_3O^+ \leftrightarrows BH^- + H_2O$; $BH^- + H_3O^+ \leftrightarrows BH_2 + H_2O$). The construction of appropriate tangents makes it possible to define an *excess absorbance triangle*, distance relationships within which provide the concentrations of M, MB, and MB_2. This information in turn permits stability constants to be ascertained, as demonstrated in the case daphnetin/$FeCl_3$.

Literature

1. C. J. Pedersen, *J. Amer. Chem. Soc.* 89 (1967) 2495.
2. C. J. Pedersen, *J. Org. Chem.* 36 (1971) 254.
3. E. Weber, F. Vögtle, *Topics Curr. Chem.* 98 (1981) 1.
4. F. Vögtle, E. Weber, *Angew. Chem.* 86 (1974) 896; 91 (1979) 813.
5. J.-M. Lehn, J. Simon, *Helv. Chim. Acta* 60 (1977) 141.
6. J.-M. Lehn, J. P. Sauvage, *J. Amer. Chem. Soc.* 97 (1975) 6700.
7. B. Dietrich, J.-M. Lehn, J. Simon, *Angew. Chem.* 86(1974) 443.
8. J.-M. Lehn, M. E. Stubbs, *J. Amer. Chem. Soc.* 96 (1974) 4011.
9. F. Vögtle, H. Sieger, *Angew. Chem.* 89(1977) 410.
10. B. Dietrich, J.-M. Lehn, *Tetrahedron Lett.* 1973, 1225.
11. C. L. Liotta, H. P. Harris, *J. Amer. Chem. Soc.* 96 (1974) 2250.
12. M. Cinquini, F. Montanari, P. Tundo, *J. C. S. Chem. Commun.* 1975, 393.
13. D. C. Cornish, *Chimia* 29 (1975) 398.
14. K. Cammann, *Das Arbeiten mit ionenselektiven Elektroden.* Springer-Verlag, Berlin-Heidelberg-New York 1973.
15. D. Ammann, E. Pretsch, W. Simon, *Tetrahedron Lett.* 1972, 2473.
16. P. Wuhrmann, A.P. Thoma, W. Simon, *Chimia* 27 (1973) 673.
17. R. B. Fischer, *J. Chem. Educ.* 51 (1974) 387.
18. N. N. L. Kirsch, W. Simon, *Helv. Chim. Acta* 59 (1976) 357.
19. D. Ammann, E. Pretsch, W. Simon, *Analyt. Lett.* 7 (1974) 23.
20. M. Guggi, M. Dehme, E. Pretsch, W. Simon, *Helv. Chim. Acta* 59 (1976) 2417.
21. N. N. L. Kirsch, R. J. J. Funck, E. Pretsch, W. Simon, *Helv. Chim. Acta* 60 (1977) 2326.
22. J. Bjerrum, *Metal Ammine Formation in Aqueous Solution.* Haase, Copenhagen 1941.
23. M. T. Beck, *Chemistry of Complex Equilibria.* Van Nostrand Reinhold Company, London-New York-Toronto-Melbourne 1970.
24. H. L. Schläfer, *Metallkomplexbildung in Lösung.* Heidelberg, Springer 1961.
25. F. J. C. Rosotti, H. S. Rossotti, *The Determination of Stability Constants.* McGraw Hill, New York 1961.
26. T. Kaden, A. D. Zuberbühler, *Talanta* 18 (1971) 61.
27. A. Sabatini, A. Vacca, *J. Chem. Soc. Dalton* 1972, 1693.
28. G. A. Cumme, *Talanta* 20 (1973) 1009.
29. A. Sabatini, A. Vacca, P. Gans, *Talanta* 21 (1974) 53.
30. G. Ginzburg, *Talanta* 23 (1976) 149.
31. R. Karlsson, L. Kullberg, *Chem. Scr.* 9 (1976) 54.
32. A. M. Corrie, G. K. R. Makar, M. L. D. Touche, D. R. Williams, *J. Chem. Soc. Dalton* 1975, 105.
33. D. J. Legett, *Talanta* 24 (1977) 535.
34. M. Wozniak, G. Nowogrocki, *Talanta* 25 (1978) 643.
35. H. Gampp, M. Maeder, A. D. Zuberbühler, T. Kaden, *Talanta* 27 (1980) 513.
36. E. S. Johansen, O. Jons, *Acta Chem. Scand.* A 35 (1981) 233.
37. R. J. Motekaitis, A. E. Martell, *Can. J. Chem.* 60 (1982) 2403.
38. M. Meloun, J. Havel, *Computation of Solution Equilibria.* 1.Spectrometry. Folia Fac. Sci. Nat., Univ. Purkynianae Brunensis, Brno, 1983.
39. J. Havel, M. Meloun, *Talanta* 32 (1985) 171.

40. D. J. Legett, *Computational Methods for the Determination of Formation Constants.* Plenum Press, New York-London 1985.
41. H. Gampp, M. Maeder, C. J. Meyer, A. D. Zuberbühler, *Talanta* 32 (1985) 257.
42. A. Laouenan, E. Suet, *Talanta* 32 (1985) 245.
43. S. Ahrland, P. Bläuenstein, B. Tagesson, D. Tuhtar, *Acta Chem. Scand.* A 34 (1980) 265.
44. S. Ahrland, I. Persson, R. Portanova, *Acta Chem. Scand.* A 35 (1981) 49.
45. S. Ahrland, N.-O. Bjord, I. Persson, *Acta Chem. Scand.* A 35 (1981) 67.
46. B. Holmberg, G. Thome, *Acta Chem. Scand.* A 34 (1980) 421.
47. I. Granberg, S. Sjöberg, *Acta Chem. Scand.* 35 (1981) 193.
48. G. Anderegg, *Helv. Chim. Acta* 64 (1981) 1790.
49. W. Forsling, I. Granberg, S. Sjöberg, *Acta Chem. Scand.* A 35 (1981) 473.
50. L.-O. Öhmann, S. Sjöberg, *Acta Chem. Scand.* A 35 (1981) 201.
51. L.-O. Öhmann, S. Sjöberg, *Acta Chem. Scand.* A 36 (1982) 47.
52. I. Granberg, W. Forsling, S. Sjöberg, *Acta Chem. Scand.* A 36 (1982) 819.
53. Z. Szeverenyi, U. Knopp, A. D. Zuberbühler, *Helv. Chim. Acta* 65 (1982) 2529.
54. Y. Wu, Th. Kaden, *Helv. Chim. Acta* 66 (1983) 1588.
55. M. Wilgocki, J. Bjerrum, *Acta Chem. Scand.* A 37 (1983) 307.
56. H. Saarinen, M. Orama, T. Raikas, J. Korvenranta, *Acta Chem. Scand.* A 37 (1983) 631.
57. R. J. Motekaitis, A. E. Martell, J.-P. Lecomte, J.-M. Lehn, *Inorg. Chem.* 22 (1983) 609.
58. I. Yoshida, R. J. Motekaitis, A. E. Martell, *Inorg. Chem.* 22 (1983) 2795.
59. L.-O. Öhmann, S. Sjöberg, N. Ingri, *Acta Chem. Scand.* A 37 (1983) 561.
60. L.-O. Öhmann, S. Sjöberg, *Acta Chem. Scand.* A 37 (1983) 875.
61. J. Granberg, S. Sjöberg, *Acta Chem. Scand.* A 37 (1983) 415.
62. A. W. Hamburg, M. T. Nemeth, D. W. Margerum, *Inorg. Chem.* 22 (1983) 3535.
63. S. Ishigura, Y. Oka, H. Ohtaki, *Bull. Chem. Soc. Jpn.* 57 (1984).
64. S. Sjöberg, L.-O. Öhmann, N. Ingri, *Acta Chem. Scand.* A 39 (1985) 93.
65. S. Ahrland, S.-I. Ishiguro, A. Marton, I. Persson, *Acta Chem. Scand.* A 39 (1985) 227.
66. W. A. E. McBryde, *Talanta* 21 (1974) 979.
67. T. Kaden, A. D. Zuberbühler, *Talanta* 18 (1971) 61.
68. D. J. Leggett, W. A. E. McBryde, *Anal. Chem.* 47 (1975) 1065.
69. M. Meloun, J. Cernak, *Talanta* 23 (1976) 15.
70. H. Gampp, M. Maeder, A. D. Zuberbühler, *Talanta* 27 (1980) 1037.
71. F. Gaizer, A. Puskas, *Talanta* 28 (1981) 925.
72. M. Meloun, J. Cernak, *Talanta* 31 (1984) 947.
73. M. Meloun, M. Javurek, *Talanta* 31 (1984) 1083.
74. M. Meloun, M. Javurek, *Talanta* 32 (1985) 973.
75. M. Meloun, M. Javurek, *Talanta* 32 (1985) 1011.
76. H. Gampp, M. Maeder, C. J. Meyer, A. D. Zuberbühler, *Talanta* 32 (1985) 95.
77. H. Gampp, M. Maeder, C. J. Meyer, A. D. Zuberbühler, *Talanta* 32 (1985) 257.
78. H. Gampp, M. Maeder, C. J. Meyer, A. D. Zuberbühler, *Talanta* 32 (1985) 1133.
79. J. Havel, M. Meloun, *Talanta* 33 (1986) 435.
80. M. Meloun, M. Javurek, J. Havel, *Talanta* 33 (1986) 513.
81. M. Briellman, A. D. Zuberbühler, *Helv. Chim. Acta* 65 (1982) 46
82. P. Job, *Ann. Chim. Paris* 9 (1928) 113.
83. R. Winkler-Oswatitsch, M. Eigen, *Angew. Chem.* 91 (1979) 20.

Literature

84. J. Polster, *Habilitationsschrift*. Tübingen 1980.
85. J. Polster, *Z. Naturforsch.* 31 b (1976) 1621.
86. O. Mäkitio, H. Saarinen, *Suomen Kemistilehti* B 44 (1971) 209.
87. H. Saarinen, O. Mäkitie, *Acta Chem. Scand.* 24 (1970) 2877.
88. J. Dreszer, *Mathematik Handbuch für Technik und Naturwissenschaft*. Verlag Harri Deutsch, Zürich-Frankfurt/Main-Thun 1975.
89. M. Schwenk, *Diplomarbeit*. Technische Universität München, Freising-Weihenstephan 1985.
90. J. Polster, M. Schwenk, *Z. Physik. Chem. N. F.* 150 (1986) 87.
91. Roth, Daunderer, Kormann, *Giftpflanzen-Pilzgifte*. ecomed Verlagsgesellschaft, Landsberg, München 1984.
92. P. Kirch, J. Danzer, G. Krisam, H.-L. Schmidt, *Proceedings of the 4th Enzyme Engineering Conference*, Bad Neuenahr 1977.
93. H.-L. Schmidt, G. Grenner, *Eur. J. Biochem.* 67 (1976) 295.
94. B. Dolabdjian, *Dissertation*. Technische Universität München, Freising-Weihenstephan 1978.
95. P. Davies, K. Mosbach, *Biochem. Biophys. Acta* 370 (1974) 329.
96. M. Kessler, L. C. Clark Jr., D. W. Lübbers, I. A. Silver, W. Simon, *Ion and Enzyme Electrodes in Biology and Medicine*. Urban und Schwarzenberg, München 1976.
97. J. G. Schindler, M. M. Schindler, *Bioelektronische Membranelektroden*. Walter de Gruyter, Berlin, New York 1983.

14 Association Equilibria

14.1 Establishing Association Constants for Dimerism Equilibria Related to Protolysis Systems

Every *association* equilibrium might just as logically be described as a *dissociation* equilibrium:

$$B + C \underset{\text{dissociation}}{\overset{\text{association}}{\rightleftharpoons}} D$$

It has long been known that many of the ions constituting pigments show a tendency to undergo self-association, and that the amount of light absorbed by such substances is not proportional to their concentration even in highly dilute solution (cf. the self-association of acridine orange).[1–16] Association phenomena tend to be more prevalent the larger the dye molecule and the less effectively its surface is shielded by hydrophilic (hydrated) groups.[17,18]

Molecular association involving unlike molecules is also common, as in the complex formed between naphthalene and 1,3,5-trinitrobenzene.[19–25] Even small ions like the halides are found to participate in such interactions (again, for example, with 1,3,5-trinitrobenzene),[26] which have acquired the general designation *electron donor–acceptor complexes*. Several procedures have been developed for investigating these complexes.[19,13,27] The best known are that of Benesi and Hildebrand[28,29] and an alternative method involving what are called "Scatchard plots".[30] Both methods have been subjected to various modifications and further development by a number of research groups.[31–34]

Molecular associations from the realm of biochemistry have been the subject of particular attention in recent years, in part due to the wide range of substances with which biopolymers have been found to associate. It has been shown, for example, that aminophenanthridine dyes such as ethidium bromide may bind either competitively or non-competitively (through intercalation) with DNA.[34–37] Attached dye molecules may subsequently be partially displaced from these complexes by addition of neutral salts such as NaCl, KCl, or $MgCl_2$. Since most dye molecules are also protolytes, it is not surprising that their tendency to undergo association is influenced by the acidity of the medium, and that acid–base equilibria must be taken into account in the course of a proper investigation.[38,39]

We begin our consideration of association by examining a case in which two identical dyestuff ions B^- combine to give the dimer C^{2-}, with B^- again arising when the dimer dissociates. Such a system may be described by the following pair of equilibria:

$$BH \overset{K_1}{\leftrightarrows} B^- \, (H^+) \tag{14-1}$$

$$2\,B^- \overset{\beta}{\leftrightarrows} C^{2-} \tag{14-2}$$

where $K_1 = \dfrac{b \cdot h}{bh}$ and $\beta = \dfrac{c}{b^2}$

In order to determine the constants K_1 and β it is necessary that the equilibria be subjected to controlled displacement. If one wishes to avoid the dynamic methods mentioned briefly in Sec. 13.1.1 the only alternatives are to vary either the pH (the possibility we discuss first) or the concentration of the dye (cf. Sec. 14.2). If pH is chosen to be the variable, as in an acid–base titration, then the following relationship will be applicable at each titration step (where b_0 = initial molar concentration):

$$b_0 = bh + b + 2c \tag{14-3}$$

Equations 14-2 and 14-3 permit the concentrations of all components to be expressed as a function of pH, and this in turn makes it possible to construct the characteristic concentration diagram bh vs. c (cf. Fig. 14-1). In the absence of dimerization (i.e., $\beta = 0$) the only consequence of increasing the basicity is conversion of BH to B^-. All the titration points (P_1) in the diagram bh vs. c would then fall along the bh axis. If dimer formation occurs, then the diagram consists of a curve whose nature is dependent on the magnitude of β (cf. the point P in Fig. 14-1). Regardless of the shape of the curve, the bh axis remains tangent to the point BH.

If in the diagram bh vs. c pairs of points P_1 and P that share a common pH are joined by straight lines, all the resulting lines must intersect at a common point (P_0). The truth of this assertion becomes apparent upon combining equations 14-2 and 14-3:

$$bh = \frac{h}{K_1 + h}(b_0 - 2c) \tag{14-4}$$

In other words, given constant values of K_1 and h, any set of concentrations bh and c will produce a point somewhere along a single straight line regardless of the value of β, and that line will intersect the abscissa at a specific point P_0 where $c = b_0/2$ (cf. Fig. 14-1). Therefore, allowing h to vary results in a pencil of rays emanating from the point P_0, which is subject to the conditions:

$$bh(P_0) = b(P_0) = 0 \quad \text{and} \quad c(P_0) = \frac{b_0}{2} \tag{14-5}$$

14.1 Establishing Association Constants for Dimerism Equilibria

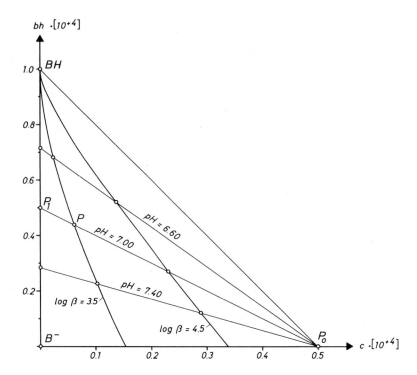

Fig. 14–1. Simulated characteristic concentration diagrams bh vs. c for the system shown as equation 14–1, constructed as a function of the parameter β (log β = 3.5 and 4.5; $b_0 = 10^{-4}$ M; $pK_1 = 7.0$.) The point P represents a typical titration point. In the event that no dimer is formed (i.e., if $\beta = 0$) then the titration is equivalent to that of a single-step protolyte (cf. the point P_1).

The same relationships are also applicable within an *A-diagram*. Thus, by developing the equation

$$A_{\lambda_i} = l(\varepsilon_{\lambda_i BH}\, bh + \varepsilon_{\lambda_i B}\, b + \varepsilon_{\lambda_i C}\, c) \tag{14–6}$$

for the wavelengths λ_i ($i = 1, 2$) and imposing the conditions laid down by equation 14–3, one obtains the following result:

$$\begin{pmatrix} A_1 \\ A_2 \end{pmatrix} = \begin{pmatrix} q_{11} & q_{12} \\ q_{21} & q_{22} \end{pmatrix} \begin{pmatrix} bh \\ c \end{pmatrix} + lb_0 \begin{pmatrix} \varepsilon_{1B} \\ \varepsilon_{2B} \end{pmatrix} \tag{14–7}$$

where $q_{\lambda_i 1} = l(\varepsilon_{\lambda_i BH} - \varepsilon_{\lambda_i B})$ and $q_{\lambda_i 2} = l(\varepsilon_{\lambda_i C} - 2\varepsilon_{\lambda_i B})$

This relationship affirms the fact that an affine transformation may be utilized to convert the characteristic concentration diagram into the A-diagram A_1 vs. A_2 (cf. Sec. 7.6).

Evaluation of this titration system with the aid of an A-diagram entails construction of a pencil of rays analogous to that joining the points P_1 and P in Fig. 14–1, but doing so requires that one know the locations of the appropriate points. A direct determination of the points P_1 is possible only if the value of the association constant β is relatively small, because only then will dilution suppress the dimerization phenomenon to an extent sufficient to permit separate titration and analysis of the acid–base equilibrium. On the other hand, if one knows in advance the magnitude of pK_1 together with the molar absorptivities $\varepsilon_{\lambda_i\mathrm{BH}}$ and $\varepsilon_{\lambda_i\mathrm{B}}$, then the points P_1 can be calculated directly from measurements taken on concentrated solutions, thus defining the desired pencil of rays. From a triangle corresponding to that shown in Fig. 14–1 (with vertices BH, B$^-$, and P_0) it is then possible to determine by methods previously discussed the concentrations of all components. This in turn provides the information necessary for calculating the desired quantity β (cf. Sec. 7.3.3).

The above-described methods are also applicable in cases where only protonated species (BH) in the titration system form dimers; i.e.:

$$2\,\mathrm{BH} \leftrightarrows \mathrm{C}$$
$$\mathrm{BH} \leftrightarrows \mathrm{B}^- \;(\mathrm{H}^+) \tag{14–8}$$

Similarly, the procedure may be extended to cover three-dimensional A-diagrams, as required for the following situation (cf. Sec. 8.3.4):

$$\mathrm{BH}_2 \leftrightarrows \mathrm{BH}^- \;(\mathrm{H}^+)$$
$$\mathrm{BH}^- \leftrightarrows \mathrm{B}^{2-} \;(\mathrm{H}^+)$$
$$2\,\mathrm{B}^{2-} \leftrightarrows \mathrm{C}^{4-}$$

Thus, we may assert in general that the *pencil of rays methods* developed in the context of protolysis and metal-complex systems is also appropriate for the analysis of association equilibria so long as these involve protolytes.[40]

14.2 Determining Association Constants for the System $2\,\mathrm{B} \leftrightarrows \mathrm{C}$ by Successively Increasing the Dye Concentration

14.2.1 Absorbance Diagrams

If in the case of a titration system like 14–1 or 14–8 one were to select a pH at which the protolyte B is present almost exclusively in the form BH or B$^-$, then the only spectrometrically active equilibrium would be that for the dimerization itself:

14.2 Determining Association Constants for the System $2B \leftrightarrows C$

$$2B \leftrightarrows C \tag{14-9}$$

This equilibrium (from which charges have been omitted) must be distinguished from the related case $B \leftrightarrows C$, which is of particular importance in the context of tautomerism (cf. Sec. 12.1) as well as in thermo- and photochromism.[41–47] The equilibrium $2B \leftrightarrows C$ is conveniently investigated by successively increasing the total concentration b_0 of dye present in the titration solution (i.e., by conducting a b_0 titration). Throughout such a process it follows that

$$b_0 = b + 2c \tag{14-10}$$

where the total amount (b_0) of dye present at any given point is assumed to be known. The corresponding association constant may be described by the expression

$$\beta = \frac{c}{b^2} \tag{14-11}$$

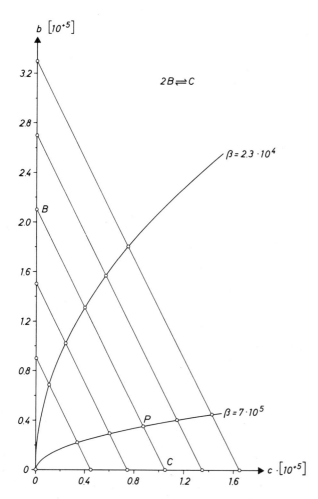

Fig. 14–2. Characteristic concentration diagrams for the system $2B \leftrightarrows C$ ($\beta = 2.3 \cdot 10^4$ M^{-1} and $7 \cdot 10^5$ M^{-1}). Point P represents one of the titration points. The variable in this case is the quantity b_0 ($0 < b_0 < 6 \cdot 10^{-5}$ M). At the point B, $\beta = 0$ ($c = 0$), while at the point C, $\beta \to \infty$ ($b = 0$).

Using equations 14–10 and 14–11 it is possible to calculate concentrations b and c at each step in the titration, permitting construction of the diagram b vs. c (cf. Fig. 14–2). If b_0 is constant, then regardless of the value of β it follows directly from equation 14–10 that all concentrations must fall along a single straight line (cf. point P in Fig. 14–2). If b_0 is permitted to change, all the resulting lines must have the same slope. In the case $\beta = 0$ only the species B is present, while in the limiting case $\beta \to \infty$ the only species remaining is C (cf. the points B and C in Fig. 14–2).

The concentrations b and c may be regarded here as independent variables, because b_0 is altered at each step in the titration. Therefore (cf. Sec. 7.6), a corresponding A-diagram must be an affine distortion of the characteristic concentration diagram b vs. c due to the relationship

$$A_\lambda \;=\; l(\varepsilon_{\lambda B} \cdot b \;+\; \varepsilon_{\lambda C} \cdot c) \tag{14–12}$$

It follows that all relationships established with respect to the concentration diagram are equally valid for the A-diagram (cf. Fig. 14–3).

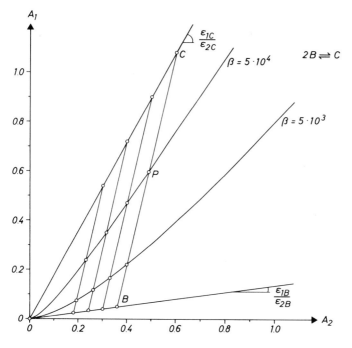

Fig. 14–3. A-diagrams for the system $2\,\text{B} \leftrightarrows \text{C}$ ($\beta = 5 \cdot 10^3\;\text{M}^{-1}$ and $5 \cdot 10^4\;\text{M}^{-1}$), where P is one of the titration points. Molar absorptivities λ_i ($i = 1, 2$): $\varepsilon_{\lambda_i B} = 2\,000$ and $15\,000$; $\varepsilon_{\lambda_i C} = 90\,000$ and $50\,000$ (units: $\text{cm}^{-1}\,\text{M}^{-1}$). Titration here entails varying the quantity b_0 ($0 < b_0 < 6 \cdot 10^{-5}\;\text{M}$). At the point B, $\beta = 0$, while at the point C, $\beta \to \infty$.

14.2 Determining Association Constants for the System 2B ⇆ C

The molar absorptivities ε_{1B} and ε_{2B} can be determined directly only if dimerization can be suppressed almost completely by extensive dilution of the dye mixture. This in turn requires that β have a relatively low value. Similarly, establishing molar absorptivities ε_{1C} and ε_{2C} for the dimer requires that approximations be made based on measurements with concentrated solutions, and this in turn presupposes the absence of higher-order complexes (i.e., trimers, etc.).

As will be demonstrated below, non-linear regression techniques make it possible to determine the quotients $\varepsilon_{1B}/\varepsilon_{2B}$ and $\varepsilon_{1C}/\varepsilon_{2C}$ on the basis of A-diagrams alone.[40,48] These quotients provide the slopes of reference lines where absorbances are due only to the "pure" monomer and dimer, respectively (cf. the points B and C in Fig. 14–3). If the molar absorptivity of one of the components (B or C) is known, together with the quantity b_0 applicable to each of the data points, then the location of the point B (or C) corresponding to each P may be calculated, permitting construction of a set of parallel lines $\overline{BP}$ (or $\overline{PC}$). These lines need only be extended to their points of intersection with the reference line for C (or B) in order to ascertain unambiguously the locations of the missing points C (B).

The quotients $\varepsilon_{1B}/\varepsilon_{2B}$ and $\varepsilon_{1C}/\varepsilon_{2C}$ may be established from the coefficients of the function describing the curve in an A-diagram. This may be demonstrated by developing equation 14–12 for the pair of wavelengths λ_i ($i = 1, 2$) and eliminating from the resulting system of equations the concentrations b and c (taking into account the relationship shown as equation 14–11). This leads to the following expression:

$$A_1^2 + \alpha_1 A_2^2 + \alpha_2 A_1 A_2 + \alpha_3 A_1 + \alpha_4 A_2 = 0 \tag{14–13}$$

where $\alpha_1 = \left(\dfrac{\varepsilon_{1C}}{\varepsilon_{2C}}\right)^2$, $\alpha_2 = -2\left(\dfrac{\varepsilon_{1C}}{\varepsilon_{2C}}\right)$,

$$\alpha_3 = -\dfrac{1}{\beta}\left(\dfrac{\varepsilon_{2B}}{\varepsilon_{2C}^2} \cdot l\right)(\varepsilon_{2B} \cdot \varepsilon_{1C} - \varepsilon_{1B} \cdot \varepsilon_{2C}), \text{ and}$$

$$\alpha_4 = \dfrac{\varepsilon_{1B} \cdot l}{\varepsilon_{2C}^2 \cdot \beta}(\varepsilon_{2B} \cdot \varepsilon_{1C} - \varepsilon_{1B} \cdot \varepsilon_{2C})$$

A comparison of the coefficients shows that

$$\alpha_1 = \dfrac{\alpha_2^2}{4} \quad \text{and} \tag{14–14a}$$

$$\dfrac{\alpha_3}{\alpha_4} = -\dfrac{\varepsilon_{2B}}{\varepsilon_{1B}} \tag{14–14b}$$

Equation 14–14a may thus be used to transform equation 14–13 into

$$\left(A_1 + \frac{\alpha_2}{2} A_2\right)^2 + \alpha_3 A_1 + \alpha_4 A_2 = 0 \tag{14-15}$$

Equations 14–13 and 14–15 provide functional representations of the absorbance curves in an A-diagram, and the associated coefficients may be readily evaluated with the help of non-linear regression techniques (cf. Sec. 7.4.2). Thus, the quotients $\varepsilon_{1B}/\varepsilon_{2B}$ and $\varepsilon_{1C}/\varepsilon_{2C}$ required to construct the points B and C are accessible directly from α_2 and α_3/α_4.

The concentrations b and c may be obtained graphically using simple distance relationships provided one knows the locations of the points P, B, and C in Fig. 14–3. This requires visualizing lines through point P in Fig. 14–2 that lie parallel to the coordinate axes b and c of the concentration diagram. One of these lines intersects the b-axis at B', the other intersects the c-axis at C' (intercepts that are not explicitly shown in Fig. 14–2). The following relationships apply to the concentrations b and c, where "O" represents the coordinate origin:

$$b = \frac{\overline{B'O}}{\overline{BO}} \cdot b_0 \quad \text{and} \quad c = \frac{\overline{C'O}}{\overline{CO}} \cdot \frac{b_0}{2}$$

Using the geometric relationship

$$\frac{\overline{B'O}}{\overline{BO}} = \frac{\overline{PC}}{\overline{BC}} \quad \text{and} \quad \frac{\overline{C'O}}{\overline{CO}} = \frac{\overline{BP}}{\overline{BC}}$$

one obtains finally the expressions

$$b = \frac{\overline{PC}}{\overline{BC}} \cdot b_0 \quad \text{and} \quad c = \frac{\overline{BP}}{\overline{BC}} \cdot \frac{b_0}{2} \tag{14-16}$$

Since these distance relationships are apparently preserved during an affine transformation (cf. Sec. 7.6), equation 14–16 is also applicable to an A-diagram (cf. Fig. 14–3), permitting a value of β to be derived from each titration point.

14.2.2 A Practical Example Based on the Eosine System

Dimerization equilibria such as $2\,B \leftrightarrows C$ are generally examined with the aid of concentration-dependent changes in UV–VIS absorption spectra. Most evaluations of the corresponding data rely on non-linear regression procedures.[49–51] Despite the considerable computational effort that has been expended in carrying out such analyses, the association constants and molar absorptivities reported by various research groups often differ markedly.[49–53] This observation underscores the importance of utilizing as many techniques as possible when attempting to convert spectrometric measurements into equilibrium constants.[53]

14.2 Determining Association Constants for the System 2B ⇆ C

Figure 14–4 shows spectra of the compound eosine recorded at high concentration ($2.5 \cdot 10^{-3}$–$1 \cdot 10^{-2}$ M) in 0.1 N NaOH using a cuvette with a pathlength $l = 0.001$ cm. As the eosine concentration increases (i.e., during a b_0 titration) a shoulder in the region 480–510 nm becomes increasingly apparent alongside the absorbance maximum at 520 nm attributed essentially to the monomer. This shoulder represents an absorbance maximum for the eosine dimer.[12] An A-diagram for eosine prepared using various initial concentrations b_0 is depicted in Fig. 14–5. The line joining the data points P is distinctly curved. Points labeled B in this figure indicate the absorbances one would expect if points P were completely free of dimer. The locations of these points were ascertained by performing calculations on the basis of molar absorp-tivities determined at low b_0 values ($b_0 = 1 \cdot 10^{-5}$ M), where it is safe to assume that $c = 0$. All the simulated lines $\overline{BP}$ in the region of high b_0 are essentially parallel, confirming the predictions of Sec. 14.2.1 (cf. Fig. 14–3).

The following equation served as the starting point for establishing coefficients for the experimental curve (cf. equation 14–15):

$$A_1 = \frac{\varepsilon_{1B}}{\varepsilon_{2B}} A_2 - \frac{1}{\alpha_3}\left(A_1 - \frac{\varepsilon_{1C}}{\varepsilon_{2C}} A_2\right)^2$$

The coefficients themselves resulted from a non-linear regression analysis using the SPSS–8 statistical package.[54] The molar absorptivities ε_{1B} and ε_{2B} were introduced as known quantities; only α_3 and the quotient $\varepsilon_{1C}/\varepsilon_{2C}$ were the subject of approximation.

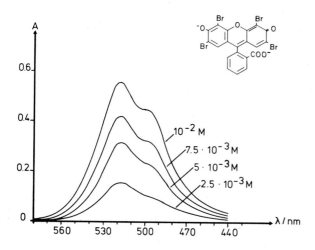

Fig. 14–4. Absorption spectra of eosine determined in 0.1 N NaOH ($l = 0.001$ cm, room temperature, concentration range $2.5 \cdot 10^{-3}$ M–$1 \cdot 10^{-2}$ M).

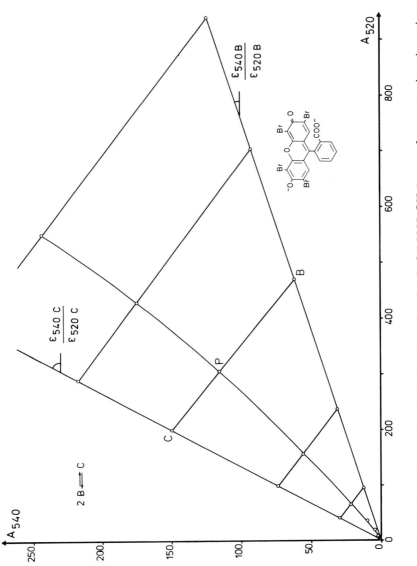

Fig. 14-5. A-diagram for the b_0 titration of eosine in 0.1 N NaOH (range of concentrations investigated: $1 \cdot 10^{-5}$ M–$1 \cdot 10^{-2}$ M; cuvette pathlengths: 1, 0.1, 0.01, 0.001 cm; room temperature).

14.2 Determining Association Constants for the System $2B \leftrightarrows C$

Non-linear regression analysis makes it possible to introduce into the A-diagram a reference line with the slope $\varepsilon_{1C}/\varepsilon_{2C}$ (cf. Fig. 14–5). Its points of intersection with the lines $\overline{BP}$ provide the locations of the points C, from which b and c can be calculated with the aid of equation 14–16. Finally, a value for β was computed using equation 14–11 (as a mean of several individual results):[48]

$$\beta \approx 400 \text{ L/mol}$$

An identical value has been reported by Th. Förster and E. König.[12]

14.2.3 Excess Absorbance and Reduced Absorbance Diagrams

It has been demonstrated in Sections 14.2.1 and 14.2.2 that the experimentally determined A-diagram for any system

$$2B \leftrightarrows C \tag{14–9}$$

traces a curved path as the substrate concentration increases (i.e., in a b_0 titration) assuming both components absorb. However, it is not immediately apparent from such a diagram whether or not the system in question is actually restricted to a single equilibrium. Fortunately, an unambiguous answer to the question may be obtained using the "excess absorbance" concept, or with the aid of a "reduced absorbance diagram".

In analogy to the way the term was defined previously (in the context of the Job method; cf. Sec. 13.3), "excess absorbance" A_λ^E refers here to the difference

$$A_\lambda^E = A_\lambda - l\, b_0\, \varepsilon_{\lambda B} \tag{14–17}$$

an important assumption being that $\varepsilon_{\lambda B}$ is a known quantity. Equations 14–10 and 14–12 provide an alternative formulation of A_λ^E applicable to the present case:

$$A_\lambda^E = l(\varepsilon_{\lambda C} - 2\varepsilon_{\lambda B})\, c \tag{14–18}$$

Developing the latter equation for the pair of wavelengths λ_i ($i = 1, 2$), and dividing one of the resulting expressions by the other, leads to the conclusion that

$$A_1^E = \frac{\varepsilon_{1C} - 2\,\varepsilon_{1B}}{\varepsilon_{2C} - 2\,\varepsilon_{2B}}\, A_2^E \tag{14–19}$$

Thus, in an *excess absorbance (EA-) diagram* A_1^E vs. A_2^E, all the data points that result from titrating the system $2B \leftrightarrows C$ must fall along a single reference line. The actual diagram obtained with our example eosine is illustrated in Fig. 14–6 (cf. also Fig. 14–5).

A knowledge of the quantity $\varepsilon_{\lambda B}$ is prerequisite to the construction of an EA-diagram, but this restriction may be avoided by working with *reduced quantities*. For example, dividing both sides of equation 14–10 by b_0 and then rearranging terms reveals that

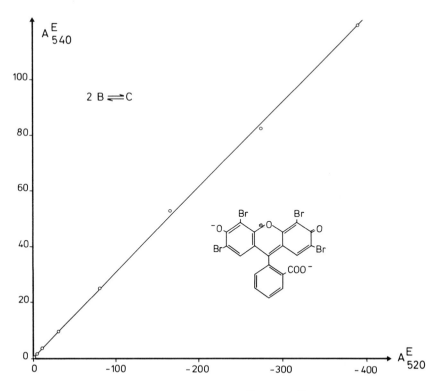

Fig. 14–6. Excess absorbance diagram A_{540}^E vs. A_{520}^E for eosine in 0.1 N NaOH. A_λ^E is governed by the relationship $A_\lambda^E = A_\lambda - lb_0\varepsilon_{\lambda B}$.

$$\frac{b}{b_0} = 1 - 2\frac{c}{b_0} \tag{14–20}$$

According to equation 14–10, b and c are linearly independent quantities (because b_0 is a variable). By contrast, the "reduced concentrations" b/b_0 and c/b_0 are linearly dependent upon each other. A similar relationship applies to the corresponding absorbances. Dividing equation 14–12 by b_0 and taking into account the relationship established by equation 14–20 provides the following expression:

$$\frac{A_\lambda}{l \cdot b_0} = \varepsilon_{\lambda B} + (\varepsilon_{\lambda C} - 2\varepsilon_{\lambda B})\frac{c}{b_0} \tag{14–21}$$

If this is elaborated in terms of the pair of wavelengths λ_i ($i = 1, 2$), and if c/b_0 is subsequently eliminated, what emerges is

$$\frac{A_1}{l \cdot b_0} = \gamma_1 \frac{A_2}{l \cdot b_0} + \gamma_2 \tag{14–22}$$

14.2 Determining Association Constants for the System $2B \leftrightarrows C$

where $\gamma_1 = \dfrac{\varepsilon_{1C} - 2\varepsilon_{1B}}{\varepsilon_{2C} - 2\varepsilon_{2B}}$ and $\gamma_2 = \dfrac{\varepsilon_{2B}\varepsilon_{1C} - \varepsilon_{1B}\varepsilon_{2C}}{2\varepsilon_{2B} - \varepsilon_{2C}}$

The same equation may also be obtained directly by rearranging terms in equation 14–19. It will be noted that the slopes predicted by the two equations are identical.

A *reduced absorbance diagram* A_1/lb_0 vs. A_2/lb_0 may be used just like an EA-diagram to establish graphically whether or not a titration system subjected to a b_0 titration (where the titrant is spectrometrically observable) consists only of a single equilibrium (e.g., $2B \leftrightarrows C$). Such reduced absorbance diagrams were first introduced by H. Mauser to assist in the kinetic analysis of chemical reactions.[55]

In order to determine the points corresponding to the pure monomer ($b_0 \to 0$) and the pure dimer ($b_0 \to \infty$) in a reduced absorbance diagram it is necessary to extrapolate the reduced concentrations b/b_0 and c/b_0 to their limiting values. Equations 14–10 and 14–11 provide the following expression for the quotient b/b_0:

$$\frac{b}{b_0} = \frac{-1 + \sqrt{1 + 8\beta b_0}}{4\beta b_0} \tag{14–23}$$

The Bernoulli–De L'Hôpital method[56,57] gives a value for the limit $b_0 \to 0$:

$$\left.\frac{b}{b_0}\right|_{b_0 \to 0} = 1 \tag{14–24}$$

Similarly, for the limit $b_0 \to \infty$:

$$\left.\frac{b}{b_0}\right|_{b_0 \to \infty} = 0 \tag{14–25}$$

Introducing these results into equation 14–20 shows that

$$\left.\frac{c}{b_0}\right|_{b_0 \to 0} = 0 \quad \text{and} \quad \left.\frac{c}{b_0}\right|_{b_0 \to \infty} = \frac{1}{2} \tag{14–26}$$

Finally, from equation 14–21 it follows that the reduced absorbances of the monomer and dimer must be

$$\left.\frac{A_\lambda}{l \cdot b_0}\right|_{b_0 \to \infty} = \varepsilon_{\lambda B} \quad \text{or} \quad \left.\frac{A_\lambda}{l \cdot b_0}\right|_{b_0 \to \infty} = \frac{\varepsilon_{\lambda C}}{2} \tag{14–27}$$

Thus, all titration points in a reduced A-diagram fall along the straight line connecting these two points. The corresponding diagram for eosine is presented as Fig. 14–7. Experience has shown that diagrams such as this, because they are based on quotients, show more scatter than is normally associated with an excess absorbance diagram (cf. Fig. 14–6).

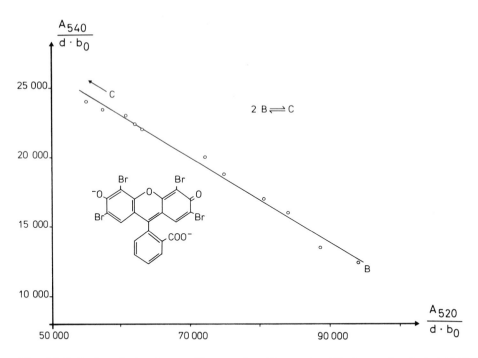

Fig. 14–7. Reduced absorbance diagram A_{540}/lb_0 vs. A_{520}/lb_0 for eosine in 0.1 N NaOH.

14.3 Reduced Absorbance Difference Quotient Diagrams and Excess Absorbance Quotient Diagrams (2B ⇆ C, C + B ⇆ E)

Sometimes the self-association of dye molecules results not only in dimers, but also in trimers and higher multimers.[58-63] Just as A-, AD-, ADQ-, ... ADQz-diagrams provide useful information with respect to a wide variety of acid–base equilibria, so too can a whole series of diagrams be applied to the problems raised by association equilibria. In principle, this provides the key to ascertaining the number of spectrometrically observable steps in a system of association equilibria.

As an illustration, consider the following combination of a dimer equilibrium and a trimer equilibrium:

$$2B \leftrightarrows C \quad \text{and} \quad C + B \leftrightarrows E \tag{14–28}$$

In this system, the mass balance conditions may be summarized as

$$b_0 = b + 2c + 3e \tag{14–29}$$

14.3 Reduced Absorbance Difference Quotient Diagrams

The corresponding expression for the absorbance A_λ is

$$A_\lambda = l(\varepsilon_{\lambda B} \cdot b + \varepsilon_{\lambda C} \cdot c + \varepsilon_{\lambda E} \cdot e) \tag{14--30}$$

We assume once again that the dye concentration b_0 will be successively increased in the course of a b_0 titration. If so, equations 14–29 and 14–30 reveal that

$$\frac{A_\lambda}{lb_0} - \varepsilon_{\lambda B} = (\varepsilon_{\lambda C} - 2\varepsilon_{\lambda B})\frac{c}{b_0} + (\varepsilon_{\lambda E} - 3\varepsilon_{\lambda B})\frac{e}{b_0} \tag{14--31}$$

The quantities c/b_0 and e/b_0 may be eliminated by developing the equation in terms of three different wavelengths; thus:

$$\frac{A_1}{lb_0} - \varepsilon_{1B} = \tau_1\left(\frac{A_2}{lb_0} - \varepsilon_{2B}\right) + \tau_2\left(\frac{A_3}{lb_0} - \varepsilon_{3B}\right) \tag{14--32a}$$

where
$$\tau_1 = \frac{\begin{vmatrix}(\varepsilon_{1C} - 2\varepsilon_{1B}) & (\varepsilon_{1E} - 3\varepsilon_{1B}) \\ (\varepsilon_{3C} - 2\varepsilon_{3B}) & (\varepsilon_{3E} - 3\varepsilon_{3B})\end{vmatrix}}{|Q|},$$

$$\tau_2 = \frac{\begin{vmatrix}(\varepsilon_{1C} - 2\varepsilon_{1B}) & (\varepsilon_{1E} - 3\varepsilon_{1B}) \\ (\varepsilon_{2C} - 2\varepsilon_{2B}) & (\varepsilon_{2E} - 3\varepsilon_{2B})\end{vmatrix}}{|Q|}, \text{ and}$$

$$|Q| = \begin{vmatrix}(\varepsilon_{2C} - 2\varepsilon_{2B}) & (\varepsilon_{2E} - 3\varepsilon_{2B}) \\ (\varepsilon_{3C} - 2\varepsilon_{3B}) & (\varepsilon_{3E} - 3\varepsilon_{3B})\end{vmatrix} \neq 0$$

Rearranging terms one obtains

$$\frac{A_1}{lb_0} = \tau_1\frac{A_2}{lb_0} + \tau_2\frac{A_3}{lb_0} + \tau_3 \tag{14--32b}$$

with $\tau_3 = \varepsilon_{1B} - \tau_1\varepsilon_{2B} - \tau_2\varepsilon_{3B}$

Every titration point referring to both equilibria of equation 14–28 must fulfill these conditions. Consequently, the coefficient τ_3 may be eliminated by constructing differences based on one of the titration points P' (with A'_λ/lb_0 chosen at random):

$$\frac{A_1}{lb_0} - \frac{A'_1}{lb'_0} = \tau_1\left(\frac{A_2}{lb_0} - \frac{A'_2}{lb'_0}\right) + \tau_2\left(\frac{A_3}{lb_0} - \frac{A'_3}{lb'_0}\right) \tag{14--33}$$

Division by the quantity $A_3/lb_0 - A'_3/lb'_0$ provides the quotients required for a *reduced absorbance difference quotient diagram*.

Assuming the titrant to be responsible for a measurable level of absorbance, a diagram of this type makes it possible to perform a direct graphical matrix-rank analysis. Alternatively, it is also possible to accomplish the same end using excess absorbances. Thus, multiplying both sides of equation 14–32a by lb_0 and rearranging terms leads to the following expression for an *excess absorbance quotient diagram*:

$$\frac{A_1^E}{A_2^E} = \tau_1 + \tau_2\frac{A_3^E}{A_2^E} \tag{14--34}$$

where $A_\lambda^E = A_\lambda - lb_0\varepsilon_{\lambda B}$

14.4 Determining Association Constants of Charge-Transfer Complexes

Interaction of an electron donor (B) with an electron acceptor (C) often leads to formation of what is known as an EDA- or charge-transfer complex, a phenomenon described by the following equilibrium:[19,23]

$$B + C \leftrightarrows BC \tag{14-35}$$

The corresponding association constant takes the form

$$\beta = \frac{bc}{b \cdot c} \tag{14-36}$$

Values are normally assigned to such constants on the basis of spectrometric analysis.[19,23] The UV–VIS spectrum of a charge-transfer complex generally reveals a bathochromic shift relative to the spectra of the starting materials, frequently permitting isolated analysis of the complex itself.

The Lambert–Beer–Bouguer law provides the following general expression for absorbance with the equilibrium $B + C \leftrightarrows BC$:

$$A_\lambda = l(\varepsilon_{\lambda B} \cdot b + \varepsilon_{\lambda C} \cdot c + \varepsilon_{\lambda BC} \cdot bc) \tag{14-37}$$

The appropriate mass-balance equations are

$$b_0 = b + bc \quad \text{and} \quad c_0 = c + bc \tag{14-38}$$

where b_0 and c_0 are the initial molar concentrations of B and C, respectively.

It follows from these considerations that

$$A_\lambda - l\varepsilon_{\lambda B} b_0 - l\varepsilon_{\lambda C} c_0 = l(\varepsilon_{\lambda BC} - \varepsilon_{\lambda B} - \varepsilon_{\lambda C})\, bc \tag{14-39}$$

The left side of equation 14–39 may be regarded as equivalent to an excess absorbance (cf. Sec. 13-3):

$$A_\lambda^E := A_\lambda - l\varepsilon_{\lambda B} b_0 - l\varepsilon_{\lambda C} c_0 \tag{14-40}$$

Introduction of the further definition

$$q_\lambda = l(\varepsilon_{\lambda BC} - \varepsilon_{\lambda B} - \varepsilon_{\lambda C}) \tag{14-41}$$

results in the following condensed expression for bc:

$$bc = \frac{A_\lambda^E}{q_\lambda} \tag{14-42}$$

The quantity A_λ^E may now be related to the overall association constant β:

$$\beta = \frac{bc}{b \cdot c} = \frac{bc}{(b_0 - bc)(c_0 - bc)} = \frac{A_\lambda^E \cdot q_\lambda}{(q_\lambda b_0 - A_\lambda^E)(q_\lambda c_0 - A_\lambda^E)}$$

Rearrangement of terms produces the equation[19]

14.4 Determining Association Constants of Charge-Transfer Complexes

$$\frac{b_0 c_0}{A_\lambda^E} + \frac{A_\lambda^E}{q_\lambda^2} = \frac{1}{q_\lambda \beta} + \frac{b_0 + c_0}{q_\lambda} \qquad (14\text{-}43)$$

Assuming that $b_0 \gg c_0$ and $b_0 c_0 / A_\lambda^E \gg A_\lambda^E / q_\lambda^2$, this equation may be simplified to the following form, as first noted by Scott:[64]

$$\frac{b_0 c_0}{A_\lambda^E} = \frac{1}{q_\lambda \beta} + \frac{b_0}{q_\lambda} \qquad (14\text{-}44)$$

According to Benesi–Hildebrand,[28,19,23] this equation may be divided by b_0, producing an expression that suggests using a variation approach (i.e., performing a b_0 titration, for example) in order to evaluate both β and q_λ:

$$\frac{c_0}{A_\lambda^E} = \frac{1}{q_\lambda \beta} \cdot \frac{1}{b_0} + \frac{1}{q_\lambda} \qquad (14\text{-}45)$$

The equilibrium $B + C \leftrightarrows BC$ may also serve as a useful model for investigating the interaction between an anion and an enzyme, an especially important observation since approximately 2/3 of all enzymes play host to anionic substrates or cofactors.

Similarly, $B + C \leftrightarrows BC$ may be regarded as describing the entrapment of an organic anion within the hydrophobic cavity of a macrocyclic host. For example, Schmidtchen and others have examined the complexation of the dinitrophenolate anion (DNP) with macrotricyclic ammonium salts in aqueous solution:[65–68]

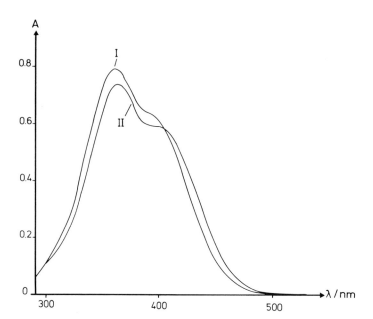

Fig. 14–8. Absorption spectra in TAPS buffer, pH 8.85, of dinitrophenolate anion (Curve I; $1 \cdot 10^{-3}$ M; $l = 1$ mm) and the complex formed between dinitrophenolate anion (DNP) and a host macrotricyclic ammonium salt "$N_4C_8(CH_3)_4F_4$". The relevant structures are shown in the text (Curve II; DNP: $1 \cdot 10^{-3}$ M; ammonium salt: $2.5 \cdot 10^{-3}$ M; $l = 1$ mm). These spectra were kindly placed at our disposal by F. P. Schmidtchen from the Institute for Organic Chemistry and Biochemistry at the Technische Universität München (Garching).

As may be seen in Fig. 14–8, formation of the above host–guest complex leads to a bathochromic shift in the dinitrophenolate spectrum. Evaluation of experimental data for this system on the basis of equation 14–43 (assuming $b_0c_0/A_\lambda^E > A_\lambda^E/q_\lambda^2$) is illustrated in Fig. 14–9, and it leads to the results:[66]

$$\beta = 220 \text{ M}^{-1} \quad \text{and} \quad q_\lambda = 1520 \text{ M}^{-1}$$

As in the case of the system $2\,B \leftrightarrows C$ it is possible here as well to take advantage of a series of special diagrams in order to ascertain how many spectrometrically observable equilibrium processes are at work. Thus, development of equation 14–42 for two different wavelengths leads directly to the function required for constructing an EA-diagram for the system $B + C \leftrightarrows BC$:

$$A_1^E = \frac{q_1}{q_2} A_2^E \tag{14–46}$$

It is also possible to create a "reduced excess absorbance diagram" for the system $B + C \leftrightarrows BC$ (cf. ref. 55), analogous to the reduced A-diagram of the system $2B \leftrightarrows C$ (cf. Sec. 14.2.3). In the place of equation 14–22 it is here necessary that one introduce the relationship

14.4 Determining Association Constants of Charge-Transfer Complexes

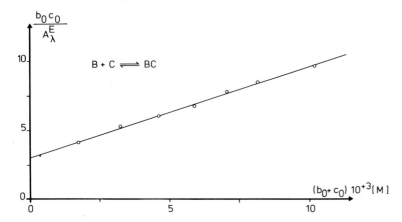

Fig. 14–9. Evaluation using equation 14–43 of a host–guest equilibrium B + C ⇌ BC (B = dinitrophenolate anion, C = the macrocyclic ammonium salt "$N_4C_8(CH_3)_4F_4$"), with the assumption $b_0c_0/A_\lambda^E > A_\lambda^E/q_\lambda^2$. Concentrations were varied over the following ranges: (b) 1.14–$1.44 \cdot 10^{-4}$ M; (c) 1.6–$10.06 \cdot 10^{-3}$ M [TAPS buffer, pH 8.85; ionic strength 0.5 M (KF); T = 299 K; l = 0.5 cm]. This diagram was kindly provided by F. P. Schmidtchen, Technische Universität München/Garching (cf. Fig. 14–8).

$$\frac{A_1^E}{lb_0} = \delta_1 \frac{A_2^E}{lb_0} + \delta_2 \tag{14-47}$$

where $A_\lambda^E = A_\lambda - l\varepsilon_{\lambda C} c_0$, $\delta_1 = \dfrac{\varepsilon_{1B} + \varepsilon_{1C} - \varepsilon_{1BC}}{\varepsilon_{2B} + \varepsilon_{2C} - \varepsilon_{2BC}}$, and

$$\delta_2 = \frac{\varepsilon_{2B}(\varepsilon_{1BC} - \varepsilon_{1C}) + \varepsilon_{1B}(\varepsilon_{2C} - \varepsilon_{2BC})}{\varepsilon_{2B} + \varepsilon_{2C} - \varepsilon_{2BC}}$$

This equation may be derived by rearranging the terms in equation 14–46 (though it should be noted that this results in the symbol A_λ^E acquiring a different meaning).

14.5 Summary

The association constant β for the titration system

$$BH + H_2O \leftrightarrows B^- + H_3O^+$$

$$2\,B^- \overset{\beta}{\leftrightarrows} C^{2-}$$

may be determined using a *pencil of rays method* analogous to that applied to protolysis and metal-complex equilibrium systems. Data points derived from a pH titration permit construction of a (two-dimensional) A-diagram in which a pencil of rays may be used to define an *absorbance triangle*. The concentrations of the various components may then be established on the basis of simple distance relationships, leading to β.

As has been demonstrated with the example eosine, the self-association of dye molecules (viewed as a $2\,B \leftrightarrows C$ system) may be analyzed with the aid of a set of parallel lines in a (two-dimensional) A-diagram. The requisite titration involves successive increases in the dye concentration (a b_0 titration). Non-linear regression makes it possible to establish either of the quotients $\varepsilon_{1C}/\varepsilon_{2C}$ or $\varepsilon_{1B}/\varepsilon_{2B}$. With this information it becomes possible to use distance relationships to establish the concentrations of B and C, permitting subsequent computation of the association constant β. However, this procedure does presuppose a knowledge of one of the quantities $\varepsilon_{\lambda B}$ or $\varepsilon_{\lambda C}$.

Verification of the presence or absence of equilibria resulting in the formation of trimers and higher multimers in addition to dimers may be established either with the aid of *excess absorbance* and *excess absorbance quotient diagrams* or by working with *reduced absorbance* and *reduced absorbance difference quotient diagrams* (or extensions thereof). The procedure is analogous to the use of A-, AD-, and ADQ-diagrams in problems associated with protolysis systems. The only limitation is the requirement that the titrant be subject to spectrometric determination.

Charge-transfer equilibria are normally evaluated on the basis of the Benesi–Hildebrand equation or the Scott equation. A host–guest complex has been used to demonstrate how the association constant for such a $B + C \leftrightarrows BC$ system may be established.

Literature

1. G. Kortüm, *Z. Physik. Chem.* B 33 (1936) 1.
2. G. Kortüm, *Z. Physik. Chem.* B 34 (1936) 255.
3. G. Scheibe, *Angew. Chem.* 50 (1937) 212.
4. G. Scheibe, *Angew. Chem.* 52 (1939) 631.
5. G. Scheibe, *Z. Elektrochem.* 47 (1941) 79.
6. G. Scheibe, *Z. Elektrochem.* 52 (1948) 283.
7. E. Rabinowitch, L. Epstein, *J.Amer. Chem. Soc.* 63 (1941) 69.
8. T. Förster, *Naturwissenschaften* 33 (1946) 166.
9. V. Zanker, *Z. Physik. Chem.* 199 (1952) 259.
10. V. Zanker, *Z. Physik. Chem.* 200 (1952) 250.
11. H. Zimmermann, G. Scheibe, *Z. Elektrochem.* 60 (1956) 566.
12. T. Förster, E. König, *Z. Elektrochem.* 61 (1957) 344.
13. G. Kortüm, G. Weber, *Z. Elektrochem.* 64 (1960) 642.
14. J. E. Selwyn, J. I. Steinfeld, *J. Phys. Chem.* 76 (1972) 764.
15. K.Conrow, G.D.Johnson, R. E.Bowen, *J. Amer. Chem. Soc.* 86 (1964) 1025.
16. J. T. Bulmer, H. F. Shurvell, *J. Phys. Chem.* 77 (1973) 256.
17. G. Kortüm, *Lehrbuch der Elektrochemie.* Verlag Chemie, Weinheim 1972.
18. H. A. Staab, *Einführung in die theoretische organische Chemie.* 4. Aufl. Verlag Chemie, Weinheim 1966.
19. G. Briegleb, *Elektronen-Donator-Acceptor-Komplexe.* Springer Verlag, Berlin 1961.
20. G. Briegleb, J. Czekalla, *Z. Elektrochem.* 59 (1955) 184.
21. R. R. Foster, *Organic Charge-Transfer Complexes.* Academic Press, London 1969.
22. R. S. Mulliken, W. B. Person, *Molecular Complexes.* Wiley. Interscience, New York 1969.
23. M. A. Slifkin, *Charge Transfer Interactions of Biomolecules.* Academic Press, London, New York 1971.
24. G. Briegleb, W. Liptay, M. Cantner, *Z. Physik. Chem. N. F.* 26 (1960) 55.
25. G. Briegleb, J. Czekalla, G. Reuss, *Z. Physik. Chem. N. F.* 30 (1961) 333.
26. G. Briegleb, W. Liptay, R. Fick, *Z. Elektrochem.* 66 (1962) 851.
27. W. Liptay, *Z. Elektrochem.* 65 (1961) 375.
28. H. A. Benesi, J. H. Hildebrand, *J. Amer. Chem. Soc.* 71 (1949) 2703.
29. R. M. Keefer, L. J. Andrews, *J. Amer. Chem. Soc.* 74 (1952) 1891.
30. G. Scatchard, *Ann. N. Y. Acad. Sci.* 51 (1949) 660.
31. R. F. Steiner, J. Roth, J. Robbins, *J. Biol . Chem.* 241 (1966) 560.
32. D. E. V. Schmeckel, D. M. Crothers, *Biopolymers* 10 (1971) 465.
33. H. J. Li, D. M. Crothers, *J. Mol. Biol.* 39 (1969) 461.
34. J. Pauluhn, H. W. Zimmermann, *Ber. Bunsenges. Phys. Chem.* 82 (1978) 1264.
35. J. Pauluhn, H. W. Zimmermann, *Ber. Bunsenges. Phys. Chem.* 83 (1979) 76.
36. J. Pauluhn, R. Grezes, H. W. Zimmermann, *Ber. Bunsenges. Phys. Chem.* 83 (1979) 708.
37. J. L. Bresloff, D. M. Crothers, *J. Mol. Biol.* 95 (1975) 103.
38. I. Zimmermann, H. W. Zimmermann, *Ber. Bunsenges. Phys. Chem.* 81 (1977) 81.
39. I. Zimmermann, H. W. Zimmermann, *Z. Naturforsch.* 31 c (1976) 656.
40. J. Polster, *Habilitationsschrift.* Tübingen 1980.
41. G. Kortüm, *Angew. Chem.* 70 (1958) 14.
42. W. Theilacker, G. Kortüm, H. Elliehausen, *Z. Naturforsch.* 96 (1954) 167.

43. W. Theilacker, G. Kortüm, H. Wilski, H. Elliehausen, *Chem. Ber.* 89 (1956) 1578.
44. W. T. Grubb, G. B. Kistiakowsky, *J. Amer. Chem. Soc.* 724 (1950) 419.
45. G. Kortüm, W. Theilacker, V. Braun, *Z. Physik. Chem. N. F.* 2 (1954) 179.
46. Y. Hirschberg, E. Fischer, *J. Chem. Soc. London* 1952, 4523; 1954, 297, 3129.
47. R. Suhrmann, H.-H.Percampus, *Z. Elektrochem. Ber. Bunsenges. Physik. Chem.* 56 (1952) 743.
48. J. Polster, *Z. Naturforsch.* 42 a (1987) 636.
49. M. E. Lamm, M. Neville, *J. Phys. Chem.* 69 (1965) 3872.
50. T. Kurucsev, U. P. Strauss, *J. Phys. Chem.* 74 (1970) 3081.
51. R. E. Ballard, C. H. Park, *J. Chem. Soc.* A 1970, 1340.
52. C.A.Brignoli, H.Devoe, *J. Phys. Chem.* 82 (1978) 2570.
53. P. Hampe, D. Fassler, *Z. Phys. Chem. Leipzig* 263 (1982) 111.
54. N. H. Nie, H. Hull, *SPSS 8, Statistik-Programm-System für die Sozialwissenschaften.* Gustav-Fischer-Verlag, Stuttgart, New York 1980.
55. H. Mauser, *Formale Kinetik.* Bertelsmann Universitätsverlag, Düsseldorf 1974.
56. J. Dreszer, *Mathematik Handbuch für Technik, Naturwissenschaft.* Verlag Harry Deutsch, Zürich, Frankfurt/Main, Thun 1975.
57. H. Netz, *Formeln der Mathematik.* Carl Hauser Verlag, München, Wien 1977.
58. P. Mukerjee, A. K. Gosh, *J. Amer. Chem. Soc.* 92 (1970) 6403.
59. A. K. Gosh, P. Mukerjee, *J. Amer. Chem. Soc.* 92 (1970) 6408.
60. A. K. Gosh, P. Mukerjee, *J. Amer. Chem. Soc.* 92 (1970) 6413.
61. A. K. Gosh, *J. Amer. Chem. Soc.* 92 (1970) 6415.
62. P. Mukerjee, A. K. Gosh, *J. Amer. Chem. Soc.* 92 (1970) 6419.
63. U. Ahrens, H. Kuhn, *Z. Physik. Chem. N. F.* 37 (1963) 1.
64. R. L. Scott, *Rec.Trav. Chim.* 75 (1965) 787.
65. F. P. Schmidtchen, *Angew. Chem.* 89 (1977) 751.
66. F. P. Schmidtchen, *Chem. Ber.* 114 (1981) 597.
67. F. P. Schmidtchen, *Chem. Ber.* 117 (1984) 725.
68. F. P. Schmidtchen, *Chem. Ber.* 117 (1984) 1287.

15 Redox Equilibria

15.1 Determining Standard Electrode Potentials

An oxidation–reduction (*redox*) reaction may be envisioned as comprising two redox pairs that differ in their electron affinities. In the course of such a reaction the pair with the lower electron affinity donates n electrons,

$$\text{Red}_1 \leftrightarrows \text{Ox}_1 + n\text{e} \qquad (15\text{--}1)$$

and these are accepted by the pair with the greater electron affinity:

$$\text{Ox}_2 + n\text{e} \leftrightarrows \text{Red}_2 \qquad (15\text{--}2)$$

Two such redox pairs may be treated as *half-cells* and linked so that they form a complete galvanic cell, causing electrons to flow from one half-cell to the other. An arrangement of this type is illustrated in Fig. 15–1, where electrons are assumed to flow from the half-cell on the left to that on the right with the aid of the two chemically inert (platinum) electrodes shown inserted into the respective solutions. In order to preserve electrical neutrality it is also necessary that the two solutions be linked by a second pathway that permits the flow of ions (typically a tube containing an agar gel and a concentrated solution of KCl).

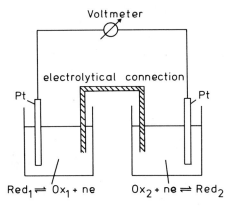

Fig. 15–1. Experimental system for the measurement of a redox potential.

Current will continue to flow in such a system until the overall free energy reaches a minimum, corresponding to chemical equilibrium. The resulting electrical work may be maximized by maintaining the system in a constant state of reversibility (i.e., by restricting the electron transfer phenomenon to a series of infinitesimally small steps). The magnitude of the electromotive force (emf) or *electrical potential* developed between the pair of electrodes is an indication of the extent to which equilibration has been achieved.[1-3] Electrical potentials may be determined with the aid either of a Poggendorf compensatory circuit[1,4] (Wheatstone bridge) or a high-resistance voltmeter (cf. Fig. 15–1).

The corresponding overall chemical reaction, i.e.,

$$Red_1 + Ox_2 \leftrightarrows Ox_1 + Red_2 \tag{15-3}$$

is subject to the condition[1-5]

$$E = E_0 - \frac{RT}{nF} \ln \frac{a_{Ox_1} \cdot a_{Red_2}}{a_{Red_1} \cdot a_{Ox_2}} \tag{15-4}$$

where E = emf of the cell
 E_0 = $E_{0h}(Ox_2/Red_2) - E_{0h}(Ox_1/Red_1)$; i.e., the difference between the "standard potentials" (see below) of the two half-cells (Ox_2/Red_2) and (Ox_1/Red_1)
 R = ideal gas constant (8.3144 J K^{-1} mol^{-1})
 T = temperature measured in Kelvin
 F = Faraday's constant (96 496 A s/g equivalent)
 a_{Ox_1} = activity of Ox_1 (and similarly for a_{Red_1}, a_{Red_2}, and a_{Ox_2})

The thermodynamic equilibrium constant for this reaction may be readily calculated once the standard potential E_0 is known. When such a galvanic cell reaches equilibrium, $E = 0$ and no more current flows; therefore, from equation 15–4:

$$\ln \frac{a^e_{Ox_1} \cdot a^e_{Red_2}}{a^e_{Red_1} \cdot a^e_{Ox_2}} = \frac{nF}{RT} E_0 \tag{15-5}$$

where the superscript "e" specifies an activity corresponding to the state of equilibrium. The term on the right may now be related directly to the logarithm of the thermodynamic equilibrium constant K:

$$\ln K = \frac{nF}{RT} E_0 \tag{15-6}$$

Consequently, the equilibrium constant K may be expressed as

$$K = \frac{a^e_{Ox_1} \cdot a^e_{Red_2}}{a^e_{Red_1} \cdot a^e_{Ox_2}} = e^{\frac{nF}{RT} \cdot E_0} \tag{15-7}$$

and K may be calculated directly from E_0.

15.1 Determining Standard Electrode Potentials

The *standard electrode potential* E_{0h} of a half-cell is a potential defined relative to what is known as a *normal hydrogen electrode* (NHE), a device within which the following reaction takes place:

$$H_2 + 2 H_2O \leftrightarrows 2 H_3O^+ + 2 e \qquad (15\text{--}8)$$

Thus, if a normal hydrogen electrode is substituted for the half-cell on the left in Fig. 15–1, the resulting overall chemical reaction (with $n = 2$) is:

$$H_2 + Ox_2 + 2 H_2O \leftrightarrows 2 H_3O^+ + Red_2 \qquad (15\text{--}9)$$

Equation 15–4 provides the following expression for emf in this galvanic cell:

$$E = E_0 - \frac{RT}{2F} \ln \frac{a_{H_3O^+}^2 \cdot a_{Red_2}}{a_{H_2} \cdot a_{Ox_2}} \qquad (15\text{--}10)$$

A normal hydrogen electrode operates under standard conditions; that is, the hydronium ion activity is maintained at unity ($a_{H_3O^+} = 1$), and at 298.15 K the hydrogen gas pressure is 1.013 bar. Thus, since $a_{H_2} = 1$, equation 15–10 may be rewritten as:

$$E = E_{0h} + \frac{RT}{2F} \ln \frac{a_{Ox_2}}{a_{Red_2}} \qquad (15\text{--}11)$$

where

$$E_{0h} = E_{0h}(Ox_2/Red_2) - E_{0h}(NHE)$$

and

$$E_{0h}(NHE) := 0$$

If a set of redox half-reactions is arranged in order of decreasing E_{0h}, the resulting list constitutes what is known as an *electromotive force series*. The equilibrium constant for a redox equilibrium like 15–3 may be calculated by subtracting the appropriate pair of E_{0h} values and introducing the resulting difference into equation 15–7 in place of E_0 [i.e., $E_{0h}(Ox_2/Red_2) - E_{0h}(Ox_1/Red_1) = E_0$].

A value for E_0 may also be computed on the basis of equation 15–11, but the process is greatly facilitated by replacing activities with concentrations (that is, setting $a_{Ox_2} = ox_2$ and $a_{Red_2} = red_2$), since this permits use of the alternative formulation

$$E = E_{0h} + \frac{RT}{2F} \ln \frac{ox_2}{red_2} \qquad (15\text{--}12)$$

The latter equation bears a convenient resemblance to an important relationship discussed previously in the context of the single-step acid–base equilibrium

$$BH \overset{K_1}{\leftrightarrows} B^- (H^+)$$

namely (cf. Chapter 4, equation 4–5b):

$$pH = pK_1 + \log \frac{b}{bh} \qquad (6\text{--}7)$$

This resemblance makes it useful to construct diagrams ox_2 vs. E and red_2 vs. E in analogy to the diagram bh vs. pH (or β vs. pH cf. Fig. 6–11 in Sec. 6.3.4), since a point of inflection in such a diagram can be used to assign a standard electrode potential E_{0h}. The required experimental data may be obtained from a titration, one in which the ratio ox_2/red_2 in the appropriate half-cell of a complete galvanic cell is altered by the addition of a strong oxidizing or reducing agent (avoiding significant dilution of the system). The titration itself may be monitored spectrometrically provided the components Ox_2 and Red_2 are spectroscopically active. Given the relationships

$$r_0 = ox_2 + red_2 = \text{constant} \tag{15-13}$$

$$A_\lambda = l(\varepsilon_{\lambda Ox_2} \cdot ox_2 + \varepsilon_{\lambda Red_2} \cdot red_2) \tag{15-14}$$

and $K = e^{\frac{2F}{RT} \cdot E_{0h}}$

it follows (with equation 15–12) in analogy to equation 6–14a that[6]

$$(A_\lambda - l r_0 \varepsilon_{\lambda Ox_2}) e^{\frac{2F}{RT} \cdot E} = l r_0 \varepsilon_{\lambda Red_2} e^{\frac{2F}{RT} \cdot E_{0h}} - e^{\frac{2F}{RT} \cdot E_{0h}} A_\lambda \tag{15-15}$$

Thus, if the term on the left is plotted against A_λ, both E_{0h} and $\varepsilon_{\lambda Red_2}$ may be evaluated. It should be noted, however, that the value obtained for E_{0h} is a function of the conditions (pH, ionic strength, etc.) under which the measurements have been made. Furthermore, the thermodynamic equilibrium constant $\mathbf{K}$ is transformed into the classical equilibrium constant K whenever activities are replaced by concentrations.

To date, spectrometric investigations have played a relatively minor role in the analysis of redox equilibria. Such systems are most often investigated by polarographic means, including pulse- or differential-polarography and cyclic voltammetry.[7] These methods are both rapid and highly precise, so they are rarely subject to serious competition.

15.2 The Redox System $Red_1 + Ox_2 \leftrightarrows Ox_1 + Red_2$

15.2.1 Evaluating the Equilibrium Constant by Successively Increasing the Concentration of One Component (First Method)

Certain sets of half-cells (redox pairs) appear incapable of combining in such a way as to produce a stable redox potential across two electrodes.[8-10] Interestingly, many of the known cases involve half-cells that react quite reversibly in other combinations. This behavior occasionally presents serious problems with respect to electrometric determination of equilibrium constants, and spectrometric techniques can prove a valuable alternative.

We consider first the system

$$Red_1 + Ox_2 \overset{K}{\leftrightarrows} Ox_1 + Red_2 \tag{15-3}$$

for which the equilibrium constant K is to be determined. Let us assume that the component Red_1 is to be introduced first into the titration vessel, and that the position of equilibrium is to be adjusted by the stepwise addition of specific amounts of the component Ox_2. The following mass balance equations must then be applicable:

$$r_0 = red_1 + ox_1 \tag{15-16a}$$

and

$$o_0 = red_2 + ox_2 \tag{15-16b}$$

Since o_0 is the quantity that is to be varied, r_0 may be regarded as constant provided volume changes during the titration are negligible. Equal amounts of the products Ox_1 and Red_2 must be formed; i.e.:

$$ox_1 = red_2 \tag{15-17}$$

so an appropriate expression for the equilibrium constant is:

$$K = \frac{ox_1 \cdot red_2}{red_1 \cdot ox_2} = \frac{ox_1^2}{red_1 \cdot ox_2} \tag{15-18}$$

In order to ascertain K by means of an A-diagram we must first consider the corresponding characteristic concentration diagram, shown for the case o_0 vs. ox_1 in Fig. 15-2. It is clear from equations 15-16a through 15-18 that the analytical expression for this curve must take the form:

$$o_0 = ox_1 + \frac{ox_1^2}{K(r_0 - ox_1)} \tag{15-19}$$

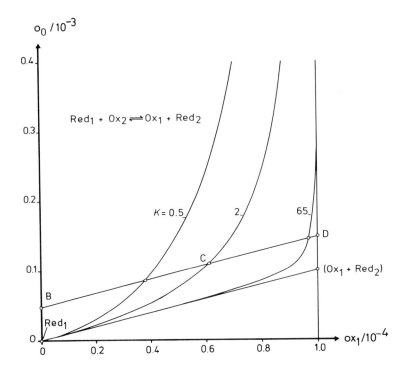

Fig. 15–2. Characteristic concentration diagrams o_0 vs. ox_1 for the system shown as equation 15–3, plotted as a function of the parameter K ($K = 0.5$, 2, and 65). The parallel to the line $\overline{\text{Red}_1\,(\text{Ox}_1 + \text{Red}_2)}$ passing through point C intersects the o_0-axis at B and intersects the parallel to the o_0-axis through $(\text{Ox}_1 + \text{Red}_2)$ at D.

The slopes at the limits $ox_1 \to 0$ and $ox_1 \to r_0$ must therefore be given by:

$$\left.\frac{do_0}{dox_1}\right|_{ox_1 \to 0} = 1 \quad \text{and} \quad \left.\frac{do_0}{dox_1}\right|_{ox_1 \to r_0} = \infty \tag{15-20}$$

At the limit $ox_1 \to 0$ the only species present is Red_1 ($o_0 = ox_1 = 0$), so the corresponding data point (Red_1) must lie at the coordinate origin. At the limit $ox_1 \to r_0$ all the Red_1 has been converted to Ox_1. Depending on the magnitude of K, the limit $ox_1 \to r_0$ may be reached at large or small values of o_0; if K is very large (i.e., $K \to \infty$), then in the region $0 < o_0 < r_0$ virtually all the added Ox_2 will be converted into Red_2. In this case, the curve traverses the line $\overline{\text{Red}_1\,(\text{Ox}_1 + \text{Red}_2)}$ (cf. Fig. 15–2). For very small values of K the curve will begin to deviate somewhere in the region $0 < o_0 < r_0$ from the line $\overline{\text{Red}_1\,(\text{Ox}_1 + \text{Red}_2)}$, in preparation for assuming a path parallel to the o_0-axis at high o_0 values.

If lines are constructed tangent to the limiting points $ox_1 \to 0$ and $ox_1 \to r_0$ these will be found to intersect at the point $(\text{Ox}_1 + \text{Red}_2)$, as indicated in Fig. 15–2. The same

15.2 The Redox System $Red_1 + Ox_2 \leftrightarrows Ox_1 + Red_2$

is true within an A-diagram (cf. Fig. 15–3), which is an affine distortion of the characteristic concentration diagram (cf. Sec. 7.6). Thus:

$$A_\lambda = l(\varepsilon_{\lambda Red_1} \cdot red_1 + \varepsilon_{\lambda Ox_2} \cdot ox_2 + \varepsilon_{\lambda Ox_1} \cdot ox_1 + \varepsilon_{\lambda Red_2} \cdot red_2)$$

$$= l\left[\varepsilon_{\lambda Red_1} \cdot r_0 + \varepsilon_{\lambda Ox_2} \cdot o_0 + \left(\varepsilon_{\lambda Ox_1} + \varepsilon_{\lambda Red_2} - \varepsilon_{\lambda Ox_2} - \varepsilon_{\lambda Red_1}\right) ox_1\right] \quad (15\text{–}21)$$

The reference line through the origin in Fig. 15–3 corresponds to the o_0-axis in Fig. 15–2, and it contains all the titration points that would result from varying o_0 if $r_0 = 0$. Therefore, this line must be parallel to the line $\overline{(Ox_1 + Red_2)\,D}$, which is tangent to the curve at high o_0 values.

Given a knowledge of the intersection point of the tangents ($Ox_1 + Red_2$) and of the o_0 values corresponding to the titration points (C), the concentrations of all components may be established in a straightforward way. Thus, it is clear that the lines $\overline{(Ox_1 + Red_2)\,D}$ and $\overline{Red_1\,B}$ must be parallel, as must also the lines $\overline{Red_1\,(Ox_1 + Red_2)}$ and $\overline{B\,C}$, which means that the following relationships must apply:

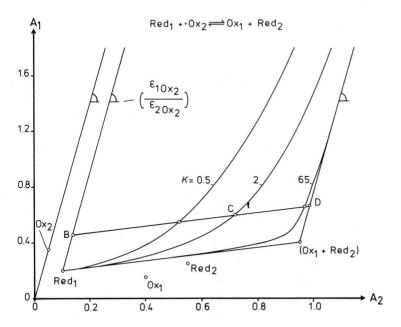

Fig. 15–3. A-diagrams for the system $Red_1 + Ox_2 \leftrightarrows Ox_1 + Red_2$ as a function of the parameter K ($K = 0.5$, 2, and 65; $r_0 = 10^{-4}$ M). Molar absorptivities λ_i ($i = 1, 2$): $\varepsilon_{\lambda_i Red_1} = 2000$ and 1000; $\varepsilon_{\lambda_i Ox_2} = 3500$ and 500; $\varepsilon_{\lambda_i Ox_1} = 1500$ and 4000; $\varepsilon_{\lambda_i Red_2} = 2500$ and 5500 (units: cm^{-1}M^{-1}). For the points B and D cf. Fig. 15–2.

$$red_1 = \frac{\overline{C\ D}}{Red_1\ (Ox_1\ +\ Red_2)}\ r_0 \quad \text{and}$$

$$ox_1 = red_2 = \frac{\overline{B\ C}}{Red_1\ (Ox_1\ +\ Red_2)}\ r_0 \tag{15-22}$$

The missing concentration (ox_2) may be established with the aid of equation 15–16b, and this completes the set of concentration data required for calculating K. It should be noted that the procedure just described does not presuppose a knowledge of the molar absorptivities for Ox_1 and Red_2.

Once K has been determined, a standard half-cell potential E_{0h} may be computed for the redox pair Red_1/Ox_1 under the conditions defined by the experiment—irrespective of whether this redox pair is actually capable of producing a stable redox potential at the electrode of a galvanic cell. The calculation itself entails evaluating in terms of E_0 the expression

$$K = e^{\frac{nF}{RT} \cdot E_0}$$

Combining this result with the known value $E_{0h}(Ox_2/Red_2)$ for the second redox pair leads to the desired value of $E_{0h}(Ox_1/Red_1)$:

$$E_{0h}(Ox_1/Red_1) = E_{0h}(Ox_2/Red_2) - E_0 \tag{15-23}$$

15.2.2 Determining the Equilibrium Constant by the Method of Continuous Variation (Second Method)

An equilibrium constant for the system

$$Red_1 + Ox_2 \leftrightarrows Ox_1 + Red_2 \tag{15-3}$$

may also be determined quite easily by the method of continuous variation, a procedure already encountered in Sec. 13.3. As in the previous case, the initial concentrations of the starting materials (Red_1 and Ox_2) are varied in such a way that their sum (c_0) remains constant. It is obvious from equations 15–16a, 15–16b, and 15–17 that

$$c_0 = r_0 + o_0 = red_1 + ox_2 + 2\ ox_1 \tag{15-24}$$

Since c_0 is a constant, such a system is characterized by only two linearly independent concentrations (e.g., red_1 and ox_2), and these in turn may be used to define a characteristic concentration diagram (cf. Fig. 15–4). An implicit representation of the curve constituting the diagram red_1 vs. ox_2 may be constructed as follows (on the

15.2 The Redox System $Red_1 + Ox_2 \leftrightarrows Ox_1 + Red_2$

basis of equation 15–18):
$$F = red_1^2 + red_1(2\,ox_2 - 4K\,ox_2 - 2c_0) + ox_2^2 - 2\,ox_2\,c_0 + c_0^2 = 0 \tag{15-25}$$

The slope $dred_1/dox_1$ for this curve must be (cf. Sections 7.3.3 and 13.3):[11]

$$\frac{d\,red_1}{d\,ox_2} = -\frac{\dfrac{\partial F}{\partial\,ox_2}}{\dfrac{\partial F}{\partial\,red_1}} = -\frac{red_1(2 - 4K) + 2\,ox_2 - 2c_0}{ox_2(2 - 4K) + 2\,red_1 - 2c_0}$$

It follows for the limiting values $ox_2 = 0$ $(c_0 = red_1)$ and $ox_2 = c_0$ $(red_1 = 0)$ that

$$\left.\frac{d\,red_1}{d\,ox_2}\right|_{ox_2=0} = -\infty \quad \text{and} \quad \left.\frac{d\,red_1}{d\,ox_2}\right|_{ox_2=c_0} = 0 \tag{15-26}$$

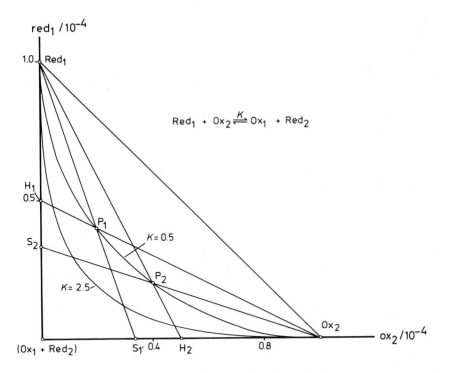

Fig. 15-4. Characteristic concentration diagrams for the titration system $Red_1 + Ox_2 \leftrightarrows Ox_1 + Red_2$ as a function of K (using the method of continuous variation). The point ($Ox_1 + Red_2$) is at the intersection of tangents to the points Red_1 and Ox_2. The points S_1 and S_2 may be used with equation 15–28 to establish a value for K based exclusively on distance relationships.

Hence, the coordinate axes are tangents to the points Red$_1$ and Ox$_2$ (cf. Fig. 15–4) or, to put it another way, the coordinate origin is the pole of the polar $\overline{\text{Red}_1 \text{Ox}_2}$ (for definitions of "pole" and "polar" see Sec. 7.3.3). This means that at the coordinate origin $red_1 = ox_2 = 0$, and according to equations 15–24 and 15–17:

$$ox_1 = red_2 = \frac{c_0}{2} \tag{15–27}$$

The point (Ox$_1$ + Red$_2$), at which only the species Ox$_1$ and Red$_2$ are present, apparently coincides with the coordinate origin.[12] The triangle defined by the vertices Red$_1$, Ox$_2$, and (Ox$_1$ + Red$_2$) may be evaluated in the same way as the comparable triangle for a two-step acid–base system (cf. Sec. 7.5.4). Consequently, lines $\overline{\text{H}_1 \text{Ox}_2}$ and $\overline{\text{H}_2 \text{Red}_1}$ have been introduced into Fig. 15–4 bisecting the two key sides of the triangle, and points P$_1$ and P$_2$ are singled out where these lines intersect the experimental curve. Points P$_1$ and P$_2$ may be used to construct the additional lines

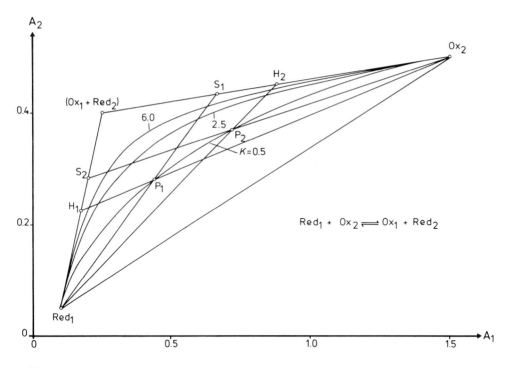

Fig. 15–5. Simulated absorbance curves for the titration system Red$_1$ + Ox$_2$ ⇌ Ox$_1$ + Red$_2$ shown as a function of K (based on the method of continuous variation), with $c_0 = 1 \cdot 10^{-4}$ M and $l = 1$ cm; $\varepsilon_{1\text{Red}_1} = 1\,000$, $\varepsilon_{1\text{Red}_2} = 3\,500$, $\varepsilon_{1\text{Ox}_1} = 1\,500$, $\varepsilon_{1\text{Ox}_2} = 15\,000$; $\varepsilon_{2\text{Red}_1} = 500$, $\varepsilon_{2\text{Red}_2} = 4\,000$, $\varepsilon_{2\text{Ox}_1} = 4\,000$, $\varepsilon_{2\text{Ox}_2} = 5\,000$ (units: cm^{-1}M^{-1}). The point (Ox$_1$ + Red$_2$) is at the intersection of the tangents to Red$_1$ and Ox$_2$. The points S$_1$ and S$_2$ may be used with equation 15–28 to establish a value of K based exclusively on distance relationships.

15.2 The Redox System $Red_1 + Ox_2 \leftrightarrows Ox_1 + Red_2$

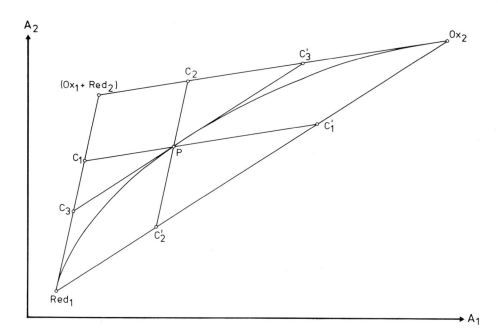

Fig. 15–6. Determination of a set of concentrations from an A-diagram (cf. equation 15–30).

$\overline{Red_1 P_1}$ and $\overline{Ox_2 P_2}$, whose points of intersection (S_1 and S_2) with the sides $\overline{(Ox_1 + Red_2) Ox_2}$ and $\overline{(Ox_1 + Red_2) Red_1}$ may in turn be established. Finally, the points S_1 and S_2 make it possible to determine the equilibrium constant on the basis of simple distance relationships (cf. Sec. 7.5.4):[12]

$$4K = \frac{\overline{S_1 \, Ox_2}}{\overline{S_1 \, (Ox_1 + Red_2)}} = \frac{\overline{S_2 \, Red_1}}{\overline{S_2 \, (Ox_1 + Red_2)}} \qquad (15\text{–}28)$$

Corresponding geometric relationships also apply within an A-diagram, since this is an affine distortion of the characteristic concentration diagram (cf. Sec. 7.6). Thus, the point $(Ox_1 + Red_2)$ in Fig. 15–5 is the point of intersection of tangents to the curve at Red_1 and Ox_2, at which the following relationship must hold true:

$$A_\lambda(Ox_1 + Red_2) = l(\varepsilon_{\lambda Ox_1} + \varepsilon_{\lambda Red_2}) \frac{c_0}{2} \qquad (15\text{–}29)$$

Once identified, the absorbance triangle may be used to determine K by application of equation 15–28. The concentrations of the various components present may be determined through distance relationships as well. In analogy to the conclusions drawn with respect to Figures 7–13, 13–3, and 13–13, the following relationships apply within Fig. 15–6:[12]

$$red_1 = \frac{\overline{C_1 (Ox_1 + Red_2)}}{Red_1 (Ox_1 + Red_2)} c_0 = \frac{\overline{C_1' Ox_2}}{Red_1 Ox_2} c_0$$

$$ox_2 = \frac{\overline{C_2 (Ox_1 + Red_2)}}{Ox_2 (Ox_1 + Red_2)} c_0 = \frac{\overline{C_2' Red_1}}{Red_1 Ox_2} c_0 \quad \text{and}$$

$$ox_1 + red_2 = \frac{\overline{C_3 Red_1}}{Red_1 (Ox_1 + Red_2)} c_0 = \frac{\overline{C_3' Ox_2}}{Ox_2 (Ox_1 + Red_2)} c_0 \quad (15\text{--}30)$$

This permits K to be computed using equation 15–18. Just as with equation 15–28, no need exists for the molar absorptivities of Ox_1 and Red_2 to be known in advance. However, it should be noted that a determination of K through equation 15–28 is subject to severe limitation if K is either very large or very small. In such cases the computed distance ratios will at best represent rough approximations. Nevertheless, equation 15–30 still makes it possible to determine an equilibrium constant in such cases through the use of data points that do not fall along the sides of the absorbance triangle (cf. Fig. 15–5).

15.2.3 Determining the Equilibrium Constant for an NADH-Dependent Dehydrogenase Reaction

Reactions of the type $A + B \leftrightarrows C + D$ are extremely common throughout chemistry, including biochemistry. For example, many reversible redox reactions involve the transfer of a pair of electrons in the course of two successive steps.[13-21] A case in point relates to so-called "vinylogous compounds", structures that may be viewed as elaborations on a vinyl group, and whose reduced forms contain unshared pairs of electrons (or extra π-electrons);[21-26] e.g.:

15.2 The Redox System $Red_1 + Ox_2 \leftrightarrows Ox_1 + Red_2$

Such a system may be represented in general by the formulation

$$Red \underset{+e}{\overset{-e}{\leftrightarrows}} S \underset{+e}{\overset{-e}{\leftrightarrows}} Ox$$

where "S" stands for "semiquinone".

In principle, a semiquinone-type structure might also arise as a result of disproportionation of the corresponding oxidized and reduced forms:

$$Red + Ox \leftrightarrows 2\,S$$

This equilibrium may be regarded as equivalent to the system described by equation 15–3.

Equilibria of the type $A + B \leftrightarrows C + D$ are frequently encountered in the study of metal complexes, as for example with metalloporphyrins. Thus, it is often observed that one neutrally bound metal ion (M_1) within a porphyrin will undergo exchange with another (M_2), giving rise to the reaction:[27,28]

$$M_1P + M_2 \leftrightarrows M_2P + M_1$$

The study of enzyme-catalyzed biochemical reactions also furnishes innumerable examples of equilibria in this category. NADH-dependent oxidoreductases are typical; thus, D-3-hydroxybutyrate: NAD^+ oxidoreductase (EC 1.1.1.30, also known as 3-hydroxybutyrate dehydrogenase, HBDH) catalyzes equilibration in the following system:

$$\underset{\underset{Ox_2}{\text{acetoacetate}}}{CH_3-\underset{\underset{O}{\|}}{C}-CH_2-COO^-} + \underset{Red_1}{NADH} + H^+ \overset{HBDH}{\rightleftharpoons} \underset{\underset{Red_2}{\text{D-3-hydroxybutyrate}}}{CH_3-\underset{\underset{OH}{|}}{\overset{\overset{H}{|}}{C}}-CH_2-COO^-} + \underset{Ox_1}{NAD^+}$$

(15–31)

The corresponding equilibrium constant K may be expressed as (cf. equation 15–18):

$$K \cdot 10^{-pH} = \frac{ox_1^2}{red_1 \cdot ox_2} \tag{15-32}$$

The constant K may in this case be determined conveniently by application of the method of continuous variation (cf. Sec. 15.2.2). An A-diagram for the system is shown in Fig. 15–7.[12] Evaluation on the basis of equation 15–28 leads to the value $K \cdot 10^{-pH} = 0.58$. A calculation based on the concentrations of the various components (cf. equation 15–30) suggests that $K \cdot 10^{-pH} = 0.62$ (representing a mean value derived from five data points). Taking the average of these two results ($K \cdot 10^{-pH} = 0.60$) and introducing $pH = 9.15$ gives the following value for the equilibrium constant itself:[12]

$$K = 0.60 \cdot 10^{+9.15}\ M^{-1} = 10^{+8.93}\ M^{-1} \quad (25\ °C)$$

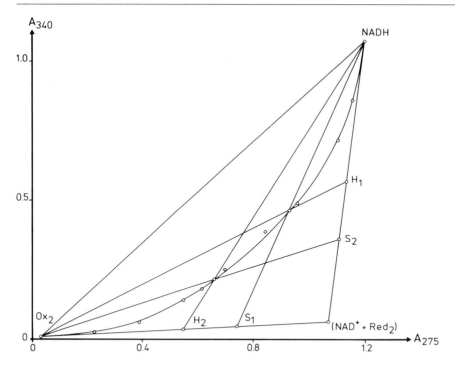

Fig. 15–7. Use of the method of continuous variation to establish the value of K for reaction 15–31. To separate 3 mL portions of glycine buffer (pH 9.15) were added the following amounts of a $1.2 \cdot 10^{-2}$ M solution of NADH (Red$_1$): 50 µL, 45 µL, 40 µL, 35 µL, 30 µL, 25 µL, 20 µL, 15 µL, 10 µL, and 5 µL. The solutions were then supplemented with corresponding amounts (i.e., 0 µL, 5 µL, 10 µL, ...,50 µL) of a lithium acetoacetate solution (Ox$_2$) having the same molar concentration. Equilibration was initiated by introducing 10 µL of a suspension of HBDH at 25.0 °C, equilibrium being reached after a period of ca. 0.5–1.5 h. Tangents to the points NADH and Ox$_2$ intersect at the point (NAD$^+$ + Red$_2$). Lines bisecting the appropriate sides of the absorbance triangle permit the construction of additional lines for locating points S$_1$ and S$_2$, from which K may be computed with the aid of equation 15–28. Concentrations of the individual components may also be used to calculate a value for K (cf. equation 15–30).

Krebs et al.[29] report the value:

$$K = \frac{1}{1.42 \cdot 10^{-9}\,\text{M}} = 10^{+8.85}\,\text{M}^{-1}\ (25\,°\text{C})$$

Once K has been established, a standard electrode potential E_{0h} corresponding to pH 7 may be calculated for the half-cell 3-hydroxybutyrate/acetoacetate with the aid of equations 15–7 and 5–23 and an appropriate E_{0h} value for NADH/NAD$^+$. Assuming E_{0h} for NADH/NAD$^+$ to be -0.320 V at pH 7,[30] and with $K = 10^{+8.93}$ M^{-1}, the final result (cf. ref. 29) is:

$$E_{0h} = -0.263\,\text{V}$$

15.3 Summary

A-Diagrams provide two alternative methods for establishing the equilibrium constant of the system $Red_1 + Ox_2 \leftrightarrows Ox_1 + Red_2$, which consists of a combination of the two redox pairs Red_1/Ox_1 and Ox_2/Red_2. The *first method* involves starting with the component Red_1 and performing a titration by adding successive portions of Ox_2, the appropriate component from the second redox pair. Interpretation of the resulting A-diagram with the aid of tangents provides information about the concentrations of all components, from which the equilibrium constant may in turn be computed. The *second method* is based on the continuous variation approach. Initial concentrations of Red_1 and Ox_2 are varied in such a way that their sum remains constant, leading to an A-diagram within which an absorbance triangle may be defined. Lines bisecting the sides of this triangle make it possible to establish the equilibrium constant on the basis of simple distance relationships. Similar distance relationships provide the concentrations of all components. Both procedures presuppose a knowledge of molar absorptivities only for Red_1 and Ox_2.

Use of the second method has been illustrated with data from the following equilibrium, which is catalyzed by 3-hydroxybutyrate dehydrogenase:

acetoacetate + NADH + H^+ $\leftrightarrows$ D-3-hydroxybutyrate + NAD^+

Literature

1. G. Kortüm, *Lehrbuch der Elektrochemie*. Verlag Chemie, Weinheim 1966.
2. E. A. Moelwyn-Hughes, *Physikalische Chemie*. Georg Thieme Verlag, Stuttgart 1970.
3. G. Wedler, *Lehrbuch der Physikalischen Chemie*. Verlag Chemie, Weinheim 1982.
4. H. Rau, *Kurze Einführung in die Physikalische Chemie*. Vieweg-Verlag, Wiesbaden 1977.
5. J. Gareth Morris, *Physikalische Chemie für Biologen*. Verlag Chemie, Weinheim 1976.
6. J. Polster, *Habilitationsschrift*. Tübingen 1980.
7. H. Schmidt, M. von Stachelberg, *Neuartige polarographische Methoden*. Verlag Chemie, Weinheim 1962.
8. H. Mauser, B. Nickel, *Angew. Chem.* 77, 378 (1965).
9. B. Nickel, H. Mauser, U. Hezel, *Z. Physik. Chem. N. F.* 54 (1967) 196.
10. B. Nickel, H. Mauser, U. Hezel, *Z. Physik. Chem. N. F.* 54 (1967) 214.
11. J. Dreszer, *Mathematik Handbuch für Technik, Naturwissenschaft*. Verlag Harry Deutsch, Zürich-Frankfurt/Main-Thun 1975.
12. J. Konstanczak, J. Polster, *Z. Naturforsch.* 41 a (1986) 1352.
13. U. Nickel, E. Haase, W. Jaenicke, *Ber. Bunsenges. Phys. Chem.* 81 (1977) 849.
14. U. Nickel, G. Loos, W. Jaenicke, *Ber. Bunsenges. Phys. Chem.* 78 (1974) 1271.

15. S. Hünig, P. Schilling, *Liebigs Ann. Chem.* 1967, 1103.
16. S. Hünig, M. Horner, P. Schilling, *Angew. Chem.* 87 (1975) 548; *Angew. Chem. Int. Ed. Engl.* 14 (1975) 556.
17. S. Hünig, G. Kießlich, H. Quast, D. Scheutzow, *Liebigs Ann. Chem.* 1973, 310.
18. S. Hünig, G. Ruider, *Liebigs Ann. Chem.* 1974, 1415.
19. S. Hünig, J. Groß, E. Lier, H. Quast, *Liebigs Ann. Chem.* 1973, 339.
20. S. Hünig, H.-C. Steinmetzer, *Liebigs Ann. Chem.* 1976, 1090.
21. K. Deuchert, S. Hünig, *Angew. Chem.* 90 (1978) 927.
22. S. Hünig, *Pure Appl. Chem.* 15 (1967) 109.
23. S. Hünig, F. Linhart, D. Scheutzow, *Liebigs Ann. Chem.* 1975, 2089.
24. S. Hünig, D. Scheutzow, H. Schlaf, A. Schott, *Liebigs Ann. Chem.* 1974, 1423.
25. S. Hünig, D. Scheutzow, H. Schlaf, H. Pütter, *Liebigs Ann. Chem.* 1974, 1436.
26. S. Hünig, F. Linhart, *Liebigs Ann. Chem.* 1975, 2116.
27. J. E. Falk, *Porphyrins and Metalloporphyrins*. Elsevier Publishing Company, Amsterdam, London, New York 1964.
28. P. Hambright, *Coord. Chem. Rev.* 6 (1971) 247-268.
29. H. A. Krebs, J. Mellanby, D. H. Williamson, *Biochem. J.* 82 (1962) 96.
30. W. Hoppe, W. Lohmann, H. Markl, H. Ziegler, *Biophysik*. Springer-Verlag, Berlin, Heidelberg, New York 1982.

Part III

Instrumental Methods

16 Suitable Apparatus for Spectrometric Titration

As noted in Sec. 1.2, spectrometric titration is nearly always conducted as a series of discrete steps, and it thus classes as a *discontinuous process*. Our focus of attention in this chapter is *absorption spectrometric acid–base titration*, in which a *step* is defined by the establishment of a specific pH. The goal is to record at each titration step either a complete absorption spectrum, or else a series of absorbances A_λ measured at a discrete set of wavelengths. Only when this end has been accomplished is the pH altered to produce a new set of conditions corresponding to the next titration step.

Discontinuous spectrometric titration can be conducted in either of two ways, as discussed in the two sections that follow.

16.1 m Experiments, Each Performed in a Different Buffer Medium

The first approach entails preparing separate buffer solutions for each of the pH values of interest. Samples of the subject protolyte are then added to the buffer solutions either by weight or as measured amounts of a concentrated stock solution.

Examples of titrations of this type are widely reported in the literature.[1-3] The principal *advantages* of such a "batch" approach include the following:

(1) No special titration apparatus is required. Once a test sample has been prepared it is merely transferred to a cuvette, which is then placed directly in the spectrometer.

(2) The dilution errors that of necessity accompany a standard titration are here completely avoided.

(3) Ionic strength may be maintained constant, at least over the pH range 2–12, by adjusting the concentrations of the buffer components.

Despite these advantages, a large number of *disadvantages* must also be considered:

(1) The procedure is very time-consuming, since m separate buffer solutions must be prepared. All initial concentrations of protolyte c_0 should be identical, so m precise sample transfers are required. For this reason, the number of titration steps investigated is normally quite small (typically $\leq$ 5–7 steps per proton transfer).[1,3]

(2) Reproducibility of the initial concentration c_0 is much lower than is the case with the methods to be described in Sec. 16.2. Most of the scatter found in A_λ vs. pH curves and in A-diagrams prepared in this way is attributable to changes in c_0. Isosbestic points also tend to be rather ill-defined.

(3) Photometric reproducibility is typically poorer than in experiments involving permanently installed titration cuvettes [cf. methods 16.2.1(b) and 16.2.2 (a) and (b)] due to the need for removing the cuvette from the spectrometer and then re-inserting it at each titration step.

(4) Carrying out m separate experiments consumes a relatively large amount of sample.

(5) It is very difficult to perform all the requisite operations under an inert atmosphere, thus virtually ruling out the investigation of easily oxidizable substances. Moreover, problems are likely to arise at short wavelength because of absorption corresponding to CO_2 and other substances from the air.

(6) If studies are to be conducted over a wide pH range (e.g., pH 2–12) then it is necessary either to change buffers several times[3] or else to use a so-called "universal buffer" (e.g., citrate/phosphate/borate).[4]

(7) Absorption spectra often differ somewhat in different buffers, especially in the short wavelength region (< 250 nm). Even universal buffers lead to spectra that change as a function of pH.

(8) The attempt to compensate for buffer effects by introducing a reference cuvette results in a significant increase in slit width (and a corresponding increase in spectral bandwidth), especially in the short-wavelength region (cf. Chapter 18).

(9) Protolytes sometimes undergo spectroscopically observable (concentration-dependent) interactions with certain buffer species, giving rise to what are known as "medium effects".[5]

(10) Chemical reactions may occur between protolytes and certain buffer species, and such reactivity is often a function of pH (as in the case of vitamin-B_6 aldehydes, which react with amine-containing buffers[3,6]).

(11) Automation of a procedure of this type is nearly impossible (in contrast to the methods described in Sec. 16.2).

16.2 Discontinuous Spectrometric Titration in the Strictest Sense

Because of the many disadvantages associated with the batch procedure, essentially all the titrations described in the present book were performed using the methods described in the section that follows.

16.2 Discontinuous Spectrometric Titration in the Strictest Sense

The procedures considered here are all based on a single sample of protolyte made up to a concentration c_0 in an appropriate titration vessel or titration apparatus. The pH is successively adjusted to the desired value for each titration step by addition of a measured amount of acid or base. At each step, a portion of the solution is introduced into the spectrometer, appropriate measurements are recorded, and then the pH of the stock solution is re-adjusted until it corresponds to the conditions appropriate to the next step.[7] Any particular measurement may easily be repeated later simply by shifting the pH up or down as required—a circumstance that further allows one to ascertain whether or not titration is accompanied by irreversible chemical reactions. Several variations are possible within the context of the basic procedure, as described below.

16.2.1 Titration Outside the Spectrometer

(a) The *simplest possible arrangement* for such a titration is illustrated in Fig. 16–1. Using a buret and a pH meter, an appropriate initial pH is first established within a beaker equipped with a magnetic stirrer. After addition of a measured amount of protolyte, a pipet is used to transfer a portion of the resulting solution to a standard cuvette. This initial sample is then returned immediately to the beaker, in which continuous stirring is maintained. Repeating the procedure twice more before recording any spectral data ensures that the contents of the cuvette and the beaker will be identical, and it also avoids the need for rinsing or drying the cuvette between measurements. Once spectral measurement is complete, the sample is returned to the original container and conditions appropriate to the next titration step are established.

Note that this procedure has been so conceived that the only dilution results from addition of titrant. The procedure is extremely simple, and associated concentration data can be assumed to be highly reliable. All the disadvantages listed in connection with the method of Sec. 16.1 are eliminated except for points (3), (5), and (11).

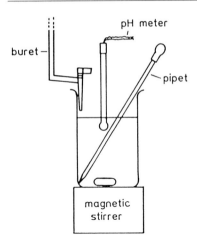

Fig. 16–1. Simple spectrometric titration arrangement consisting of a burette, a pH electrode, a magnetic stirrer, and a pipet.

(b) The apparatus depicted in Fig. 16–2 is designed to facilitate *precision spectrometric titration*.[8,9] A commercial 500-mL titration vessel (e.g., Metrohm model EA 880–500), labeled "TV", is equipped with a lid containing several ground-glass standard taper (NS 14) openings. For purposes of temperature control the vessel is surrounded by a plexiglass jacket so arranged that it does not interfere with the magnetic stirrer (MS). Gas-tight joints permit insertion of a glass electrode (E), a buret tip (B), a Bunsen valve (BV), and either a thermometer or a thermosensor (not illustrated). Absorption (or fluorescence) spectrometric measurements are conducted on samples that have been transferred to a modified Thunberg cuvette C (e.g., Hellma No. 190 or 191), which is equipped with a special flow insert. This cuvette is connected to the inlet tube (IT) by means of a piece of Teflon tubing (Tf).

Titration is initiated by filling the titration vessel with 500 mL of doubly-distilled water. The cuvette, titration vessel, and all tubing are then thoroughly flushed with nitrogen issuing from a gas inlet tube (GI) and regulated by valves V1 and V2. Protolyte is introduced rapidly, usually to a concentration c_0 of 10^{-4}–10^{-5} M. A nitrogen stream is maintained across the top of the solution throughout the course of titration with the aid of valve V2.

As soon as the appropriate pH has been established (using a buret and pH meter), a syringe (S) is used to draw solution through the inert transfer tubing (Tf) and into the cuvette. The cuvette is then flushed three times with sample solution before any spectral measurements are made. The chamber behind the plunger is kept filled with nitrogen at all times to prevent any introduction of air through the syringe.

> Since impurities in the nitrogen stream would be likely to accumulate over the course of a titration, it is advisable in the case of precision work either to employ exceptionally pure nitrogen or else to incorporate provisions for removal of residual CO_2 and/or O_2. Use of a membrane filter is also recommended.

16.2 Discontinuous Spectrometric Titration in the Strictest Sense

A titration apparatus of this type is quite portable, and it is well-suited to measurements of various types, including UV–VIS absorption, fluorescence, and CD/ORD. The only spectrometer modification that may be necessary is adaptation of the sample chamber cover so that it permits free passage of the requisite connecting tubes. Installing several cuvettes in a parallel arrangement even makes it possible to employ a variety of spectral techniques simultaneously. Additional sensors may be introduced into the titration vessel as desired so that not only pH but also other electrochemical parameters may be monitored (i.e., other potential differences, conductivity values, polarographic information, etc.). With suitable modifications it is possible to use coulometric techniques to generate certain types of titrants directly within the titration vessel.

The large volume (500 mL) specified for the titration vessel ensures that dilution accompanying a titration will be nearly negligible (e.g., < 0.5% with introduction of 1–2 mL of titrant). This is an important consideration, because it is not always practical to compensate mathematically for dilution effects, especially with regard to

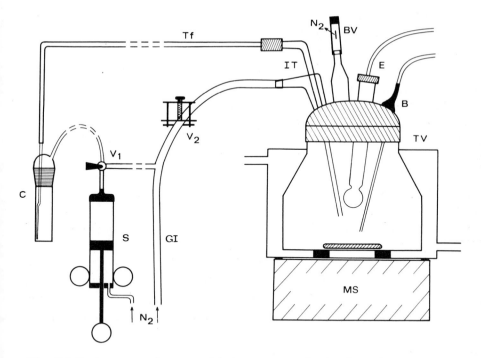

Fig. 16–2. Apparatus for a precision spectrometric titration. Components include: glass electrode (E), buret tip (B), Bunsen valve (BV), Thunberg cuvette (C), Teflon tubing (Tf), inlet tube (IT), gas inlet (GI), stopcock (V1), pinch clamp (V2), syringe (S), titration vessel (TV), magnetic stirrer (MS)

spectra recorded on an analog device. Dilution errors of the order of magnitude involved here are virtually impossible to detect, and even isosbestic points may be expected to show a high degree of definition (cf. Figures 6–1 and 18–2). Nevertheless, should there be a compelling reason to do so, the titration volume V_0 may be reduced to 50–100 mL provided burets of sufficient precision are available (see below). The lower limit with respect to solution volume is determined largely by the dead-volume of the cuvette and its connecting tubes, since the pH electrode must at all times remain immersed in solution.

The reproducibility associated with solution transport into the cuvette for such an apparatus is better than the reproducibility of absorption spectrometry itself (< 0.02%). A series of spectra recorded after a single filling is virtually impossible to distinguish from a similar set of spectra obtained by first emptying and then re-filling the cuvette before each run. However, this guarantee of reproducibility applies only to spectra recorded in the course of a single experiment, during which the cuvette must remain firmly in place within a thermostated sample holder maintained at the same temperature as the titration vessel (e.g., 25.0 ± 0.1 °C).

> One potential source of error in the procedure described is gas bubbles that may become trapped along the walls of the cuvette. Traces of grease or oil contribute significantly to this problem, so it is advisable to clean the entire apparatus periodically with acetone. Teflon sleeves rather than grease should always be used for sealing the standard taper joints.

The above titration procedure eliminates all the problems ascribed to the method of Sec. 16.1. Dilution effects will be present, but where they pose a significant problem they can easily be eliminated mathematically. Essentially constant ionic strength may be maintained throughout the pH range ca. 2–12 by addition of an inert salt (e.g., 0.1 M KCl). In contrast to the procedure of Sec. 16.1, all ionic strength effects will be associated with a consistent set of ionic species, permitting more straightforward calculation than is possible in the presence of a complex buffer system. Nearly all the spectrometric titrations described in the present book were performed using this procedure, and a similar approach has been shown to be successful in the hands of others.[10]

The fact that it is easy to pump solution back and forth between the titration vessel and the cuvette facilitates rapid, quantitative sample exchange. Three cycles normally suffice, and these can be completed within 15–30 seconds. Nevertheless, the process would be somewhat difficult to automate. Indeed, if automation is the goal one should consider using a *flow system* (cf. Chapter 19), though it should be noted that this approach requires separate inlet and outlet lines, both inert. Thus, for a given tubing diameter and flow rate, considerably more time is required to ensure complete replacement of sample.

(c) In a third variant on the basic procedure, a *fiber-optics device* is used to establish direct contact between the titration vessel and the spectrometer. The technique is illustrated schematically in Fig. 16–3, and it altogether eliminates the need for a

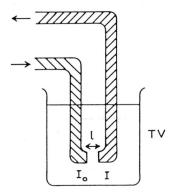

Fig. 16–3. Spectrometric titration setup utilizing a fiber-optics device with pathlength l.

cuvette. Monochromatic radiation of undiminished intensity (I_0) passes into the titration vessel (TV) through a fiber-optics link. There it is allowed to traverse a predetermined portion of solution (equivalent to a pathlength l) before returning to the spectrometer (with the intensity I) through a second set of fiber optics. The literature describes procedures that are applicable to both absorbance and fluorescence measurements,[11–13] and systems are available commercially (e.g., Metrohm model E 616) that combine photometers with the integrated fiber-optics devices required for photometric titration in the visible region.

A fiber-optics system offers the following *advantages*:

(1) Any desired titration vessel may be utilized.
(2) The pathlength l may be varied continuously with relatively little effort.
(3) No liquid whatsoever enters the spectrometer, so problems associated with thermostated sample compartments are avoided.

There are also *disadvantages*, however:

(1) The use of glass fiber-optics devices is limited to the visible region of the spectrum (i.e., above ca. 400 nm). Even with expensive quartz systems, transmission is found to decrease significantly below about 250–300 nm.
(2) Any fiber-optics system must be fully integrated into the associated spectrometer. Since it is extremely difficult to construct a reasonably symmetrical double-beam instrument based on this principle, most work to date has been restricted to single-beam mode, with all the accompanying disadvantages (cf. Chapter 18).
(3) That portion of the titration vessel subject to spectral measurement must be carefully shielded from external stray light. This requirement severely restricts the possibility of exercising visual control over the process (ensuring, for example, that no gas bubbles have entered the light path).

(4) It is difficult and costly to construct a fiber-optics probe that is completely inert to the effects of strong acids and bases, organic solvents, etc.

16.2.2 Titration Inside the Spectrometer

One way to avoid the need for a direct connection between the titration vessel and the spectrometer is to perform the entire titration inside the sample chamber of the spectrometer. This may be accomplished in either of two ways:

(a) The sample chamber may be outfitted with a *large, custom titration vessel*, a portion of which is equipped with parallel planar quartz windows that allow it to function as a cuvette.[14]

The following *disadvantages* are associated with this approach:

(1) Many spectrometers are so designed as to virtually exclude the possibility of introducing a titration vessel of sufficient size.

(2) Any such apparatus is certain to be expensive and relatively inflexible.

(3) It is unlikely that pathlength variability could be achieved in any convenient way.

(4) Adequate thermostatic control is problematic, especially since in most instruments only the walls of the sample compartment are subject to temperature regulation.

(b) Alternatively, one might carry out a *microtitration within a standard cuvette*.[15] In this case there is a different set of disadvantages to be confronted:

(1) Precision microburets are required, if only because of the need to compensate adequately for dilution effects (see below).

(2) Stirring is necessary within the cuvette, and this entails construction of an appropriate cuvette housing.

(3) It is difficult to prevent diffusion of titrant through the buret tip (see below).

(4) Gas bubbles have a tendency to accumulate along the windows of the cuvette.

(5) Not all the spectroscopic techniques referred to in Section 16.2.1(b) lend themselves readily to miniaturization.

(6) Pathlength changes are difficult (and costly) to implement.

16.3 Optimal Sample Size for Spectrometric Titration

Microscale titration techniques have a more limited impact on sample requirements than might at first be supposed. Table 16–1 provides an informative comparison of methods 16.2.1(b) and 16.2.2(b) as they apply to a case in which the maximum molar absorptivity ε_λ is assumed to be 10 000 cm^{-1}M^{-1} and the corresponding absorbance A_λ is expected to have the value 1.

Table 16–1. Sample requirements for absorption spectrometric titrations conducted inside and outside the spectrometer (assuming $A_{max} = 1$ for $\varepsilon_{max} = 10\,000$ cm^{-1}M^{-1}).

Method	Pathlength l	Concentration c_0	Total Volume V_0	Amount of sample required
Titration outside the spectrometer, cf. Sec. 16.2.1(b)	1 cm	10^{-4} M	500 mL	$5 \cdot 10^{-5}$ mol
	10 cm	10^{-5} M	50 mL	$5 \cdot 10^{-7}$ mol
Microtitration within the spectrometer, cf. Sec.16.2.2(b)	1 cm	10^{-4} M	≈ 3 mL	$3 \cdot 10^{-7}$ mol
	10 cm	10^{-5} M	≈ 10 mL	$1 \cdot 10^{-7}$ mol

Under optimum conditions for a small sample ($l = 10$ cm, $V_0 = 50$ mL, $c_0 = 10^{-5}$ M), titration outside the spectrometer using method 16.2.1(b) requires $5 \cdot 10^{-7}$ mol of protolyte. Microtitration directly in the cuvette by method 16.2.2(b) requires $3 \cdot 10^{-7}$ mol of protolyte with a 1 cm pathlength, an amount that may be reduced to ca. $1 \cdot 10^{-7}$ mol by providing a 10 cm optical path. It would therefore appear that confronting the extra problems accompanying miniaturization is warranted only if scale reduction can be combined with full automation.

It should be noted that the entries in Table 16–1 refer only to absorption spectrometry, and the situation may be somewhat different in the case of other techniques (cf. Sec. 18.2).

16.4 Introducing the Titrant: Burets and Other Alternatives

Most of the titrations discussed in this book are of the acid–base type, in which the independent variable is a function of some titrant. In fact, the variable itself plays no direct role in the evaluation of the titration function $A(\lambda,\mathrm{pH})$ except to the degree that it produces a dilution effect. Table 16–2 provides an overview of the level of reproducibility and accuracy that can be achieved using modern, high-quality burets. The acid–base titrations that have served as examples throughout the book were conducted almost without exception in 500-mL titration vessels, so it is safe to assume that buret errors would have a negligible effect on the results.

Table 16–2. Reproducibility and accuracy attainable with precision burets.

Buret type	Cylinder volume	Reproducibility	Accuracy
Manual microburet (e.g., Metrohm model E 457)	0.5 mL	0.0002 mL	0.005 mL
	5.0 mL	0.005 mL	0.02 mL
Multiburet with interchangeable reservoirs (manual or motorized; e.g., Metrohm models E 415, E 485)	5 mL	0.005 mL	0.02 mL
	10 mL	0.005 mL	0.02 mL
	20 mL	0.010 mL	0.03 mL

In general, (statistical) buret errors are likely to be overshadowed by systematic errors resulting from *diffusion of titrant from the buret tip*. This conclusion should come as no great surprise; after all, during a 50-step titration the buret tip is apt to remain immersed in the titration solution for several hours, which provides ample time for a considerable amount of diffusion. The extent of the problem may be reduced significantly by introducing a piece of platinum wire into the buret tip, thereby transforming the actual opening into a tiny coaxial ring. An even better solution is to obtain a buret tip equipped with a microvalve (e.g., Metrohm model EA 1118–32), an expedient that can reduce titrant outflow to a matter of nanoliters per hour. Yet another option is to remove the buret tip from the solution after each addition of titrant, although doing so is a major inconvenience.

The amount of titrant required near the extremes of the pH scale may be minimized by making use of acid and base solutions of various concentrations. Thus, within the pH range 3–11 it is reasonable to work with 0.1 M solutions of HCl and NaOH (assuming an initial volume of 100–500 mL), but outside this range it is more appropriate to employ 1 M (or even 5 M) solutions. An entire titration may be performed with 1 M acid and base, but in this case it is absolutely essential that diffusion-resistant buret tips be utilized. If titration frequently requires the use of several different acid and base solutions it is wise to invest in a "multiburet", a device equipped with a set of interchangeable titrant reservoirs.

All alkaline titration solutions should be prepared using CO_2-free water. The corresponding buret reservoirs should be carefully flushed with N_2 and then protected against subsequent air contamination either with a calcium chloride tube or with a liquid barrier. These precautions will ensure absorption spectra that are free of artifacts due to CO_2 in the region below 250 nm.

In some cases it is possible to avoid liquid titrants altogether. For example, a strongly alkaline titration limit may be approached with essentially no dilution errors by adding a *solid titration reagent*: NaOH pellets. In the case of redox titrations it may be feasible to use a *gaseous titrant*, such as O_2 supplied at a variety of partial pressures.[16] Alternatively, *coulometric generation* of a titrant may represent a viable option.[17]

Literature

1. R. E. Benesch, R. Benesch, *J. Amer. Chem. Soc.* 77 (1955) 5877.
2. D. E. Metzler, E. E: Snell, *J. Amer. Chem. Soc.* 89 (1967) 2895.
3. R. J. Jonson, D. E. Metzler, *Meth. Enzymol.* 18 A (1970) 433.
4. *Handbook of Biochemistry*, 2nd ed. The Chemical Rubber Co., Cleveland 1970. Section J 243.
5. G. Kortüm, *Lehrbuch der Elektrochemie*. Verlag Chemie, Weinheim 1972. Chap. XI.
6. W. P. Jencks, *Catalysis in Chemistry and Enzymology*. McGraw-Hill, New York 1969. Chap. 2.
7. W. T. Jenkins, L. D'Ari, *J. Biol. Chem.* 241 (1966) 5667.
8. H. Lachmann, *Dissertation*. Tübingen 1973.
9. R. Blume, H. Lachmann, H. Mauser, F. Schneider, *Z. Naturforsch.* 29 b (1974) 500.
10. M. Meloun. M. Javurek, *Talanta* 31 (1984) 1083.
11. A. Jacobsen, H. Dieslich, *Angew. Chem.* 85 (1973) 468.
12. A. Jacobsen, *Feinwerktechnik und Meßtechnik* 83 (1975) 117.
13. A. M. Gary, J. P. Schwing, *Bull. Soc. Chim. Fr.* 1972, 3654.
14. B. Nickel, H. Mauser, U. B. Hetzel, *Z. Phys. Chem. N. F.* 54 (1967) 199.
15. G. P. Foust, D. B. Burleigh jr., S. G. Mayhew, C. H. William jr., V. Massey, *Anal. Biochem.* 27 (1969) 530.
16. L. A. Kiesow, J. W. Bless, J. B. Shelton, *Science* 179 (1973) 1236.
17. S. Ebel, W. Parzefall, *Experimentelle Einführung in die Potentiometrie*. Verlag Chemie, Weinheim 1975.

17 Electrochemical Methods

The electrochemical titration variable upon which we focus our attention in this chapter is *pH*, the parameter that plays the most prominent role in subjects dealt with in the present book. Other electrochemical quantities such as ionic activity (or p_{Ion}) and redox potentials are treated elsewhere and in appropriate contexts (i.e., in Sections 13.5 and 15.1, respectively).

17.1 Problems Associated with the Conventional pH Scale

Spectrochemical titration relies heavily on the ability to measure accurately either the hydronium ion activity $a_{H_3O^+}$ (:= h) of a solution, or else the closely related quantity pH:

$$\text{pH} := -\log h = -\log c_{H_3O^+} \cdot f_{H_3O^+} \tag{17-1}$$

The requisite information is normally obtained by electrometric means; that is, with the aid of a pH meter and an appropriate series of electrodes. Unfortunately, the activity (or activity coefficient) of an individual ion such as H_3O^+ is not subject to precise thermodynamic definition; in fact, it is only ascertainable in certain specific situations (e.g., by using bridged concentration cells operating within the range of validity of the Debye–Hückel law[1,2]).

In general, the most viable alternative is to work with what is known as the *conventional pH scale*, in which the "true" pH scale is normalized so that it conforms to a particular type of measurement.[2–5] Thus, the electrical potential E of the galvanic cell

$$\text{Pt (H}_2) \mid \text{solution } X \mid \text{conc. KCl solution} \mid \text{reference electrode} \tag{17-2}$$
$$\text{(salt bridge)}$$

is given by the equation

$$E = E_0' - \frac{RT}{F} \cdot \ln h + \sum \varepsilon_{\text{diff.}} \tag{17-3}$$

Here the standard potential E_0' includes contributions due to the reference electrode, but these may be regarded as constant so long as all measurements are made with the same system (e.g., a calomel or silver/silver chloride electrode) and provided that the system is one that leads to highly reproducible results. The expression $\sum \varepsilon_{\text{diff.}}$ incorporates several different diffusion potentials (*junction potentials*) $\varepsilon_{\text{diff}}$ reflecting diffusion-controlled potential jumps that occur at phase boundaries within the electrochemical measuring sequence.

Accepting the conventional pH scale as valid is equivalent to assuming that the expression $\sum \varepsilon_{\text{diff.}}$ in equation 17-3 is a constant; i.e., the sum of the diffusion potentials is taken to be independent of the nature and concentrations of all species present in the solution X. There can be little doubt that this assumption represents an approximation, but in a great many practical situations it is quite acceptable, especially when comparisons are being made between solutions of similar pH and similar ionic strength. Putting the matter mathematically, one assumes that

$$E' := E_0' + \varepsilon_{\text{diff.}} = \text{constant} \tag{17-4}$$

To the extent this is true, it follows from equation 17-3 at 25 °C that

$$E = E' - \frac{RT}{F} \cdot \ln h = E' + 0.05916 \cdot \text{pH} \tag{17-5}$$

Any measured E' value is a function not only of the corresponding reference electrode, but also of the salt bridge and all phase boundaries; therefore such a measurement may *not* be regarded as a standard reference value even within the conventional pH scale. Instead, one is obliged to make frequent use of what are known as *standard buffer solutions*, solutions whose pH values have been determined as precisely as possible within guidelines established on a national or international basis (e.g., NBS, DIN 19266,[6] IUPAC[7]). Every electrochemical measuring system must be calibrated with the aid of reliable standard buffers; only then is it possible to assume that a linear relationship (i.e., equation 17-5) has been established between E and pH throughout the range over which a set of measurements is to be made.

The considerations above pertain not only to the single-electrode case exemplified by the cell Pt (H_2) | solution X; indeed, provided appropriate precautions are taken, they are equally applicable to the more complicated system defined by the typical glass electrode, a device that is especially convenient and easy to operate (see below).

17.2 Calibrating a pH Meter Using Standard Buffers

Table 17–1 describes the makeup of the most important standard buffers, and it also reveals how the pH of each buffer varies as a function of temperature over the range 0–40 °C. All the buffers listed are commercially available either in the form of concentrates or as ready-to-use solutions. They may also be prepared in the laboratory as needed, but doing so requires careful adherence to precise sets of instructions.[2,6,8]

Buffers may be used to calibrate pH meters in a variety of ways:

(a) *Single-point calibration*: in this approach the pH meter is carefully adjusted so that the pH registered in the presence of one of the standard buffers matches that reported in the literature. This method is certainly the most convenient, but it is *inappropriate* for spectrometric titration purposes, because only pH values in the immediate vicinity of the standard in question may be relied upon with any certainty.

(b) *Two-point calibration*: Using both the electrical zero-point adjustment and the slope control potentiometer, the pH meter is set to report accurate pH values for *two* standard buffers. This method is suitable for titrations demanding only moderate precision, but titration should be confined to the interval between the pH values of the two standard buffers. Over this interval one may assume to a fair approximation that the pH value displayed on the meter corresponds directly with the conventional pH scale; no further correction is required.

(c) *Regression calibration*: In this most accurate approach the pH meter is first adjusted so that it matches two (arbitrarily chosen) standard buffers, again using the zero-point and slope controls as in the preceding method. (An even better procedure calls for employing a standard voltage source to set the zero-point for the electrode sequence and to match a theoretical value for the slope, such as 59.159 mV/pH at 25 °C.[2,8]) The next step is to measure as many standard buffers as possible, widely distributed over the pH range of interest. The results of these measurements are then plotted as a graph of pH (measured) vs. pH (standard), as shown in Fig. 17–1. Finally, linear regression techniques are used to produce a *calibration curve* for application in subsequent titrations. Any deviations observed between calculated and theoretical values for the electrode zero-point and slope may be used as indications of the condition of the electrodes, which are commonly subject to "aging".

Table 17-1. The pH values exhibited by various standard buffers, reported as a function of temperature over the range 0–40 °C, together with a description of the composition and stability of each buffer.

t (°C)	Potassium tetraoxalate	Potassium hydrogen tartrate	Potassium dihydrogen citrate	Potassium hydrogen phthalate	Phosphate	Phosphate	Borate	Sodium carbonate/ sodium hydrogen carbonate	Calcium hydroxide
0	1.666	–	3.863	4.003	6.984	7.534	9.464	10.317	13.423
5	1.668	–	3.840	3.999	6.951	7.500	9.395	10.245	13.207
10	1.670	–	3.820	3.998	6.923	7.472	9.332	10.179	13.003
15	1.672	–	3.802	3.999	6.900	7.448	9.276	10.118	12.810
20	1.675	–	3.788	4.002	6.881	7.429	9.225	10.062	12.627
25	1.679	3.557	3.776	4.008	6.865	7.413	9.180	10.012	12.454
30	1.683	3.552	3.766	4.015	6.853	7.400	9.139	9.966	12.289
35	1.688	3.549	3.759	4.024	6.844	7.389	9.102	9.925	12.133
40	1.694	3.547	3.750	4.035	6.838	7.380	9.063	9.856	11.984
Composition:	0.05 M $C_4H_3KO_8 \cdot 2\,H_2O$	Sat'd. soln. of $C_4H_5KO_6$ at 25 °C	0.05 M $C_6H_7KO_7$	0.05 M $C_8H_5KO_4$	0.025 M each of KH_2PO_4 and Na_2HPO_4	1.179 g KH_2PO_4 and 4.30 g Na_2HPO_4 per L soln. (25 °C)	0.01 M $Na_2B_4O_7 \cdot 10\,H_2O$	0.025 M each of Na_2CO_3 and $NaHCO_3$	Sat'd. soln. of $Ca(OH)_2$ at 25 °C
Stability:	Stable for ca. two months	Tendency toward fungal growth	Tendency toward fungal growth	Tendency toward fungal growth	Stable for ca. two months	Stable for ca. two months	Sensitive to CO_2	Sensitive to CO_2	Very sensitive to CO_2

17.2 Calibrating a pH Meter Using Standard Buffers

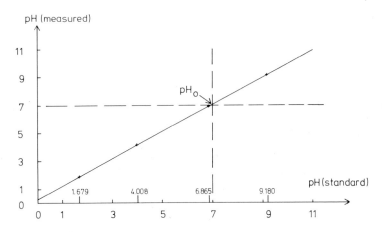

Fig. 17–1. Calibration curve for a pH meter (regression line).

Method (c) is the method of choice for precision titrations throughout the pH range 1.5–12.5. Its use requires that a "true" pH relative to the conventional pH scale be computed for each titration point by adjusting the apparent (measured) value with the aid of the following equation:

$$\text{pH (conventional)} = \frac{\text{pH (measured)} - 7 + S \cdot \text{pH}_0}{S} \qquad (17\text{–}6)$$

where pH_0 = zero-point potential at (measured) pH 7.0
S = slope of the calibration line

This equation also serves as the basis for what is sometimes known as the "electrode slope",[8] equal to $59.16 \cdot S$ mV/pH. Implementation of method (c) in a titration is greatly facilitated if a computer is available for evaluating the data.

Any calibration procedure involving standard buffers should be performed at the temperature that is expected to apply to the titration solution of interest (e.g., 25.0 ± 0.1 °C). To ensure consistency, equilibration time and even the stirring rate should also be carefully monitored.

Neutral and alkaline buffer solutions that are to be used repeatedly and over a prolonged period of time should be carefully flushed with nitrogen. Detailed information concerning the lifetimes of typical buffer solutions may be found in references 2 and 8, as well as in literature supplied by buffer manufacturers.

17.3 pH Measurements

Accurate measurement of any electrical potential presupposes that the system in question is in a state of reversible thermodynamic equilibrium. This means that current flow within the measuring circuit must be negligible, a condition that is met by modern pH meters with the aid of an extremely high input resistance ($> 10^{12}$ ohm). Reproducibility and accuracy with a precision instrument may be expected to fall within the range ± 0.001–0.003 pH units. This specification should apply to the entire region from pH 0 to pH 14 and should be valid over a time period of more than 10 hours.

The standard buffers used for performing calibrations with respect to the conventional pH scale are appropriate for work requiring this degree of precision, but only if measurements are made in the absence of liquid junctions (in cells without transference).[1,2,9,10]

In fact, however, most spectrometric titrations are carried out using *glass electrodes* in measuring circuits that *do* contain liquid junctions (cells *with* transference). Two arrangements are especially common:

(a) *Single-probe circuits* (cf. Fig. 17–2): In a system of this type the reference electrode R is surrounded concentrically by a glass electrode (normally of the calomel or silver/silver chloride variety). The latter maintains contact with the sample solution S through a tiny diaphragm D. Single-probe electrodes are quite convenient to use and they occupy very little space, an especially important consideration in the case of a microtitration. The most critical element is the diaphragm, since this may easily become clogged, and it can also be a source of interference potentials (see below).

(b) *Two-probe circuits*: In this case the reference electrode, together with its diaphragm, constitutes a wholly separate probe. The corresponding diaphragm is typically much larger than with a single-probe system. In order to avoid diaphragm problems, including clogging or "poisoning", some suppliers choose to equip their reference electrodes instead with a capillary tip through which a small amount of electrolyte solution (usually KCl) is constantly allowed to flow. This expedient has the disadvantage that it results in a gradual increase in ionic strength within the titration vessel.

The *calibration curve* for a glass electrode is very nearly linear throughout the pH range 1.5–12; that is, such an electrode performs in accordance with equation 17–5. Outside this range, however, significant deviations must be anticipated (known as "acidic error" and "alkaline error"), the precise nature of which will depend on the type of glass utilized. Appropriate correction curves are available from the electrode manufacturers.

Prior to recording pH data during a precision titration it is important that one allow for an *equilibration period*—under nitrogen—that is both consistent and sufficient (i.e.,

17.3 pH Measurements

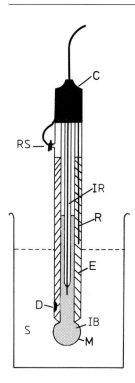

Fig. 17–2. Single-probe pH electrode (schematic representation). Components include: internal reference electrode, Ag/AgCl (IR); internal buffer solution (IB); reference electrode, silver wire (R); reference electrolyte, e.g. KCl, 3 mol/L (E); diaphragm junction between sample solution and reference electrolyte (D); rubber stopper, opened to perform measurements (RS); pH-sensitive glass membrane (M); connector (C); sample solution (S).

a minimum of 4–5 minutes). The time required for equilibration typically increases as an electrode ages.

The *accuracy* of any pH measurement established with a system that involves liquid junctions (i.e., cells with transference) is limited by the concentration dependencies of various diffusion potentials. Below 40 °C the systematic error to be expected from this source is less than 0.02 pH units (cf. ref. 11, for example), but even this means that it is inappropriate to report spectrometrically determined pK values to an accuracy greater than 0.02 pH units regardless of the fact that the reproducibility of the associated measurements may suggest precision that is an order of magnitude higher. Only in those few special cases in which it is possible to avoid an electrolyte bridge and a diaphragm (i.e., using cells without transference, where measurements are made in the absence of liquid junctions) can one take full advantage of the accuracy inherent in a modern pH meter calibrated with reliable standard buffers (cf. ref. 8).

Certain *types* of results may be achieved with greater accuracy, however, including pK determinations based solely on photometric data (cf., for example, Sec. 6.3.6), or ΔpK values (cf. Sections 7.4 and 8.4). The methods described in Chapter 11, in which simultaneous titration is used as a way of establishing *relative* pK values, also circumvent dependence upon pH data, in that the pK of a reference protolyte is treated as a known quantity comparable to the pH of a standard buffer.

Literature

1. G. Kortüm, *Lehrbuch der Elektrochemie*. Verlag Chemie, Weinheim 1972. Chap. X.
2. K. Schwabe, *pH-Meßtechnik*. 4. Aufl. Verlag Th. Steinkopff, Dresden 1976.
3. G. Kortüm, *Lehrbuch der Elektrochemie*. Verlag Chemie, Weinheim 1972. Chap. IX.
4. C. H. Hamann, W. Vielstich, *Elektrochemie I*. Verlag Chemie, Weinheim 1975.
5. E. J. King, *Acid-Base Equilibria*. Pergamon Press, London 1965.
6. *DIN 19 266*, Beuth-Vertrieb, Köln 1971.
7. *IUPAC Handbuch der Symbole und der Terminologie physikochemischer Größen und Einheiten*. Deutsche Übersetzung. Verlag Chemie, Weinheim 1973.
8. S. Ebel, W. Parzefall, *Experimentelle Einführung in die Potentiometrie*. Verlag Chemie, Weinheim 1975.
9. G. Kortüm, W. Vogel, K. Andrussow, *Pure Appl. Chem.* 1 (1960) 190.
10. R. G. Bates, *Determination of pH*. Wiley, New York 1964.
11. *Puffersubstanzen, Pufferlösungen, Puffer-Titrisol®-Konzentrate*. Publication of Firma Merck, Darmstadt (West Germany), No. 23/6344/5/480R.

18 Spectrometric Methods

18.1 UV–VIS Absorption

18.1.1 Spectrometric Data Collection

Multiple-wavelength UV–VIS spectrometric titration can be conducted with the aid of either a *single-beam* or a *double-beam* spectrophotometer (cf. Figures 18–1a and 18–1b). Solution in a reference cuvette R is used to define the 100% transmission level at each wavelength of interest. With a single-beam instrument this means that each measurement involving the sample cuvette S must be preceded by a calibration with the reference cuvette R. Using ordinary analytical cuvettes and a very simple type of titration setup, such as that described in Sec. 16.2.1(a), the procedure is quite straightforward, but any more sophisticated arrangement is likely to demand frequent readjustments in several connecting tubes (or fiber-optics devices), some of which may prove quite inflexible. Only by the use of very sturdy cuvette holders can accidental spillage of the cuvette contents be avoided.

A single-beam apparatus is of course quite unsatisfactory for recording analog titration spectra, so a strong case can be made for working exclusively with a *recording, double-beam spectrophotometer*. The *optimal* instrument is one capable not only of generating complete titration spectra but also of pausing at 5–10 preselected wavelengths to permit the display or storage of a set of precise absorbance values A_λ (cf. Chapter 19). The principal *advantages* of this type of system are as follows:

Fig. 18–1. (a) Optical arrangement of a single-beam absorption spectrophotometer with a light source L, monochromator M, sample cell S, reference cell R, and photomultiplier PM; (b) Optical arrangement of a double-beam spectrophotometer equipped with a mechanical chopper Ch.

(1) Recorded titration spectra are entirely compatible with the corresponding A_λ (pH) data, since both sets of information are derived from the same experiment. This equivalence of data makes it easier to discover the presence of systematic errors, and to verify the effectiveness of corrective measures that may need to be taken.

(2) Interrupting the scan program momentarily at specific wavelengths ensures that such points will be defined just as precisely as they would be with a research-quality single-beam instrument, eliminating the "overshoot errors" that frequently accompany rapid scanning.

(3) Damping factors (integration times) may be optimized separately for titration spectra and individual A_λ (pH) measurements, thereby minimizing the overall time required for a complete spectral analysis.

Most of the spectrometric titrations described in this book were performed using an instrument with these characteristics (a Zeiss model DMR 21 WL). Total *elapsed time* for a single analysis (a spectrum recorded over the range 500–200 nm, scanned at a rate of 120 nm/min, together with separate measurements at 8–10 fixed wavelengths, requiring 5 seconds each) averaged 4–5 minutes, including time for resetting the monochromator to the initial wavelength. In other words, the entire procedure can be completed in the time required for equilibrating the pH electrode. It usually takes about one additional minute to establish the conditions corresponding to the next titration step, so roughly 10–15 sets of data can be accumulated in the course of an hour. A useful "rule of thumb" suggests allocating about 15–20 titrations steps to each stage of equilibrium in a precision titration of a complex reaction system. Several hours of work may thus be necessary, but with modern instrumentation the task can still be completed before instrumental drift becomes appreciable.

Most instruments today utilize a single *detector* throughout the entire UV–VIS spectral region of ca. 200–800 nm. Typically this detector is a photomultiplier with enhanced NIR-sensitivity. Two *light sources* are required, however: an incandescent lamp for the visible region and a deuterium lamp for the UV. The source must ordinarily be changed at about 350 nm, but with many instruments this change occurs automatically.

<small>Certain of the spectra reproduced in this book (e.g., Figures 6–1, 6–12, 6–15, and 6–16) were recorded over the range 200–500 nm using *only* the deuterium lamp, causing artifacts in the region 460–490 nm.[1,2]</small>

Spectrometric titration in aqueous solution normally does not demand a high degree of resolution, so a system with a *single monochromator* may well suffice. Prism monochromators provide the highest resolution in the UV-region, while grating devices are superior in the visible. Nevertheless, *double monochromators* offer the advantage that they permit passage of much less stray light, and this difference can prove significant in a spectrometric titration involving high absorbances ($A_\lambda > 2$) in the short-wavelength region of the UV (cf. Sec. 18.1.2).

18.1 UV–VIS Absorption

With respect to multiple-wavelength measurements, *wavelength reproducibility* is much more important than high resolution. Indeed, effects due to slight shifts in wavelength are sometimes mistakenly interpreted as evidence of poor photometric reproducibility, especially if measurements have been taken along a shoulder of a strong absorption band. The best of the modern spectrometers (including the one mentioned above) provide a level of *wavelength reproducibility* (in automatic mode) exceeding ± 0.01 to 0.02 nm over the range 200–400 nm. *Photometric reproducibility* with a high-quality instrument (for $A_\lambda \leq 1$) should prove better than ± 0.001 absorbance units, perhaps even ± 0.0002 units (standard deviation for an integration time of 3 s). Nevertheless, outstanding specifications such as these will be relevant in the context of a spectrometric titration only if a fixed titration cuvette is employed, as in the titration system described in Sec. 16.2.1(b). Drift in the *zero-point setting* ($A_\lambda = 0$) of an instrument in this category is generally less than ± 0.0004 units per hour.

To demonstrate the precision and reproducibility that can be achieved in a spectrometric titration we conclude this section with one further example derived from a single-step acid–base equilibrium system. Figure 18–2 shows titration spectra reflecting dissociation of the phenolic –OH in *2-hydroxy-5-nitro-α-toluenesulfonic acid*. Six sharp isosbestic points are present within the pH range 3.5–10. The characteristics of the spectrometer used are such that the region from ca. 365 nm to 320 nm is displayed twice (at two different scale expansions).

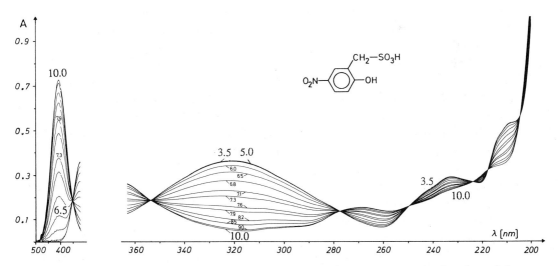

Fig. 18–2. Titration spectra obtained during an absorption spectrometric titration of 2-hydroxy-5-nitro-α-toluenesulfonic acid ($c_0 = 5 \cdot 10^{-5}$ M).

Approximately 2.5 mL of 0.1 M HCl served as the titrant in this case (pH range: 3.5–10). Since the initial volume V_0 was 500 mL, dilution errors are limited to ≤ 0.5%, with no perceptible effects on the analog titration spectra. Tiny breaks or irregularities at 400, 320, and 210 nm are a result of interrupting the recording process to permit collection of digital data for a subsequent $A_\lambda(\text{pH})$ evaluation. The A-diagrams themselves are not depicted, but all were found to be strictly linear. Self-absorption by hydroxide ion is not a problem because the pH was not allowed to exceed 10. The strongly acidic sulfonic acid group was not titrated in the interest of minimizing dilution.

The compound 2-hydroxy-5-nitro-α-toluenesulfonic acid is a useful test substance for evaluating the performance of any recording spectrophotometer because it is highly stable and its titration spectra are unusually rich in isosbestic points.[3,4] It also provides a convenient demonstration of the way spectrometric studies can reveal the effects of pH on a chemical reaction. Thus, it is not the acid itself that one weighs into the titration vessel; rather, the acid is generated *in situ* in the course of the following hydrolysis reaction:[1,4]

$$\text{O}_2\text{N}-\underset{}{\text{C}_6\text{H}_3}\underset{\text{CH}_2-\text{SO}_2-\text{O}}{\overset{}{\bigg\langle}} \quad \xrightarrow[+\text{H}_2\text{O}]{(\text{OH}^-)} \quad \text{O}_2\text{N}-\text{C}_6\text{H}_3(\text{CH}_2-\text{SO}_3\text{H})-\text{OH} \tag{18-1}$$

Hydrolysis of the starting 2-hydroxy-5-nitro-α-toluenesulfonic acid sultone is an irreversible process over the entire pH range of interest, so the final solution contains only a single stable substance that may be titrated spectrometrically without interference by extraneous species.

18.1.2 The Lambert–Beer–Bouguer Law and its Associated Error Function

The Lambert–Beer–Bouguer law itself is discussed in detail in Sec. 3.1, as are the assumptions upon which it is based. Our concern here is with certain technicalities that have important practical implications.

The quantity that one actually measures in the course of a spectrophotometric analysis is *light intensity* I_λ, not absorbance A_λ. The latter is defined in a secondary way as a part of the Lambert–Beer–Bouguer law:

$$A_\lambda := \log \frac{I_{0\lambda}}{I_\lambda} = l \cdot \varepsilon_\lambda \cdot c \tag{18-2}$$

The reason for defining absorbance A_λ is that it is much easier to work with a quantity that is directly proportional to concentration. Nevertheless, most of the error in a photometric measurement is a direct function of the more fundamental quantity I_λ, so

18.1 UV–VIS Absorption

is important to ascertain how such error is propagated in terms of A_λ. We begin by differentiating equation 18–2 with respect to I_λ:

$$\frac{dA_\lambda}{dI_\lambda} = 0 - \frac{\log e}{I_\lambda} \tag{18-3}$$

Rearranging terms and dividing by A_λ leads to[5-7]

$$\frac{dA_\lambda}{A_\lambda} = -\frac{0.4343}{A_\lambda} \cdot \frac{dI_\lambda}{I_\lambda} = -\frac{0.4343}{A_\lambda \cdot 10^{-A_\lambda}} \cdot \frac{dI_\lambda}{I_{0\lambda}} \tag{18-4}$$

Figure 18–3 is a plot of the relative error $\Delta A_\lambda/A_\lambda$ as a function of A_λ. Most of the relative error in the measured light intensity I_λ is a result of statistical noise (the Schottky effect), and it is generally constant; in Fig. 18–3 the quantity $\Delta I_\lambda/I_{0\lambda}$ has been assigned the value 0.15%.

The *photometric error function* is seen to be a trough-shaped curve with a minimum at $A_\lambda = 0.4343$. Between the limits $0.2 < A_\lambda < 0.8$ the relative error in A_λ is less than 0.5%, and it is this observation that serves as the basis for the general rule that photometric measurements should be restricted to this range (or at least to the range $0.1 < A_\lambda < 1.2$). Observing this precaution normally presents no great problem, particularly since a system like that described in Sec. 16.2.1(b) is adaptable to cuvettes with various pathlengths (e.g., from 0.5 to 10 cm). Indeed, the use of flow cells extends the range of possible pathlengths down into the region from 0.5 cm to ≤ 0.5 mm.*

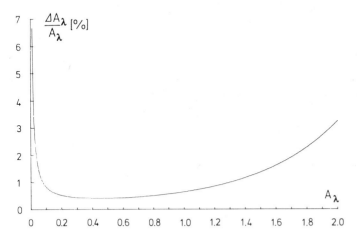

Fig. 18–3. Photometric error function associated with the Lambert–Beer–Bouguer law as a result of statistical noise (the Schottky effect).

---

* References 8 through 11 provide more extensive discussions of photometric errors, as well as suggestions regarding choice of those wavelengths best suited to spectrometric analysis.

The rapid increase in the error function below $A_\lambda = 0.1$ is due to the effect of taking a quotient $(I_\lambda/I_{0\lambda})$ of two nearly equal numbers. The only way to surmount this obstacle is to make A_λ larger (e.g., by increasing either the concentration or the pathlength). The increase above $A_\lambda = 1$ is somewhat different. It results from insufficient light reaching the photomultiplier, and some improvement may be achieved by increasing the extent of signal amplification. With modern low-noise amplifiers it is often possible to obtain reliable measurements with A_λ values as high as three. Then the limiting factor above $A_\lambda = 2$ is likely to be not the noise level of the photomultiplier but rather a systematic error attributable to *stray light* issuing from the monochromator.[7,12–14]

Reducing stray light of all kinds is important, but particularly in the short wavelength UV region it is necessary that special attention be devoted to the *spectral bandwidth* (or the *halfwidth*) $\Delta\lambda$. The Lambert–Beer–Bouguer law is a limiting relationship, and it is strictly applicable only with a truly monochromatic light source. In fact, light that emerges from a monochromator encompasses a (relatively narrow) *band* of wavelengths; the width of this band at its half-height $h/2$ (cf. Fig. 18–4) is often used as a measure of spectral purity. In the case of a grating monochromator, bandwidth is essentially independent of wavelength, but with a prism the two quantities are linked by a rather complicated functional relationship.[6,7] Bandwidth is also directly proportional to the width of the monochromator exit slit. Most modern recording instruments permit direct display of the current slit width, usually in nm. Changes in slit width are programmed into the scan sequences of nearly all double-beam spectrometers, with the slits opening more widely as the short-wavelength UV-region is approached. One must therefore remain alert to the possibility that the prerequisites of the Lambert–Beer–Bouguer law will not be adequately met at very short wavelengths. A useful generalization states that if measurements must be taken on the shoulder of an absorption peak, the corresponding bandwidth should not be allowed to exceed 1–2 nm. References 6, 15, and 16 provide comprehensive discussions of the error function associated with the Lambert–Beer–Bouguer law as it applies to both oligo- and polychromatic light sources.

A further consequence of error-function considerations is that the substance placed in the *reference cell* should be responsible for as little absorbance as possible.

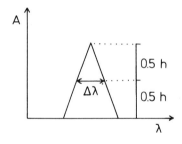

Fig. 18–4. Spectral energy distribution (slit function) $A(\lambda)$ for a monochromator with entrance and exit slits of the same width (schematic representation based on ref. 6).

18.1 UV–VIS Absorption

The ideal choice is *extremely pure water*. Use of a more strongly absorbing reference solution forces an increase in the slit width, which produces a corresponding increase in spectral bandwidth.

Apart from the fact that there may be *physical deviations* from the Lambert–Beer–Bouguer law (as a result of insufficient monochromaticity, whether from stray radiation or excessive bandwidth), *chemical deviations* may also occur due to spectroscopically observable interactions among absorbing species, or between these species and non-absorbing components of the sample solution. Chemical deviations are most easily detected either by examining a dilution series A_λ vs. c_i or else with the aid of A-diagrams[7] (cf. Sections 3.1 and 3.2). In the absence of competing association equilibria, most protolytes tend to obey the Lambert–Beer–Bouguer law reasonably well so long as their concentrations are kept below ca. 10^{-3}–10^{-4} M.[6,14]

Finally, we note once again the possibility of problems due to *absorbance by the titrant R*. Figure 18–5 shows absorption spectra for 0.1 M solutions of both HCl and NaOH. Assuming that dilution from the titrant is kept below 3%, the final concentration of HCl or NaOH in a titration solution should never be greater than $3 \cdot 10^{-3}$ M. Under these circumstances, self-absorption by NaOH[17] only begins to cause problems (in the region 200–215 nm) when the pH exceeds 12. Potential interference due to HCl is also restricted to the region below 210 nm, and it only becomes apparent under very strongly acidic conditions. It must be noted, however, that these assertions are based on the assumption that only high-quality reagents are employed, and that all sodium hydroxide solutions are carefully protected against contamination by CO_2.

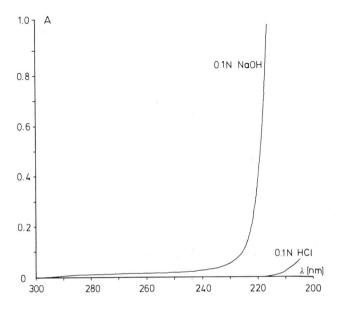

Fig. 18–5. Absorption spectra of 0.1 M NaOH and HCl.

18.2 Fluorescence

18.2.1 Spectrometric Data Collection

Fluorescence measurements are usually made with a *single-beam spectrometer* like that depicted schematically in Fig. 18–6. Provision of a light source L with quasi-continuous emission characteristics (e.g., a high-pressure xenon lamp), together with two monochromators M_{Ex} and M_{Em}, makes it possible to record either *excitation spectra* $I^F(\lambda_{Ex})$, in which emission is measured at a fixed wavelength λ_{Em}, or *emission spectra* $I^F(\lambda_{Em})$ resulting from irradiation at a fixed excitation wavelength λ_{Ex}. It is a spectrum of the latter type that is normally referred to as a *fluorescence spectrum* in the strict sense of the term.

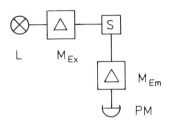

Fig. 18–6. Optical arrangement of a single-beam spectrofluorometer consisting of excitation- and emission-monochromators M_{Ex} and M_{Em}, respectively, light source L, sample cell S, and photomultiplier PM.

Sometimes the excitation source employed is one whose energy output is limited to a relatively small number of very narrow lines, as in the case of a mercury-vapor lamp. *Line sources* of this type offer the advantage of extremely intense radiation with exceptional spectral purity (i.e., minimal bandwidth). Individual emission lines are isolated with the aid of inexpensive *filters* (either colored glass or interference devices), thereby eliminating the need for a source monochromator (cf. Fig. 18–7). The disadvantages of this approach are a restricted choice of excitation wavelength and loss of the ability to record excitation spectra.

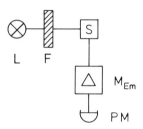

Fig. 18–7. Optical arrangement of a single-beam spectrofluorometer containing a mercury line-source L and a filter monochromator F (other labels as in Fig. 18–6).

18.2 Fluorescence

With most fluorometers, fluorescence radiation emitted by the sample is observed along a horizontal axis at an angle of 90° relative to the source. Some instruments use a vertical 90° arrangement, but this rules out magnetic stirring within the sample cuvette.

Fluorescence spectrometers generally employ *single monochromators* because of a desire to obtain the strongest possible emission signal, resulting in high sensitivity. Large spectrometric systems sometimes employ *double monochromators* on the excitation side in the interest of improving monochromaticity and reducing stray light errors. High-pressure xenon lamps are the most frequently used sources despite the fact that they are subject both to slow, continuous changes in intensity as well as to short-term intensity fluctuations caused by electrical discharge irregularities. Both phenomena further underscore the disadvantages inherent in single-beam spectrometry. In recent years there has been increasing interest in what are known as *ratio spectrometers*.[18] Within an instrument of this type a portion of the radiation leaving the source monochromator is diverted in such a way that it by-passes the sample and impinges directly upon a second photomultiplier, whose output then provides the equivalent of a reference signal (cf. Fig. 18–8). An electronically developed quotient of the signals emerging from the two photomultipliers is largely free of the effects of lamp fluctuation. Further improvements have resulted from the finding that a xenon source can be stabilized considerably by subjecting it to the influence of a magnetic field. Spectrometric titration typically requires several hours to complete, so it is most advantageous to secure access to an instrument of the ratio type, preferably one that also incorporates magnetic field stabilization.

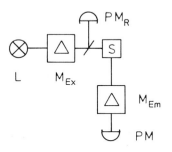

Fig. 18–8. Optical arrangement of a ratio spectrofluorometer containing a supplementary reference photomultiplier PM$_R$ (other labels as in Fig. 18–6).

On the other hand, there is relatively little to be gained by going the further step of choosing what is known as a *fully-compensated spectrofluorometer*. This is an instrument designed to overcome problems of distorted *emission spectra* caused by fluctuations in photomultiplier sensitivity and wavelength-related changes in the efficiency with which radiation is transmitted through the optical components. Because of factors such as these, uncompensated emission spectra recorded on different instruments differ in their relative peak intensities. The same is true with respect to *excitation spectra*, which are especially subject to deformations related to unique emission characteristics of a particular source. These would be serious shortcomings if one's goal were identifying or characterizing specific spectral bands, making structural assignments, or examining relationships between fluorescence spectra and absorption spectra. Indeed, a number of techniques have been developed to combat the problem, and it is the subject of extensive discussion in the technical literature.[19-21] Nevertheless, distortions of this type are essentially irrelevant to the researcher interested only in spectrofluorometric titration, including matrix-rank analysis, so long as the consequences are invariant throughout the course of a particular investigation.

Apart from factors related to source fluctuation and spectral distortion, the specifications applicable to the various mechanical, optical, and electronic aspects of modern fluorescence spectrometers compare favorably with those of absorption spectrometers.

18.2.2 Quantitative Fluorescence Spectrometry and Predictable Sources of Error

In comparison with absorption spectrometry, the fluorescence technique offers sensitivity that is enhanced by a factor of 100–1000,[22] but the method is also considerably more subject to interferences. A great many prerequisites must be fulfilled before one can be assured that there will be a linear relationship between observed fluorescence intensity and concentration. The discussion that follows constitutes a somewhat simplified overview of the situation.[19,23]

We begin by assuming that the Lambert–Beer–Bouguer law is fully applicable with respect to absorption at the excitation wavelength λ':

$$A_{\lambda'} = \log \frac{I_{0\lambda'}}{I_{\lambda'}} = l \cdot \varepsilon_{\lambda'} \cdot c \quad \text{or}$$

$$I_{\lambda'} = I_{0\lambda'} \cdot 10^{-A_{\lambda'}} = I_{0\lambda'} \cdot 10^{-l \cdot \varepsilon_{\lambda'} \cdot c} \tag{18-5}$$

Neglecting effects due to reflection, stray light, etc., it may be further assumed that any observed reduction in the intensity $I_{0\lambda'}$ of the source radiation is a consequence of absorption:

$$I_{\lambda'} = I_{0\lambda'} - I_{\lambda'}^A \tag{18-6}$$

In the event that the intensity of emitted fluorescent radiation I^F is also proportional to the intensity of absorbed radiation $I_{\lambda'}^A$ (Knoblauch's law), then

$$I^F \sim I_{\lambda'}^A = I_{0\lambda'} - I_{\lambda'} \tag{18-7}$$

From equation 18–5 it follows that

18.2 Fluorescence

$$I^F \sim I_{0\lambda'} \cdot (1 - 10^{-A\lambda'}) \qquad (18\text{-}8)$$

The expression $(1 - 10^{-A\lambda'})$ may be developed as a MacLaurin series, leading to a result of the following form:

$$1 - 10^{-A\lambda'} = A_{\lambda'} \cdot \ln 10 - \frac{1}{2!} \cdot A_{\lambda'}^2 \cdot (\ln 10)^2 + \frac{1}{3!} \cdot A_{\lambda'}^3 \cdot (\ln 10)^3 - \cdots \qquad (18\text{-}9)$$

If one now accepts the restriction that absorbance will be kept *very small* (i.e., $A_{\lambda'} \ll 0.1$), this series may be truncated after a single term, which implies a direct proportionality between I^F and c:

$$I^F \sim I_{0\lambda'} \cdot A_{\lambda'} \sim c \qquad (18\text{-}10)$$

More rigorous discussions of this proportionality relationship and its derivation are provided in references 18, 19, and 24.

A great many fluorescent substances have been found to obey the following general equation for intensity of fluorescent emission I_λ^F at a given wavelength λ, with excitation at the wavelength λ':[7,14,25-28]

$$I_\lambda^F = I_{0\lambda'} \cdot l \cdot \Gamma \cdot R(A_{\lambda'}) \cdot \frac{1 - 10^{-A\lambda'}}{A_{\lambda'}} \cdot \sum_i F(\lambda, \lambda')_i \cdot \varepsilon_{\lambda'_i} \cdot c_i \qquad (18\text{-}11)$$

where Γ is a factor that takes into account the geometry of the system
$R(A_{\lambda'})$ is a function to correct for multiple reflections within the cuvette
$F(\lambda, \lambda')$ = spectral internal fluorescence yield for the ith substance at a particular pair of excitation and emission wavelengths[24]

Since the absorbances $A_{\lambda'}$ of all species may be expected to change during the course of a spectrometric titration, evaluation on the basis of equation 18–11 with respect to several fluorescent substances is an extremely complicated process. For this reason, most fluorometric titration studies described in the literature deal only with the longest wavelength fluorescence band, which is attributed to a *single* component. Moreover, the approximation $A_{\lambda'} \ll 0.1$ is nearly always invoked.

Overall, the following list of chemical and instrumental *prerequisites* must be met if a linear relationship is to be assumed between the fluorescence intensity I_λ^F and the changing concentration c_i that characterizes a spectrometric titration:

(1) Adherence to the Lambert–Beer–Bouguer law as it applies to excitation occurring at the wavelength λ' (see above);

(2) Adequate respect for the condition $A_{\lambda'} \ll 0.1$ throughout the titration;

(3) Free passage of all emitted radiation (i.e., resorption does not occur);

(4) Absence of self-quenching (concentration quenching);

(5) Lack (or at least constancy) of quenching due to other species;

(6) Assurance that the emitted radiation does not itself cause photochemical reactions;

(7) Absence of (measurable) Rayleigh and/or Raman scattering;

(8) Exclusion of chemiluminescence and other "foreign" sources of radiation;

(9) Fixed optical geometry throughout the titration;

(10) Constant energy-output by the excitation source ($I_{0\lambda'}$ = constant).

The extent to which conditions (1) and (2) are fulfilled in a given experiment should be ascertained with the aid of an independent absorption spectrometric titration, perhaps supplemented with a dilution series A_λ vs. c_i. Table 18–1 illustrates the potential magnitude of problems associated with condition (2) by showing how systematic errors due to dropping higher terms in equation 18–9 are propagated with respect to I_λ^F (cf. also reference 19).

Table 18–1. Systematic error to be expected in fluorescence intensity correlations as a result of neglecting higher terms in the MacLaurin series (cf. equation 18–9).

$A_{\lambda'}$	$A_{\lambda'} \cdot \ln 10$	$1 - 10^{-A_{\lambda'}}$	Systematic error (%)
0.5	1.1513	0.6837	+68.4
0.2	0.4605	0.3690	+24.8
0.1	0.2302	0.2056	+12.0
0.05	0.1151	0.1087	+ 5.9
0.02	0.0461	0.0450	+ 2.3
0.01	0.0230	0.0228	+ 1.2
0.005	0.0115	0.0114	+ 0.6
0.002	0.0461	0.0459	+ 0.2
0.001	0.0023	0.0023	+ 0.1

Adequate conformity with the set of requirements (2)–(5) should always be verified by examining a spectrofluorometric dilution series I_λ^F vs. c_i. In order to ensure optimal reproducibility it is wise to perform all measurements in a fixed titration cuvette, thereby reducing the likelihood of problems related to condition (9). Since titrants are a common source of quenching, especially at high concentration, it is advisable to restrict fluorometric titrations to a limited set of relatively unproblematic acids and bases (e.g., H_2SO_4, $HClO_4$, KOH).[29,30] Addition of 0.1 mol/L of KCl has been

18.2 Fluorescence

recommended as a convenient way of establishing a fairly constant level of extraneous quenching.

Fluorescence measurements are facilitated by use of the highest possible excitation intensity $I_{0\lambda'}$, but this also increases the potential for photoreactions with respect to all species present.[18] Since the only portion of the solution subject to such reactions is that through which the source beam actually passes, continual stirring within the cuvette, or maintenance of a constant (slow) flow of solution, can reduce the severity of the problem. The sole alternative is to accept a compromise involving lower excitation intensity $I_{0\lambda'}$ coupled with increased signal amplification.

Light scattering interference often points to the fact that insufficient attention has been directed to the purity of substrates and solvents. With precision work it may be necessary to subject the titration solution to membrane filtration. Even though Raman scattering (cf. Sec. 18.5) tends to be of rather low intensity, Raman bands associated with the solvent can interfere with the measurement of low-intensity fluorescence. The locations of the most intense Raman bands of water for each of several mercury excitation lines are listed in Table 18–2. Overlap between water-induced Raman bands and fluorescence bands from a titration system can often be eliminated simply by changing the excitation wavelength.

Table 18–2. The most intense Raman bands resulting from excitation of water ($\tilde{\nu} \approx 3350$–3500 cm^{-1}, cf. ref. 31) at wavelengths corresponding to the principle lines from a mercury source.*

Hg-line (nm)	313	366	405	436
Raman band (nm)	350	418	470	513

* We wish to thank the Colora company for permitting us to obtain these results with an Aminco model SPF 500 spectrofluorometer during the course of a demonstration.

Chemiluminescence is a potentially worrisome source of extraneous light, but there are other more mundane possibilities that must be taken into account as well. In particular, all those points where tubing enters the sample compartment need to be regarded as suspect, and each must be carefully shielded. Teflon tubes of the type recommended with the titration setup of Sec. 16.2.1(b) sometimes act almost like fiber-optics devices as a result of multiple internal reflections, and they may allow a considerable amount of light to enter the sample compartment. The crucial role in equation 18–11 played by geometric factors means that the sample cuvette should always remain firmly in place throughout a titration.

As noted previously, a ratio spectrometer has the great advantage of minimizing problems arising from time-dependent changes in the source intensity $I_{0\lambda'}$. Absolute fluorescence intensity I_λ^F is usually converted to a relative quantity by comparison with some chemically and photochemically stable *fluorescence standard*.[32] In the case of a ratio instrument it may only be necessary to introduce such a standard (e.g., a quinine sulfate solution[6,23] or a piece of uranyl glass[23]) into the radiation path twice during a titration: once at the beginning and once at the end. With a single-beam instrument it is advisable to adjust the 100% setting of the apparatus with a standard prior to each measurement or spectral recording. This expedient at least helps to compensate for long-term drift, though it does go counter to the recommendation that the titration cuvette should always remain firmly fixed in its sample holder.

Quantitative spectrofluorometric analysis involves an unusually large number of assumptions, and for this reason absorption spectrometry is normally the method of choice for examining protolysis equilibria (at least in the ground electronic state; see below). Only in isolated cases will fluorescence spectrometry prove to be more advantageous. The high sensitivity of fluorescence measurements might appear to be an important factor, but the sensitivity of absorption spectrometry is itself a variable, subject to increase by appropriate changes in pathlength, initial concentration, and total titration volume. Of much more fundamental interest than sensitivity is the frequently higher *selectivity* of fluorescence. Thus, various investigators have succeeded in incorporating fluorescent chromophores into macromolecules in such a way that they serve as specific analytical "probes",[18,33] making it possible, for example, to examine the dissociation state of a functional group inaccessible to ordinary absorptiometric investigation because of competitive absorption by other parts of the molecule.

One unique opportunity that fluorescence spectrometric titration presents is the potential for investigating the relationship between dissociation equilibria and electronic excitation, the topic of the section that follows.

18.2.3 Protolysis Equilibria: The Electronic Ground State vs. Excited States

A protolyte that may be present in either the ground state S_0 or the first electronic excited state S_1 is subject to a number of possible equilibria. These are depicted schematically in Fig. 18–9 in what is known as the "Förster cycle".[18,29,34] Absorption spectrometric titration provides information only about the ground state dissociation equilibrium, including its pK value, but spectrofluorometric titration affords access to protolysis equilibria involving both the ground state *and* the first excited state.

It has long been known that many protolytes behave very differently upon promotion to the first electronic excited state. For example, excited-state phenol is much more readily deprotonated than would be predicted from ground-state data; pK^* in this case lies several pH units below pK.[30,34,35] The literature describes a number of

18.2 Fluorescence

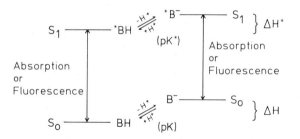

Fig. 18–9. Förster cycle representing protolysis equilibria in both the ground state (characterized by pK) and the first electronic excited state (pK^*), as well as possible transitions between the ground state system S_0 and the excited state system S_1, which entail either absorption (A) or fluorescence (F).

methods for examining equilibria comprising a Förster cycle, and several comprehensive summaries have appeared.[18,19,34,36,37] In this section we limit our attention to the question of how different electronic states manifest themselves in a set of titration spectra, and how multi-wavelength spectrometry may be employed to investigate a specific system, in this case β-*naphthol*.

Figure 18–10 shows a set of *absorptiometric titration spectra* for β-naphthol ($c_0 = 1 \cdot 10^{-4}$ M). Four sharp isosbestic points are apparent in the region 240–360 nm. Corresponding A-diagrams (not depicted) are found to be strictly linear. As expected, therefore, the titration system corresponds to the rank $s = 1$. The (ground state) pK is found to be ca. 9.5.[30]

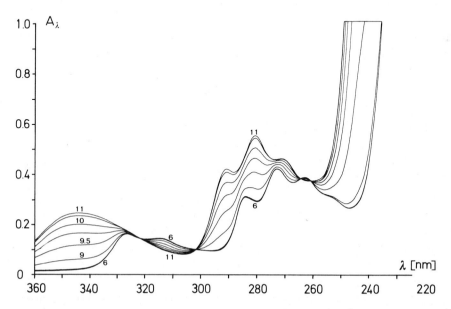

Fig. 18–10. Absorptiometric titration spectra for β-naphthol at 25.0 °C ($c_0 = 1 \cdot 10^{-4}$ M).

A fluorometric dilution series for β-naphthol, evaluated over the concentration range 0.5 to 5 · 10⁻⁵ M, is to a high degree of approximation linear, and this remains true in the presence of either 0.1 M HCl or 0.1 M NaOH. Consequently, equations 18–10 and 18–11 may be accepted as valid within this particular concentration range. Figure 18–11 shows titration spectra for a *spectrofluorometric titration* of β-naphthol ($c_0 = 5 \cdot 10^{-5}$ M). The wavelength of the radiation used for excitation purposes was 310 nm (bandwidth 2 nm). The absorbance A_λ at this wavelength varies over the course of the titration, but it never significantly exceeds 0.05 (cf. Fig. 18–10). Within each of the pH ranges 1–6 and 6–12 the titration spectra display one point of mutual intersection, known as an *isostilb*. Unfortunately, the sharpness of such points of intersection is rather difficult to evaluate due to the relatively high level of noise characterizing a fluorescence spectrum as compared to an absorption spectrum.

If a plot is now made of the measured fluorescence intensity I_λ^F at two different wavelengths (analogous to preparing an A-diagram), what results is an *intensity diagram (I-diagram)*,[25] as shown in Fig. 8–12. Combining information derived from two different emission bands (e.g., I_{420}^F and I_{360}^F) produces a line comprised of two intersecting linear segments, so the titration system must correspond to the rank $s = 2$, and ΔpK must be very large. The wavelength combination I_{420}^F vs. I_{410}^F is a trivial case in that here only a single component of the system, the excited state anion *B⁻, contributes to the fluorescence spectrum. Consequently, this particular I-diagram appears as a straight line even though the system has a rank s greater than one (cf. the analogous degeneracy in certain A-diagrams, as discussed in Sec. 7.3.1 and ref. 38).

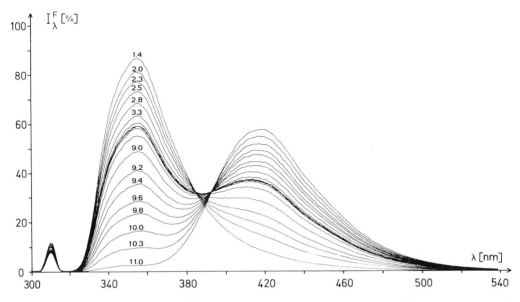

Fig. 18–11. Fluorescence titration spectra for β-naphthol at 25 °C, with excitation at 310 nm ($c_0 = 5 \cdot 10^{-5}$ M).

18.2 Fluorescence

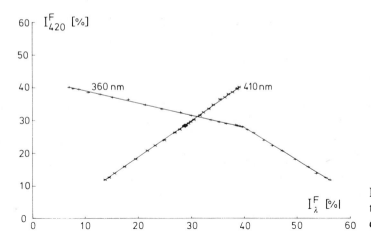

Fig. 18–12. I-Diagrams for the titration of β-naphthol described in Fig. 18–11.

Figure 18–13 shows another set of *spectrofluorometric titration spectra* for β-naphthol. These spectra were recorded under essentially the same conditions used for Fig. 18–11, but this time excitation was effected at 322 nm (bandwidth 2 nm), a wavelength corresponding to an isosbestic point in the absorption spectra. With this alteration, absorbance $A_{\lambda'}$ must remain constant throughout the titration, so any systematic error introduced by truncating the MacLaurin series (equation 18–9) is

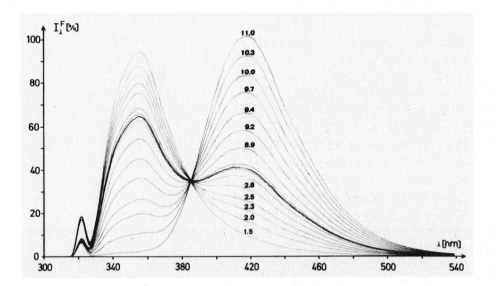

Fig. 18–13. Fluorescence titration spectra for β-naphthol at 25 °C, with excitation at 322 nm ($c_0 = 5 \cdot 10^{-5}$ M).

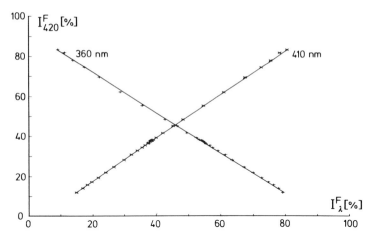

Fig. 18–14. I-Diagrams for the titration of β-naphthol described in Fig. 18–13.

also constant. Now the only species whose concentrations can affect the fluorescence spectra are the two excited state forms *BH and *B$^-$. As a result, what previously showed itself to be a two-step titration system appears to have been reduced to the rank $s = 1$, and an isostilb is present throughout the titration. Corresponding *I-diagrams* (e.g., Fig. 18–14) are linear regardless what wavelength combination is investigated.[39]

The *I-pH–diagrams* for this system at both $\lambda' = 310$ nm (Fig. 18–15) and $\lambda' = 322$ nm (Fig. 18–16) clearly reveal the presence of two non-overlapping titration steps. The pK of the ground state differs from that of the excited state (pK*) by more than six pH units, so each titration step may be evaluated individually as a single-step titration. The extremes for the two titrations lie at pH 1.5 and 11, and neither has actually been reached experimentally. For this reason, I_λ^F-pH–curves were simulated with the aid of the iterative curve-fitting program TIFIT (cf. Chapter 10). Simultaneous evaluation of four I_λ^F (pH)-series provided the set of values pK = 9.6 ± 0.1 (ground state) and pK* = 2.4 ± 0.1 (excited state). Literature values for pK* vary between 2.5 and 3.1.[19,30,37,40]

Especially in alkaline solution, β-naphthol is subject to rapid photochemical decomposition.[41] At high excitation intensity (e.g, with a mercury lamp, using the 313 nm monochromatic filter included as part of the fluorescence attachment for a Zeiss model DMR 21 spectrometer), and unless the titration solution in the cuvette is continuously agitated, fluorescence intensity diminishes by $\geq$ 20% during the time required to record a single spectrum (~2–3 min). The plots illustrated here (Figures 18–11 through 18–16) were obtained with an apparatus incorporating two monochromators (Aminco model SPF 500).† By reducing the intensity of the excitation radiation

---

† We wish to thank the Colora company for providing us the opportunity to record these spectra in the course of a demonstration.

18.2 Fluorescence

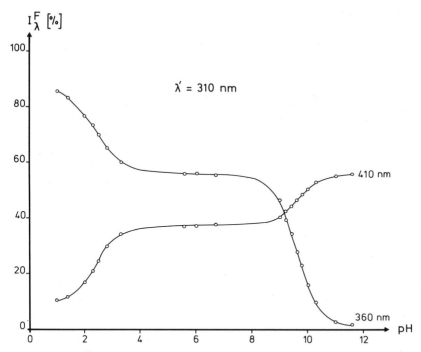

Fig. 18–15. I_λ^F-pH–Diagrams for the titration of β-naphthol (excitation at 310 nm, cf. Fig. 18–11).

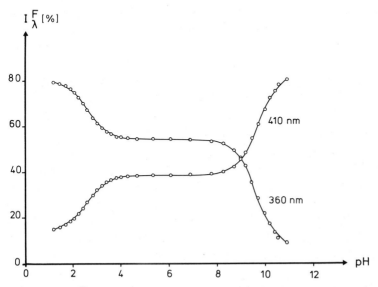

Fig. 18–16. I_λ^F-pH–Diagrams for the titration of β-naphthol (excitation at 322 nm, cf. Fig. 18–13).

to the greatest extent possible (with a bandwidth of 2 nm), signal reduction due to photochemical decomposition of the protolyte was held to < 0.5%.

A complete interpretation of the results of absorption and fluorescence spectrometric titration of β-naphthol would require a knowledge of the rate constants for all the fast photochemical and chemical reactions comprising the Förster cycle. In particular, the magnitude of the rate constants for the equilibrium *BH + H_2O ⇆ *B^- + H_3O^+ determines whether or not equilibrium is ever actually established within the lifetime of the excited state species. A complete discussion of the subject would go beyond the bounds of this book, and interested readers are advised to consult appropriate specialized literature.[30,42,43]

18.3 Circular Dichroism (CD) and Optical Rotatory Dispersion (ORD)

18.3.1 Spectrometric Data Collection

CD and ORD spectroscopy are two related techniques for investigating the interaction between chiral molecules and polarized light. The key variable in the case of circular dichroism is the *difference* in observed absorbance for left- vs. right-circularly polarized light $\Delta A_\lambda := A_{\lambda L} - A_{\lambda R}$, which may be plotted as a function of wavelength. In practice, however, the raw information is often converted into a difference in *molar absorptivities* $\Delta\varepsilon_\lambda := \varepsilon_{\lambda L} - \varepsilon_{\lambda R}$. With optical rotatory dispersion, what is measured instead is the wavelength dependence of the angle of rotation α_λ through which the vibrational plane of plane polarized light is rotated as it passes through a sample. Here again it is common to perform a data conversion, expressing findings in terms of either the corresponding *molar angle of rotation* $[\phi]_\lambda$ or the *specific rotation* $[\alpha]_\lambda$. In the case of theoretical investigations it is often desirable to work with what is known as the *ellipticity* Θ_λ, which is a function of both α_λ and $\Delta\varepsilon_\lambda$ (cf. Sec. 18.3.2).

Circular dichroism effects are only observed in regions of the spectrum where the sample is also subject to absorption (i.e., $\varepsilon_\lambda \neq 0$), while the optical rotatory dispersion phenomenon may be investigated both at wavelengths free of absorption (normal ORD) and within the envelope of an absorption band (anomalous ORD). Normal ORD effects may be described using the language of classical electrodynamics, but a full understanding of either anomalous ORD or CD requires the application of quantum theory.[44]

CD and ORD spectra are most often recorded with the aid of single-beam instruments (cf. Figures 18–17 and 18–18), although attachments have also been developed for use with double-beam absorption spectrometers.[45,46] Many of the

18.3 Circular Dichroism (CD) and Optical Rotatory Dispersion (ORD)

components of a CD or ORD spectrometer are identical to those used in absorption or fluorescence instruments. Incandescent and deuterium lamps may serve as suitable quasi-continuous *light sources*, though high-pressure xenon lamps are more common. In the past, precision quantitative polarimetry experiments at fixed wavelength were invariably carried out using mercury or sodium vapor lamps equipped with appropriate monochromatic filters,[46] but laser sources are gaining in popularity.

Most recording CD and ORD spectrometers utilize prism double-monochromators in order to minimize interference from stray light in the short-wavelength UV region (< 220 nm). In some instruments the second prism is so arranged that its optical axis functions as a *polarizer*. In most instruments, however, a separate polarizer is mounted between the monochromator and the sample (cf. Figures 18–17 and 18–18).

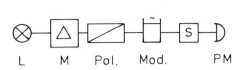

Fig. 18–17. Optical arrangement of a CD spectrometer with light source L, monochromator M, polarizer Pol, electro-optical modulator Mod, sample cell S, and photomultiplier PM.

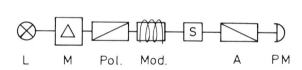

Fig. 18–18. Optical arrangement of an ORD spectrometer with light source L, monochromator M, polarizer Pol, magneto-optical modulator Mod, sample cell S, analyzer A, and photomultiplier PM.

Ordinary glass serves as a suitable cuvette material in the visible range, but quartz cuvettes have the advantage that they are applicable throughout the range 180–800 nm. Precision work should always be conducted in cuvettes of the highest possible quality, paying special attention to the absence of birefringence effects. The best cuvettes incorporate permanently-fused jackets designed to permit careful temperature control, since chiral properties tend to be highly sensitive to changes in temperature (cf. Sec. 18.3.2).

The principle *differences* between CD and ORD instruments, as well as between instruments from different manufacturers, relate to how monochromatic plane-polarized radiation is modulated and how a photomultiplier signal is converted into the quantity of interest, ΔA_λ or α_λ. Many *multi-wavelength polarimeters* utilize the *mechanical modulation* technique. Here the polarizer is caused to oscillate within a small range (ca. ± 1 degree) about its longitudinal axis, typically at a frequency of 50 Hz. The photomultiplier (PM) then serves as the source of an alternating current signal which causes a servomotor to rotate the analyzer A, located between the sample and the photomultiplier (null balance method). The angle of rotation of the analyzer is displayed digitally, providing a direct measure of the quantity α_λ.

By contrast, *ORD spectrometers* generally utilize a form of *magneto-optical modulation* based on the Faraday effect (cf. Fig. 18–18). A common design involves a quartz rod surrounded by a magnetic coil that is placed between the polarizer and the cuvette. When a 50 Hz alternating current is passed through the coil, the plane of incident plane-polarized light is caused to oscillate slightly about its null value, also with a frequency of 50 Hz. Any further optical rotation induced by the sample is then measured either with a null-balance device (see above) or else directly with the aid of a second magneto-optical modulator.[45]

A *CD spectrometer* usually incorporates an *electro-optical modulation* system based on the Kerr effect (cf. Fig. 18–17). In this case, plane-polarized light, which may be interpreted as a combination of equal amounts of left- and right-circularly polarized light, is modulated with the aid of an alternating high-voltage source (50 Hz) in what is known a Pockels cell, with the result that the ratio of left- to right-circularly polarized light changes in phase with the alternating voltage. All possible modes of elliptically-polarized light fall between the two limiting cases of purely left- and purely right-circularly polarized light. Within the confines of one of the absorption bands of an optically active sample, right- and left-circularly polarized light are found to absorb to differing extents (i.e., $\varepsilon_{\lambda L} - \varepsilon_{\lambda R} \neq 0$). Consequently, the intensity of the radiation emerging from the sample varies in phase with the alternating voltage of the Pockels cell. The signal produced by the photomultiplier can be dissected into a very large constant-potential component $V_=$ and a very small alternating-potential component $V_\sim$. A combination of these two voltages may then be used to create an electrical signal that is proportional to the experimental variable $\Delta A_\lambda := A_{\lambda L} - A_{\lambda R}$.[45;*]

CD and ORD spectrometers share many of the same optical design features, so it is not surprising that instrument manufacturers have also developed *combined CD/ORD spectrometers*.

Since the monochromators and recording devices utilized in CD and ORD instruments are quite similar to those used in UV–VIS absorption spectrometers, *wavelength reproducibility* and *wavelength precision* are equivalent in such instruments to what one would expect for an absorption spectrometer. The accuracy with which a *rotation angle* may be determined using a precision polarimeter falls in the range $\leq \pm 0.001°$ (standard deviation).[46] A CD spectrometer is required to measure a very small difference between two nearly equivalent quantities, $A_{\lambda L}$ and $A_{\lambda R}$. The degree of accuracy attainable in such a measurement is limited by fluctuations in the xenon lamp, by photomultiplier noise, and by the characteristics of the electronic circuitry; the smallest difference ΔA_λ that can currently be measured is roughly $1-3 \cdot 10^{-5}$ (for a mean absorbance $A_\lambda = 1$ and a time constant of 3–5 s).

In the case of a time-independent analysis like a spectrometric titration, the signal-to-noise ratio may often be improved markedly by establishing a very long time

---

* Note that the symbol ΔA_λ is used here to represent a quantity different from the one assigned the same symbol earlier in the context of acid–base equilibria.

18.3 Circular Dichroism (CD) and Optical Rotatory Dispersion (ORD)

assigned the same symbol earlier in the context of acid–base equilibria.
constant, or by averaging the results from a large number of measurements. Using these and other noise-reduction techniques it is sometimes possible to obtain CD or ORD results with a level of accuracy comparable to that anticipated for absorption spectrometry (cf. Sec. 18.3.2).

18.3.2 Quantitative CD and ORD Spectrometry

The quantitative relationship linking angle of rotation, pathlength, and concentration is known as *Biot's law* (1817):[47,48]

$$\alpha_\lambda = l \cdot [\phi]_\lambda^T \cdot c \tag{18-12a}$$

$$= l' \cdot [\alpha]_\lambda^T \cdot c' \tag{18-12b}$$

where α_λ = angle of rotation in degrees
l = pathlength (cm)
$[\phi]_\lambda^T$ = molar rotation at wavelength λ and temperature T
c = concentration in mol/L of solution
l' = pathlength (dm)
$[\alpha]_\lambda^T$ = specific rotation at wavelength λ and temperature T
c' = concentration in g/mL of solution

Most everyday uses of polarimetry, especially saccharimetry, continue to utilize the older formulation given as equation 18–12b.

Just like the Lambert–Beer–Bouguer law, the Biot law is a limiting relationship strictly applicable only to a very dilute solution. Its use is therefore subject to the conditions already discussed in Sections 3.1 and 18.1.2. Since polarimetry often requires the measurement of extremely small angular rotations, temperature control becomes very important. This is true not only because of the relationship that exists between temperature and molar or specific rotation, but also because any change in temperature produces at least slight changes in both pathlength and concentration.[47]

If one assumes that the Lambert–Beer–Bouguer law applies to both terms on the right-hand side of the fundamental CD equation $\Delta A_\lambda := A_{\lambda L} - A_{\lambda R}$ then the following quantitative expression can be used to describe CD-signal strength as a function of concentration:[49]

$$\Delta A_\lambda := A_{\lambda L} - A_{\lambda R} = \log \frac{I_{0\lambda}}{I_{\lambda L}} - \log \frac{I_{0\lambda}}{I_{\lambda R}} =$$
$$= l\left(\varepsilon_{\lambda L} - \varepsilon_{\lambda R}\right) \cdot c := l \cdot \Delta\varepsilon_\lambda \cdot c \tag{18-13}$$

Theoretical treatments often utilize the quantity *molar ellipticity* $[\Theta]_\lambda^T$, mentioned briefly in Sec. 18.3.1, which is proportional to $\Delta\varepsilon_{\lambda T}$:[48,50]

$$[\Theta]_\lambda^T := 3300 \cdot \Delta\varepsilon_{\lambda T} \qquad (18\text{–}14\text{a})$$

where the index "T" refers to the temperature dependencies of $\Delta\varepsilon_\lambda$ and $[\Theta]_\lambda$. The factor 3300 arises in the course of converting from rads to degrees and from natural logarithms to base-10 logarithms.[51] The ellipticity Θ is obtained by combination of equations 18–13 and 18–14a:

$$\Theta = 3300\,\Delta A_\lambda = l \cdot [\Theta]_\lambda^T \cdot c \qquad (18\text{–}14\text{b})$$

The three quantities $[\phi]_\lambda^T$, $[\Theta]_\lambda^T$, and $\Delta\varepsilon_{\lambda T}$ are related to one another through what are known as the Kronig–Kramers transformations.[44] This means that it is possible in principle to interconvert CD spectra and ORD spectra.

Since a great many substances of biological interest only display absorbance in the short-wavelength region of the UV (e.g., sugars,[52] peptides,[53,54] proteins[55,56]), efforts have intensified in recent years to extend CD and ORD spectrometry further into the region ca. 250–180 nm. Analyses of this nature are most successful if several additional precautions are observed:

(1) Temperature control must be extremely precise.

(2) Solvents must be exceptionally pure, as well as optically transparent even to short-wavelength UV light.

(3) Light scatter increases rapidly in the UV region ($\sim 1/\lambda^4$), so membrane-filtration of sample solutions is strongly recommended.

(4) Double-monochromators are required in order to compensate for the increase in stray light at short wavelength (cf. Sec. 18.3.1).

(5) Most measurements rely upon interpreting weak signals superimposed on a high absorption background ($A_\lambda > 1.5$–2). This means that photometric accuracy is an especially sensitive function of stray light errors of all types (cf. Sec. 18.1.2).

(6) The low intensity of the signal to be analyzed might tempt one to consider using a concentrated solution, but concentration must never be increased beyond the range of validity of equations 18–12 and 18–14b as established by dilution series measurements.

(7) Most CD and ORD instruments are of single-beam design, and since they normally contain high-pressure xenon lamps that are subject to significant intensity variations it is important that one be alert to possible instrumental drift. An appropriate reference cuvette should be examined repeatedly during the course of any spectrometric titration.

(8) At wavelengths below 200 nm it is necessary that a CD or ORD instrument be operated under a nitrogen atmosphere, and many instruments are designed to permit nitrogen flow during all measurements.

18.3 Circular Dichroism (CD) and Optical Rotatory Dispersion (ORD)

The following example indicates the type of information to be expected from CD and ORD spectrometric titration.

Figure 18–19 shows a set of CD spectra for the compound L-cystine. These spectra resulted from a dilution series ($10^{-4} - 2 \cdot 10^{-5}$ M) in which the solvent was 0.1 M HCl. Corresponding Θ_λ vs. c diagrams are illustrated in Fig. 18–20. Data taken at the three fixed wavelengths indicated were integrated over 16 s in order to assure a favorable signal-to-noise ratio. Figure 18–21 contains ORD results obtained from the same dilution series. These data were collected at two wavelengths using a mercury-vapor lamp (see below) and a one-second integration time. It is apparent from all the plots that the conditions imposed by equations 18–12 and 18–14b are adequately fulfilled over the concentration range of interest.

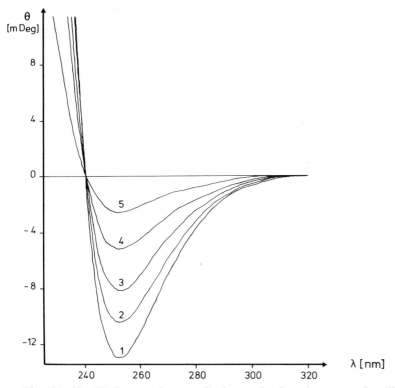

Fig. 18–19. CD Spectra from a dilution series based on L-cystine dissolved in 0.1 M HCl. (Θ in units of millidegree).

---

* The spectra presented here were recorded at the University of Tübingen on a Perkin-Elmer model 241 digital polarimeter located in the Institut für Organische Chemie and a Jasco model 20A CD/ORD-spectrometer in the Institut für Physiologische Chemie. We are grateful to Professors Dr. E. Bayer and Dr. U. Weser, respectively, for providing us access to these facilities.

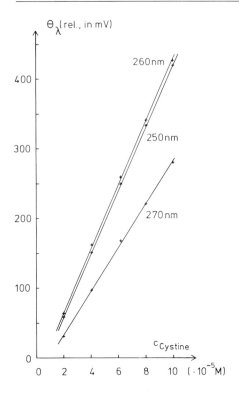

Fig. 18-20. Diagrams Θ_λ vs. c for a dilution series of L-cystine dissolved in 0.1 M HCl.

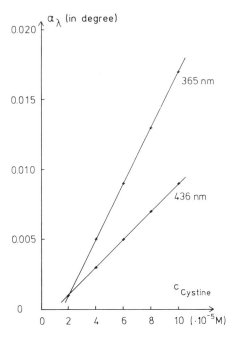

Fig. 18-21. Diagrams α_λ vs. c for a dilution series of L-cystine dissolved in 0.1 M HCl.

18.3 Circular Dichroism (CD) and Optical Rotatory Dispersion (ORD)

A set of CD spectra obtained during a titration of L-cystine ($c = 10^{-4}$ M) over the pH range 7–10 is displayed in Figure 18–22. The equivalent of an isosbestic point appears at ca. 231 nm. If, in analogy to an absorption diagram, CD-signal intensities at two different wavelengths are plotted against each other (producing an *ellipticity* or *Θ-diagram*), the result is a straight line, as shown in Fig. 18–23.

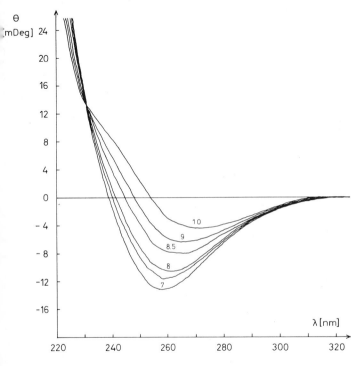

Fig. 18–22. CD spectra obtained during a pH titration of L-cystine ($c_0 = 10^{-4}$ M, Θ in units of millidegree).

A similar evaluation based on an ORD titration is presented in Fig. 18–24, which might be termed a *rotation-angle–* or *α-diagram*. Again, to a high degree of approximation all curves are linear. Actual titration curves covering the pH range 3–12 are shown in Fig. 18–25. All ORD measurements were taken using a multi-wavelength precision polarimeter equipped with sodium- and mercury-vapor lamps. Despite the fact that the integration time was much shorter than in the case of the CD titration (1 s instead of 16 s), the data contain far less noise, a result that underscores the profound difference between mercury and xenon lamps.

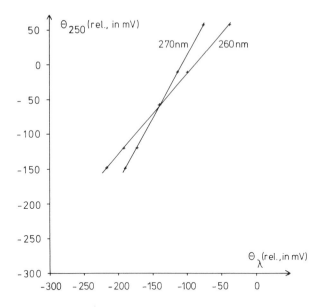

Fig. 18–23. Ellipticity diagram from the titration of L-cystine described in Fig. 18–22.

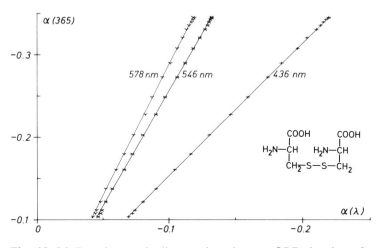

Fig. 18–24. Rotation-angle diagram based on an ORD titration of L-cystine ($c_0 = 10^{-4}$ M, α in units of degree).

It will be noted that all the titrations of L-cystine commenced at a pH greater than 3. The reason the first titration step (pK = 1.65[57]) was avoided was a desire to keep dilution errors negligibly small (< 1%).

18.4 IR Absorption

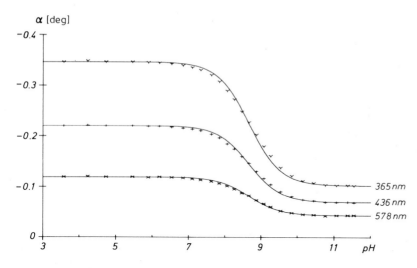

Fig. 18–25. ORD Titration curves for the titration of L-cystine described in Fig. 18–24.

18.4 IR Absorption

18.4.1 Spectrometric Data Collection

Most infrared absorption spectrometers in use today are of the *double-beam* type, and the standard optical arrangement is in many ways comparable to that employed in UV–VIS absorption spectrometers (cf. Fig. 18–1b). The quasi-continuous radiation source used for the classical IR region (ca. 2.5–25 µm or 4000–400 cm^{-1}) is ordinarily either a Nernst rod, fabricated from a blend of zirconium oxide with certain trace additives, or a device composed of silicon carbide and known as a "globar". The usual detector is a thermoelement.[58–60]

> Since in the IR region (unlike the UV) there is no need to be concerned with photochemical reactions, IR spectrometers often incorporate an "inverted" optical design. That is to say, the sample is subjected to constant irradiation by the complete spectrum issuing from the source. Only after it emerges from the sample is the light separated into its component wavelengths by passage through a monochromator.

Photometric accuracy and reproducibility as well as wavelength reproducibility for a modern IR apparatus are entirely comparable to what one would expect from UV–VIS spectrometry. Usually the quantity that is measured and recorded is *absorbance* $A_\lambda = \log(I_{0\lambda}/I_\lambda)$, though it was formerly quite common to express results in terms of terms of the related quantity *%-transmission* ($100 \cdot I_\lambda/I_{0\lambda}$). Automation and data-handling in the case of IR instruments is often rather sophisticated compared with the standard set by the usual UV–VIS equipment.

18.4.2 Quantitative Spectrometry in Aqueous Solution

Within the limitations noted in Sections 3.1 and 18.1.2, the Lambert–Beer–Bouguer law is also applicable to infrared absorption.[58,59,61] Quantitative infrared analyses are possible today with an accuracy of better than 2%, even for multi-component mixtures,[58] and reproducibility is often better than 0.1% (standard deviation).[62] It should be noted, however, that IR spectra usually contain a vast number of narrow bands, so it is particularly important to assure adequate monochromaticity of the light analyzed.

Quantitative IR analysis can sometimes be performed simply by measuring absorbances at a wavelength of maximum absorption, but it is important to ensure that the shape of the peak in question remains unchanged with changing concentration and that it is in no way influenced by other factors such as the presence of a titrant. However, in most cases it is actually preferable to measure a somewhat different quantity that is also directly proportional to c, at least given certain assumptions. This quantity is the *integral absorption*:

$$\int_{\lambda_1}^{\lambda_2} A(\lambda) d\lambda \quad \text{or} \quad \int_{\tilde{v}_1}^{\tilde{v}_2} A(\tilde{v}) d\tilde{v}$$

The limits of integration λ_1 and λ_2 (or $\tilde{v}_1$ and $\tilde{v}_2$) in the above expression are chosen such that they effectively isolate the absorption band of interest (cf. also Sec. 12.3.4). Integral absorption is primarily used for quantitative determination of single substances, but the several components of a mixture may also be analyzed in this way provided they display non-overlapping peaks. The measurements themselves are facilitated by linking the spectrometer with a computer programmed to perform the necessary numerical integration.

Unfortunately for our purposes here, quantitative IR spectrometry in *dilute aqueous solution* presents a number of serious problems:

(1) *Water* produces rather intense, non-specific background absorption.[58,63] The only way to compensate for its presence is with the use of a reference cuvette; i.e., by *difference spectroscopy*. This requires the availability of either a carefully matched set of cuvettes or else a reference cuvette with variable pathlength.[58,62] Even this solution is problematic: high absorbance on the reference side results in increased slit width at the monochromator, which in turn causes a significant increase in spectral bandwidth.

18.4 IR Absorption

(2) Most of the substances suitable as *cuvette windows* for IR spectrometry (e.g., KBr, NaCl, CsI) are water soluble. A great deal of effort has been expended in a search for potential substitutes that would be compatible with aqueous solutions. CaF_2, ZnS, AgCl, AgBr, TlBr, and TlI[58] have all been utilized, with the first two proving to be the most stable with respect to solutions of varying pH. Unfortunately, all these substances also display a high degree of (apparent) self-absorption due to reflective losses and interference phenomena.[62] Optical compensation with appropriate reference cuvettes is obviously essential, bringing with it all the familiar disadvantages.

(3) High background absorption due to solvent necessitates the use of *short pathlengths* in quantitative IR spectrometry (e.g., 0.01–0.1 cm), which in turn implies high sample concentrations (10^{-2} to ≥ 1 M). Consequently, results obtained in an IR spectrometric investigation of a dissociation equilibrium are not necessarily directly comparable with their UV–VIS counterparts (cf. Sec. 12.3.2).

As a consequence of all these problems, relatively few IR spectrometric titrations have been described in the literature, and most of those that have been reported are rather crude from a quantitative standpoint. Examples include the case of the vitamin-B_6 aldehydes discussed in Sec. 12.3.2 (performed in D_2O with AgCl windows, $l = 0.02–0.04$ mm, $c_0 \approx 0.5$ M).[64,65]

Several research groups have recently begun to re-investigate the potential for quantitative IR spectrometry with respect to aqueous solutions. In principle, the technique is of considerable interest, particularly in the context of branched equilibria (cf. Sec. 12.3.2), because IR absorption is highly selective for specific functional groups, and individual peaks can often be assigned to particular structural features with relative ease. The method certainly offers the promise of serving as an important investigative tool in the study of spectrometric titration systems, especially if it could be coupled with laser-Raman spectroscopy (cf. Sec. 18.5).

NIR Spectrometry (i.e., spectrometry conducted in the near-infrared region, ca. 0.8–2.5 µm or 12 500–4 000 cm^{-1}) is free of most of the problems discussed above. The corresponding spectrometers are ordinarily UV–VIS instruments that use incandescent lamps but offer an enhanced optical range together with a PbS cell to serve as a supplemental detector. Suitable cuvettes are made of glass or quartz (e.g., INFRASIL). The NIR region contains mainly overtone bands ($\Delta v > +1$) and combination bands related to vibrational transitions which themselves occur within the normal infrared region. Quantitative studies may be carried out in very dilute aqueous solution, and the results obtained are of nearly the same quality as UV–VIS data. Nevertheless, to date there have been almost no reports of spectrometric titration systems examined in this way.

18.5 Raman Scattering

18.5.1 Spectrometric Data Collection

The revolutionary laser light sources introduced in the 1960s heralded a veritable renaissance in Raman-spectroscopic investigations. Compared with what was possible earlier using mercury-vapor line sources, the power of the new sources led to an increase in sensitivity of several orders of magnitude,[63] permitting the design of spectrometers for routine use with recording times of only a few minutes.

The optical layout of a Raman spectrometer (cf. Fig. 18–26) resembles that employed for fluorometry in the sense that both require a 90°-offset geometry. This similarity means that strongly fluorescent substances cannot be examined using standard Raman techniques.

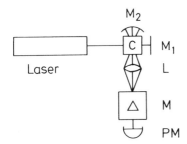

Fig. 18–26. Optical arrangement of a Raman spectrometer containing a laser, a cuvette C, mirrors M_1 and M_2, a lens L, a monochromator M, and a photomultiplier PM.

Raman spectrometers are usually *single-beam* devices, a simplification made possible by the fact that continuous wave (CW) laser sources are characterized by very constant energy output. Nevertheless, some manufacturers also offer *ratio-Raman spectrometers*. Monochromatic radiation of wavelength λ_0 (e.g., the 514.5 nm line of an argon laser) passes directly through the cuvette C and is then reflected back again by the mirror M_1, thereby increasing the effective radiation density. A portion of the new radiation that results from Raman scattering (see below) is focused by a mirror M_2 and/or a lens L onto the entrance slit of the monochromator M. Quantification of the monochromator output is accomplished with a photomultiplier. Since Raman intensities tend to be very low, other detecting aids are sometimes utilized as well, including photon counting and optical multi-channel analysis.[66]

In contrast to the IR absorption phenomenon, which is fundamentally related to changes in dipole moment within the sample, Raman scattering is a function of changes in sample polarizability. Indeed, the two techniques may be regarded as mutually supportive. The lines that constitute a normal Raman spectrum correspond to energies $h\nu_R$ somewhat lower than the energy of the source ($h\nu_0$), which is in turn

18.5 Raman Scattering

lower than the energy required to effect the longest wavelength (lowest energy) electronic transition permitted for the sample ($S_0 \to S_1$, cf. Fig. 18–27).*

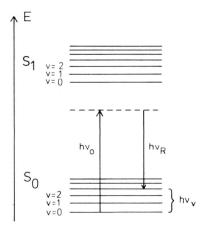

Fig. 18–27. Energy diagram illustrating Raman scattering from a system with electronic energy levels S_0 and S_1 and vibrational energy levels $v = 0, 1, 2, \ldots$.

The Raman effect is a result of inelastic collisions between light quanta emanating from the excitation source (hv_0) and molecules of the sample. Such collisions may induce changes in vibrational energy (hv_V), leading to the appearance of Raman-scattered radiation (hv_R) in the form of a *Stokes Raman spectrum*, whose lines appear at wavelengths longer than that of the excitation source:

$$hv_R = hv_0 - hv_V \tag{18-15}$$

The entire IR spectral region (e.g., from 2.5 to 25 μm or 4000–400 cm^{-1}) is incorporated within a Raman spectrum, which occupies a wavelength interval of ca. 50–100 nm somewhere in the visible region. The exact location of the spectrum is of course dependent upon the wavelength of the source (cf. Fig. 18–28).

In order to obtain a Raman spectrum with resolution comparable to that expected from an IR spectrum it is obviously necessary that the monochromator be very effective. A good monochromator is important for another reason as well, however: most of the radiation that is introduced for excitation purposes undergoes not *inelastic* collisions (leading to Raman scattering), but *elastic* collisions. These cause *Rayleigh scattering*, which is of the same wavelength λ_0 as the source radiation and several orders of magnitude more intense than the Raman scattering. In order sufficiently to block the passage of this intense "optical stray light" it is often necessary to incorporate a *double* or even a *triple monochromator*.[66] A very recent development in this regard is the *multi-channel Raman spectrometer*, a device which promises extremely high spectral resolution.[68]

---

* The reader is advised to consult more specialized sources for information regarding such Raman-related techniques as resonance-Raman spectroscopy or non-linear Raman spectroscopy.[67]

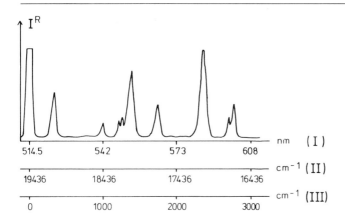

Fig. 18–28. Raman spectrum showing alternative abscissa markings; I, II: absolute scales (nm, cm^{-1}); III: cm^{-1} scale expressed relative to the frequency v of the excitation source.

18.5.2 Quantitative Spectrometry in Aqueous Solution

Water serves as an excellent solvent for Raman spectroscopy because its own Raman spectrum is very weak and restricted to only a few bands (cf. Table 18–2).[31,58,63]

Just like fluorescence, Raman emission may be regarded as directly proportional to the concentration of the sample provided certain important conditions are fulfilled. In fact, the relationship

$$I_\lambda^R \sim c \tag{18–16}$$

is found to be valid in at least some cases over a concentration range of four orders of magnitude.[63]

Raman spectra, like IR spectra, are usually quantified by measuring integral band intensities $\int I^R(\tilde{v})d\tilde{v}$, where the required numerical integration is carried out by a computer linked directly to the spectrometer.[63]

One of the principal problems encountered during laser-Raman spectrometric studies on organic molecules is *charring* of the sample due to the intensity of the laser beam. Potential damage of this type may be reduced by using a rapidly rotating cuvette,[69,70] an expedient that has the additional advantage of minimizing temperature gradients within the sample. Another common source of difficulty is the presence of dust or other *particles* in the sample. Not only will such particles evoke undesirable light scattering, but in addition they may absorb radiation at the wavelength λ_0 of the source, thereby contributing further to the problem of charring.[63] Indeed, the use of highly purified samples and solvents is even more important with Raman spectrometry than in the case of fluorescence spectrometry, and it is advisable that all particulate matter be removed by membrane filtration.

An *example* of the use of Raman spectroscopy in aqueous solution has already been presented in Sec. 12.3.2, which includes a Raman spectrum of cysteine recorded at pH 7 (cf. Fig. 12–3). Additional examples are provided by Raman analyses of

18.5 Raman Scattering

glutathione at pH 3 (cf. Fig. 18–29) and of reduced glutathione at pH 1 (cf. Fig. 18–30). These two compounds contain the thiol (–SH) and disulfide groups, respectively, both of which produce very intense Raman bands.[63,71,72] Bands of this type have made it possible to use the Raman technique to investigate conformational changes accompanying disulfide bridging in peptides and proteins[53,63] as well as various redox equilibria and the branching phenomena that complicate certain protolysis systems (cf. Sec. 12.3.2). Raman spectrometric investigations were employed as early as the 1950s to determine pK values for nitric acid and other strongly acidic species.[73–75]

The greatest disadvantage of Raman spectroscopy (as well as IR and NMR spectroscopy) from the standpoint of spectrometric titration continues to be the need for *high sample concentrations*. In the analyses of cysteine and glutathione alluded to above, the starting concentrations of substrate ranged from 0.3 to ca. 2 M, so any comparisons with results from UV–VIS absorption spectrometry must be regarded with caution. The latest generation of multi-channel Raman spectrometers promises a dramatic increase in sensitivity,[76] so it may be that more reasonable initial concentrations will soon become a reality.

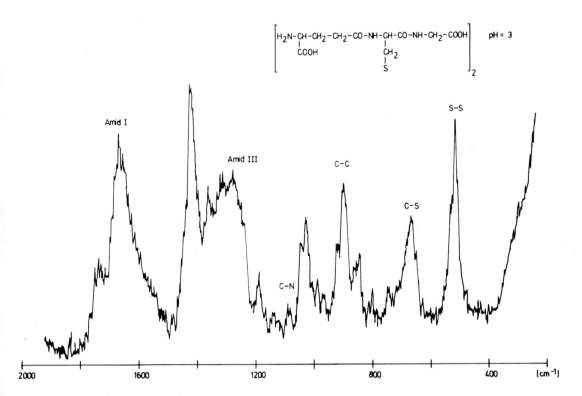

Fig. 18–29. Raman spectrum of glutathione at pH 3.0 (c_0 = 0.3 M). This spectrum was kindly placed at our disposal by Prof. G. Jung of Tübingen (Institute of Organic Chemistry).

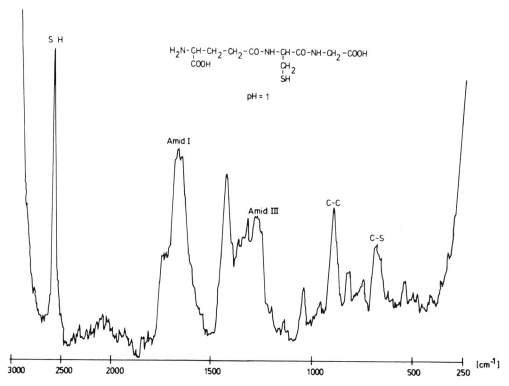

Fig. 18–30. Raman spectrum of reduced glutathione at pH 1.0 ($c_0 = 0.3$ M). This spectrum was kindly placed at our disposal by Prof. G. Jung of Tübingen (Institute of Organic Chemistry).

18.6 Nuclear Magnetic Resonance (NMR)

18.6.1 Basic Principles and Spectrometric Data Collection

In contrast to all the spectroscopic techniques discussed up to this point, nuclear magnetic resonance spectroscopy distinguishes itself by *not* being based on an optical phenomenon. Instead of optical components, an NMR spectrometer is designed around a complex set of radiofrequency devices (cf. Fig. 18–31).

There is another important characteristic that also distinguishes NMR spectroscopy from the techniques described in Sections 18.1–18.5: only when a molecule is under the influence of a strong, external magnetic field does there exist a set of

18.6 Nuclear Magnetic Resonance (NMR)

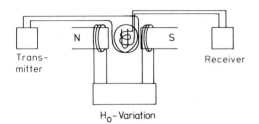

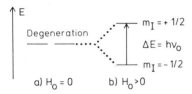

Fig. 18–31. Arrangement of the components in a nuclear magnetic resonance spectrometer.

discrete energy levels capable of serving as a basis for the spectral transitions in question.[77] We begin our discussion by reviewing the fundamental principles of NMR, starting within the somewhat limited context of *proton-magnetic resonance (PMR)*.

Every proton has associated with it a spin quantum number $I = 1/2$. Quantum theory predicts that such a species can exist in $2I + 1 = 2$ *quantum states*, each associated with its own *magnetic* quantum number $m_I = \pm 1/2$. Under normal circumstances (i.e., in the *absence* of an external magnetic field) these quantum states are of equal energy and are said to be *degenerate* (cf. Fig. 18–32a). However, the degeneracy may be lifted by a powerful, static, external magnetic field H_0, giving rise to a system characterized by two states with differing energies (Fig. 18–32b). Quantum theory supplies the following expression for the excitation energy ΔE required to induce a transition from the lower energy state to the upper energy state:

$$\Delta E = h\nu_0 = \frac{\mu_I}{I} \cdot \Delta m_I \cdot H_0 \tag{18-17}$$

where h = Planck's constant
 ν_0 = frequency of radiation (*resonance* frequency)
 μ_I = nuclear magnetic moment (= 2.793 Bohr magnetons in the case of the proton)
 I = spin quantum number (= 1/2 in the case of the proton)
 Δm_I = change in the magnetic quantum number m_I
   ($\Delta m_I = \pm 1$ for the proton)
 H_0 = magnetic field strength supplied by the static, external magnet

Fig. 18–32. Nuclear spin energy levels for the proton both in the absence (a) and in the presence (b) of an external magnetic field.

Figure 18–31 illustrates schematically one possible arrangement of components within an NMR spectrometer. The *radiation source* in this case is a radiofrequency transmitter that produces a monochromatic signal of frequency v_0, thus eliminating the need for a dispersing element (monochromator). The sample to be investigated is placed in a long cylindrical tube that is inserted between the polepieces of a magnet, which in turn supplies a static magnetic field H_0. In order to minimize effects due to magnetic field inhomogeneity the sample is spun at high velocity about its longitudinal axis, generally with the aid of an air-driven turbine. Provision is also made for close thermostatic control of the region occupied by the sample. Finally, it will be noticed that the sample is surrounded by two sets of coils. These are connected to a radio transmitter and receiver, respectively, and are mounted perpendicularly both with respect to each other and to the lines of magnetic force constituting the H_0 field. The transmitter coil furnishes the sample with monochromatic radiation hv_0. Any absorption of radiant energy by the sample causes a secondary signal to be induced simultaneously in the receiver coil, and it is this signal that is collected and quantified with the aid of a detector. An instrument such as that just described is based on a *nuclear inductance arrangement*, a design first suggested by Bloch, et al.[78–80]

In another type of NMR spectrometer the sample is surrounded by only a *single* coil, which is in turn connected to a high-frequency bridge circuit. As with a Bloch-type instrument, this design (proposed by Purcell, et al.[79,80]) also enables the user to ascertain how much radiant energy has been removed from the transmission circuit as a result of absorption by the sample.

Proton NMR spectra provide the chemist with a measure of the extent to which the native resonance frequency v_0 of a given proton (hydrogen nucleus) has been altered (*shifted*) as a result of the influence of other atoms in its vicinity. Thus, neighboring atoms are often a source of localized *shielding effects*. At the site of any particular proton the actual net, perceptible magnetic field H is typically somewhat less intense than the nominal value H_0, and the amount of the discrepancy varies from place to place within a molecule. This means that the energy $\Delta E = hv_0$ required to induce a spectral transition (cf. equation 18–17) must also vary from proton to proton. Spectroscopists describe the phenomenon in terms of *chemical shifts* imposed upon the theoretical resonance frequency v_0. The magnitude of the chemical shift is usually ascertained with the aid of a supplementary magnetic coil which can be used to alter slightly the external magnetic field and, in effect, compensate for local shielding, thereby restoring the "resonance" (absorption) condition (cf. Fig. 18–32). Therefore, the x-axis of an NMR spectrum (recorded in what is known as a "field sweep") actually provides a measure of magnetic field strength. Alternatively, an NMR instrument may be designed in such a way that it is the frequency v_0 of the transmitter that is altered in the attempt to compensate for changes due to the chemical environment, providing a "frequency-sweep" spectrum.[77]

Both frequency and field strength are extremely difficult to measure (in an absolute sense) to a high degree of accuracy, so chemical shifts are normally reported

18.6 Nuclear Magnetic Resonance (NMR)

in relative terms based on a *standard substance*. For proton-resonance spectroscopy this standard is almost always tetramethylsilane (TMS), whose twelve protons produce a single sharp absorption signal.

Chemical shifts may be described quantitatively using either the *δ-scale*

$$\delta := \frac{v_{sample} - v_{TMS}}{v_0} \qquad (18\text{-}18)$$

or the *τ-scale*, where

$$\tau := 10.0 - \delta \qquad (18\text{-}19)$$

Values on both scales are dimensionless numbers expressed in ppm (= parts per million = 10^{-6}). From equation 18–17 it is clear that the fundamental resonance frequency v_0 may be caused to assume any desired value simply by providing a magnetic field H_0 of the appropriate strength. Typical operating frequencies v_0 for proton spectroscopy include 60 MHz and 90 MHz. Figure 18–33 illustrates the relationship between chemical shifts expressed in terms of the δ and τ scales, and it provides in addition a representative selection of absolute values for chemical shift on the basis of frequency (Δv, measured in Hz) at various operating frequencies v_0.

NMR Spectrometers are essentially single-beam instruments, and since measurement times are often rather long there is every reason to be concerned with the possibility of instrumental drift. Spectra must always be calibrated with respect to a standard substance, and this constitutes one rationale for the decision to adopt a relative measurement scale. Assuming there is no danger of disruptive interactions, the preferred method of calibration calls for introducing a small amount of standard substance directly into the sample solution (*internal standardization*).[77] Should an internal standard be for some reason precluded, standard substance may instead be enclosed in a sealed coaxial capillary before inserting it into the sample tube (*external standardization*).[77] Many modern instruments also rely upon the standard (or some other substance, such as a *deuterated solvent*) to help maintain a constant magnetic field strength H_0 (or a constant ratio H_0/v_0). The process involves a special regulating circuit known as a *control* or *lock system*,[77] and it has the advantage of assuring long-term stability comparable to that of a double-beam or ratio spectrometer.

In addition to protons, there are certain *other nuclei* that also lend themselves to nuclear magnetic resonance investigation; indeed, the technique is applicable in principle to any nucleus with a nuclear spin greater than zero. Some of the elements (isotopes) falling into this category and of particular interest to organic and biochemists include ^2H, ^{13}C, ^{15}N, ^{17}O, ^{19}F, and ^{31}P. However, factors besides nuclear spin play a role in determining the practicality of nuclear magnetic resonance spectrometry, with the most important being an isotope's *relative natural abundance* and the value of its *nuclear magnetic moment*. For example, the key isotope ^{13}C has a relative natural abundance of only 1.1%, and its nuclear magnetic moment is only one-

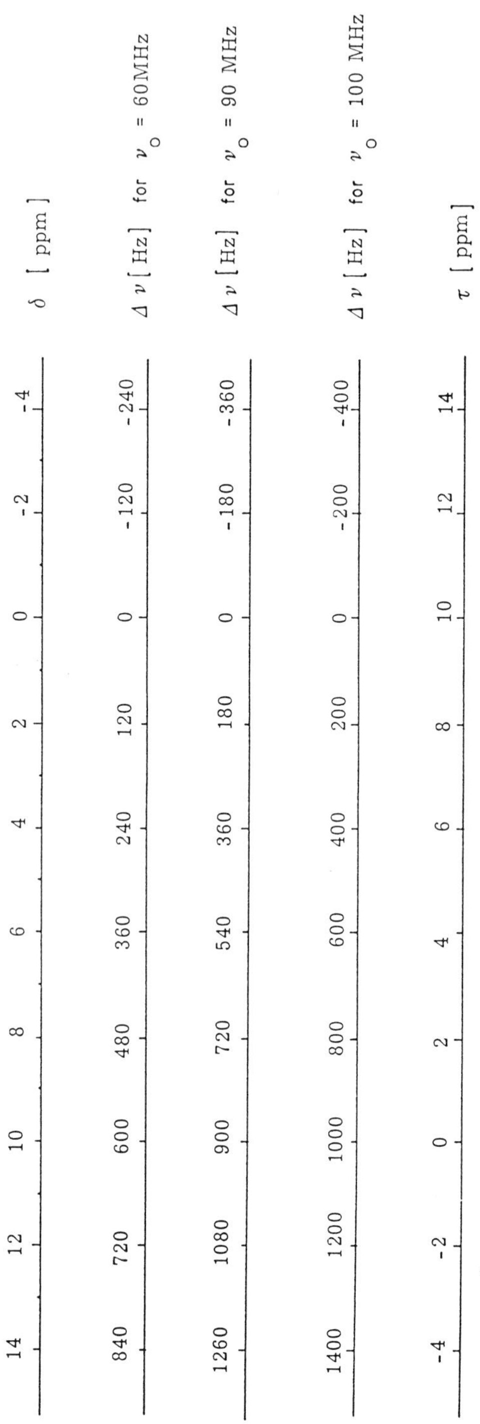

δ [ppm]	14	12	10	8	6	4	2	0	-2	-4
$\Delta\nu$ [Hz] for $\nu_0 = 60$ MHz	840	720	600	480	360	240	120	0	-120	-240
$\Delta\nu$ [Hz] for $\nu_0 = 90$ MHz	1260	1080	900	720	540	360	180	0	-180	-360
$\Delta\nu$ [Hz] for $\nu_0 = 100$ MHz	1400	1200	1000	800	600	400	200	0	-200	-400
τ [ppm]	-4	-2	0	2	4	6	8	10	12	14

Fig. 18-33. Chemical shift expressed both in relative terms (δ and τ) and directly in frequency units (Hz) for various operating frequencies ν_0.

18.6 Nuclear Magnetic Resonance (NMR)

fourth that of the proton. As a result of these differences, ^{13}C-NMR spectroscopy is less sensitive than proton NMR spectroscopy by a factor of about 5700.[81–83]

Preparative enrichment of an organic compound with ^{13}C is in most cases a very difficult undertaking, and primarily for this reason intensive, ongoing efforts have been directed toward developing new analytical techniques to facilitate natural-abundance ^{13}C-NMR spectroscopy. The most important advance has been introduction of the *pulse Fourier-transform (PFT)* spectrometer, an instrument that has expanded the scope of NMR spectrometry with respect to all nuclei, including even the proton. Fourier-transform NMR utilizes not a time-independent, constant source of monochromatic radiation of frequency v_0, but rather a brief, intense radiofrequency impulse P (t), as shown schematically in Fig. 18–34. It is possible to demonstrate[81–83] that such a pulse with a duration of only a few μs provides in addition to the carrier frequency a host of other frequencies known collectively as the "Fourier components". In other words, the transmitter in a PFT instrument emits what amounts to polychromatic radiation capable of addressing all absorption modes of the molecule simultaneously (and thus opening the way to what is known as *multiplex spectroscopy*). As a result of the synchronous absorption of energy over a wide range of frequencies there is induced in the receiver coil a complex signal known as a *pulse interferogram* (or *free induction decay, FID*), a signal that incorporates all the information required for reconstructing a normal NMR spectrum [cf. Fig. 18–34, PI (t)]. The mathematical process used to convert such an interferogram into a normal spectrum S (v) is called *Fourier transformation* [cf. Fig. 18–34, S(v)].

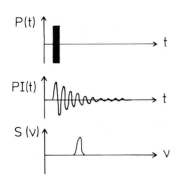

Fig. 18–34. Pulse-Fourier transform spectroscopy; schematic representation of an excitation pulse P(t), a pulse interferogram PI(t), and the frequency spectrum S(v) that results after Fourier transformation.

The PFT technique offers the following important advantages:

(1) Recording time for a spectrum is reduced to ca. 1 s.

(2) In the time required to acquire a single "classical" NMR spectrum (5-10 min) it is possible to accumulate several hundred interferograms, and if these are averaged the result represents a considerable improvement in signal-to-noise ratio.

(3) A certain amount of computation time is required to effect a Fourier transformation (usually with an integral computer system), but the delay amounts only to a matter of seconds.

(4) Compared with monochromatic radiation, brief radiofrequency pulses may be introduced at very high intensity without undue risk of "saturation" (cf. Sec. 18.6.2).

The combination of multiplex operation and increased transmitter efficiency results in a PFT-sensitivity for ^{13}C that is more than 100-fold greater than the sensitivity achieved by conventional techniques. The recent introduction of superconducting magnets has led to commercially available instruments offering magnetic field strengths H_0 as high as 117.5 kG (= 11.75 T), corresponding to a ^{1}H resonance frequency ν_0 of 500 MHz.[84] High magnetic fields provide some further increase in sensitivity, but their main advantage is a dramatic increase in resolution.

Disadvantages associated with ^{13}C-PFT spectroscopy include the following:

(1) Use of the technique requires on-line access to a rapid computer with considerable data-storage capacity.

(2) PFT Spectrometers are expensive, especially those utilizing superconducting magnets, and operating costs are also high.

(3) A high-resolution ^{13}C-NMR spectrum typically contains a multitude of absorption lines, and many of these may be further split into "multiplets" as a result of spin-spin coupling between ^{13}C- and ^{1}H-nuclei. Such a (non-decoupled) spectrum can often prove rather difficult to interpret.

(4) Spin-spin coupling ("splitting") also disperses the intensity of a given signal over a number of lines, thereby reducing the observed spectral intensity per line.

Much effort has been devoted to developing techniques for producing *decoupled* spectra in order to facilitate data interpretation and peak assignment, as well as to increase the signal strength per line. Common strategies include broadband decoupling, off-resonance decoupling, and various types of selective decoupling.[77,81–83]

Appropriate combinations of the above-described techniques make it possible to increase the ^{13}C-NMR signal-to-noise ratio by a factor of 10^3 to 10^5.[85] The relative *accuracy* of a chemical shift measurement performed on a high-quality instrument is today better than 0.1 ppm, corresponding to roughly 10^{-5}% of the operating frequency ν_0.[83,86]

18.6 Nuclear Magnetic Resonance (NMR)

18.6.2 Quantitative NMR Analysis of Titration Equilibria

The approach taken to *quantification* in an NMR-spectrometric titration is entirely different from that used with all the previously described techniques (Sections 18–1 through 18–5). Assuming titration is accompanied by very rapid equilibration (i.e., if the half-life $\tau' = 1/k < 10^{-4}$ s), as is the case with most protolysis equilibria, then the relative concentration of the various species must be ascertained from an NMR spectrum not by means of signal *intensity* measurements (associated with the y-axis of the spectrum), but rather on the basis of *chemical shift* (plotted along the x-axis). This principle is best illustrated with an example of the simplest possible system, the single-step protolysis equilibrium

$$\text{BH} \underset{k_R}{\overset{k_F}{\rightleftarrows}} \text{B}^- \; (\text{H}^+) \tag{18–20}$$

Let us therefore consider the changes observed in the ^{13}C-NMR spectrum of *pyridine* as a function of pH:[87]

Deprotonation of nitrogen causes a shift of ca. 10 ppm in the ^{13}C-NMR signal for carbon atom 4.[87] Reaction is very rapid, as suggested by the fact that the rate constants k_F and k_R are known from ultrasonic relaxation studies.[88] Repeated protonation and deprotonation causes the resonance frequency of C-4 to alternate rapidly between the two values ν_{0BH} and ν_{0B}—so rapidly, in fact, that changes occur in the shapes of the lines constituting the resonance signals. The *Bloch theory of NMR*[77,89] makes it possible to predict the line shape that will characterize a signal from a species with any given lifetime. Figure 18–35 depicts NMR spectra corresponding to a system comprised of two equilibrating species, BH and B⁻, with resonance frequencies ν_{0BH} and ν_{0B}. For simplicity it has been assumed that $bh = b$ and that the two rate constants k_F and k_R are equal,* with values ranging between 1 s⁻¹ and 10⁴ s⁻¹. If exchange is very slow (cf. the upper curve in Fig. 18–35) the spectrum will consist of two sharp lines at the expected frequencies ν_{0BH} and ν_{0B}. As the rate of equilibration increases, the two lines become broader. They also begin to merge, and at what is known as the *coalescence point* the two can no longer be

---

* The forward and reverse processes have also been treated as if both were pseudo-first-order; that is to say, it has actually been assumed that $k_F = k_R \cdot c_{H_3O^+}$, where $c_{H_3O^+}$ is regarded as a constant.

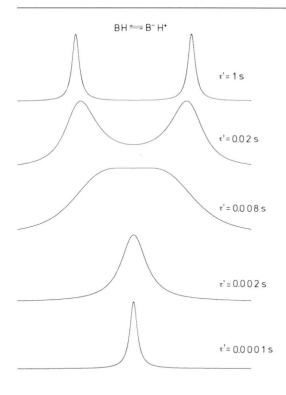

Fig. 18–35. Simulated NMR spectrum for an exchange process BH ⇌ B⁻ (H⁺) computed as a function of the average lifetime $\tau' = 1/k$.

distinguished. With further increases in rate this combined signal becomes increasingly narrow, ultimately collapsing to a single sharp line of frequency $(v_{0BH} + v_{0B})/2$ at a rate $k = 10^4 \text{ s}^{-1}$.

In Fig. 18–36 the rate of interconversion is held constant at the limiting value ($k = 10^4 \text{ s}^{-1}$), but this time the relative concentrations bh/b_0 and b/b_0 are allowed to vary just as they would in a titration. Each spectrum now consists of a single sharp line, and when $bh = b$ (i.e., when pH = pK) the spectrum obtained is identical to the last spectrum in Fig. 18–35. At the points $bh = b_0$ and $b = b_0$ the lines fall at the resonance frequencies v_{0BH} and v_{0B}, respectively. Generally speaking, in any single-step titration (cf. equation 18–20) the NMR line corresponding to any given nucleus (e.g., a particular ^{13}C nucleus) will appear somewhere between the resonance positions predicted for that nucleus in the pure species BH and B⁻. Its precise location (chemical shift) will be a direct function of the relative concentrations of the two species; i.e.:

$$v = v_{0BH} \cdot \frac{bh}{b_0} + v_{0B} \cdot \frac{b}{b_0} \tag{18-22}$$

or, transposing to the δ-scale:

$$\delta = \frac{1}{b_0} \cdot (\delta_{BH} \cdot bh + \delta_B \cdot b) \tag{18-23}$$

18.6 Nuclear Magnetic Resonance (NMR)

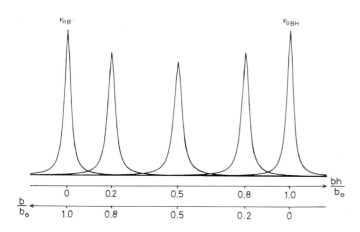

Fig. 18–36. Simulated NMR lines for a very rapid exchange process $BH \rightleftarrows B^- (H^+)$ computed as a function of the relative concentrations bh/b_0 and b/b_0.

Thus, just as in the case of the Lambert–Beer–Bouguer law, a linear relationship is found to link concentration with a spectral parameter, but here the role normally assigned to molar absorptivity ε_λ has been assumed by the chemical shift δ.

Putting the matter succinctly, the concentration-dependent experimental parameter in an NMR-spectrometric titration is not signal intensity or an absorption value (plotted along the *y*-axis of the spectrum) but rather the *position* of the signal, its frequency ν_0 (which is the *x*-variable). If one wishes on the basis of NMR to quantify the position of equilibrium in a rapid exchange phenomenon, the important variable is the chemical shift, a parameter that may be determined with an extremely high degree of accuracy (cf. Sec. 18.6.1). On the other hand, quantitative analysis of a single *stable* substance, or of a mixture of stable substances, depends in NMR as in other types of spectroscopy upon measurements made along the *y*-axis of the spectrum, and in the case of NMR these tend to be rather inexact.[79,90]

Our concrete example of a spectrometric titration in this instance is a ^{13}C-NMR titration of histidine ($c_0 = 0.3$ M).* The lower portion of Fig. 18–37 consists of a set of NMR titration curves δ vs. pH plotted in terms of various carbon atoms. In the upper portion of the same figure the chemical shifts of various pairs of atoms have been plotted against each other in the form of *chemical shift diagrams (δ-diagrams)*, which may be regarded as analogous to A-diagrams.[86,92] The ordinate in each case is δ_{COOH}, since it is this signal that proves to be the most sensitive to changes in pH. All the δ-diagrams consist of three linear segments, consistent with what would be expected for a three-step non-overlapping titration system. In each of the δ-diagrams, as well as in the δ-pH–diagrams at the bottom of the figure, it is apparent that the extent of

---

* Original data related to this study appear in ref. 91. We are grateful to Prof. G. Jung of Tübingen (Institute of Organic Chemistry) for allowing us to reproduce much of the information here.

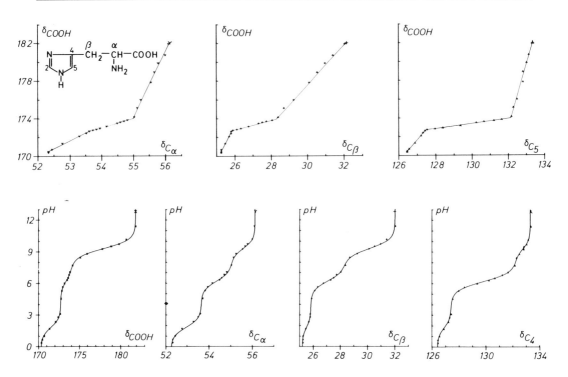

Fig. 18–37. ^{13}C-NMR Titration of histidine ($c_0 = 0.3$ M); upper portion, δ-diagrams; lower portion, δ-pH–diagrams.

dissociation of *all three* dissociable groups has an effect on each of the ^{13}C nuclei investigated (cf. also ref. 91). Similar studies on the behavior of various peptides serve as further confirmation of the utility of δ-diagrams derived from ^{13}C-NMR titrations.[86]

Recently, increasing attention has been paid to the potential for using ^{17}O- (cf. ref. 93) and ^{31}P-NMR spectrometry[94,95] as tools for investigating biochemical systems. For example, it has been shown that ^{31}P-NMR titration provides insight into the chemistry of the phosphates of vitamin-B$_6$ compounds and their derivatives,[94,95] as well as the diphosphate of vitamin-B$_1$ (thiamine pyrophosphate).[92] The results of a ^{31}P-NMR titration of thiamine pyrophosphate are illustrated in Fig. 18–38. Throughout the pH range ca. 1.5–10 the broadband proton-decoupled ^{31}P-NMR spectrum of thiamine pyrophosphate consists of two sharp doublets, which may be characterized as comprising an AB-system.[77,96] Figure 18–38 reveals the pH-dependencies of all four lines.[92] The curves δ_α(pH) and δ_β(pH) indicate the relationships between acidity and the overall chemical shifts δ_α and δ_β of the AB-system itself. Above pH 8 these chemical shifts coincide with the midpoints of the corresponding doublets, indicating that coupling here is very weak. Below pH 7.0 the signals are displaced toward the midpoint of the *pair* of doublets, a sign of stronger coupling. Figure 18–39 contains a series of δ-diagrams δ_3 vs. δ_i, each of which consists of two linear segments. This points to the presence of two non-overlapping equilibria.

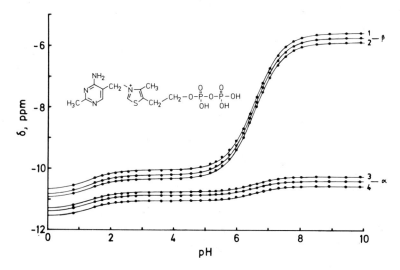

Fig. 18–38. δ-pH–Data derived from a ^{31}P titration of thiamine pyrophosphate; 1–4: observed resonance lines from the two doublets; α, β: calculated chemical shifts for a corresponding AB-system.

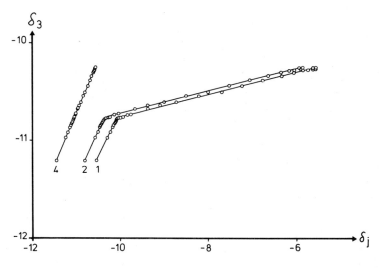

Fig. 18–39. Chemical shift diagrams for resonance lines 1–4 of the two doublets from thiamine pyrophosphate (cf. Fig. 18–38).

Table 18-3. ^{31}P-NMR chemical shifts (in ppm) for individual components of the thiamine pyrophosphate titration system.

BH$_2$	BH$^-$	B^{2-}
−11.41	−10.87	−10.40
−10.86	−10.21	− 5.75

The titration curves in Fig. 18-38 can be evaluated on the basis of equation 7-7 by introducing chemical shifts in place of absorbances. Non-linear regression analysis using the TIFIT program leads to the values pK_1 = 1.35 ± 0.1 and pK_2 = 6.57 ± 0.01, together with the set of chemical shifts reported in Table 18-3. The solid lines in Fig. 18-38 are theoretical titration curves constructed on the basis of these results.

It is of interest to compare this information with data obtained from a ^{31}P-NMR titration of pyrophosphoric acid (Fig. 18-40). In this case only a single resonance line is observed, because the two phosphorus atoms are magnetically equivalent (thus constituting an A$_2$-system),[77,97] a circumstance that rules out any use of δ-diagrams. No values could be established for the first two dissociation constants K_1 and K_2 since the data are restricted to the region above pH 3. Evaluation as a two-step titration system[92] provides the values pK_3 = 6.25 ± 0.01 and pK_4 = 8.64 ± 0.01. The chemical shifts of the pure components BH$_2$, BH$^-$, and B^{2-} were calculated to be −10.01, − 6.89, and 5.93, respectively. The solid line in Fig. 18-40 was created on the basis of these results. Literature values for pK_3 range from 5.8 (18 °C) to 6.6 (25 °C), and for pK_4 from 8.2 (18 °C) to 9.3 (25 °C).[98,99] In the case of thiamine pyrophosphate the reported values of pK_3 range from 6.5 to 6.7.[100,101]

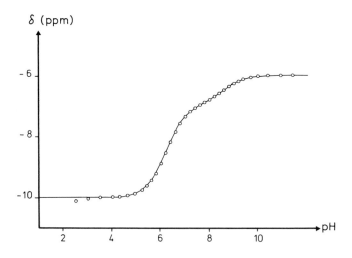

Fig. 18-40. δ-pH-Diagram for sodium pyrophosphate.

18.6 Nuclear Magnetic Resonance (NMR)

At the present time, exact quantitative evaluation of NMR-titration data remain subject to the following limitations:

(1) The only *experimental setups* so far reported involve either m samples, each prepared in a different buffer (cf. Sec. 16.1), or else the use of a separate titration vessel external to the spectrometer (cf. Sec. 16.2.1). The latter approach requires that samples be transferred to NMR tubes at each stage of titration, with appropriate checks on pH before and after each measurement. Use of inert atmosphere techniques under these conditions is extremely inconvenient. The development of suitable flow cuvettes or permanently mounted, internal titration vessels is still very much in the experimental stage. Consequently, the level of reproducibility to be expected from an NMR titration falls short of the standards set by UV–VIS, fluorometric, and CD/ORD systems.

(2) *Signal saturation* can seriously impede progress in a quantitative NMR investigation.[77-79] According to equation 18–17, the energy $\Delta E = h \cdot v_0$ required to alter a nuclear spin when the operating frequency is 60 MHz amounts to only ca. $3 \cdot 10^{-5}$ kJ mol^{-1}, or five orders of magnitude less than the thermal energy available at room temperature (ca. 2.5 kJ mol^{-1}). Using Boltzmann statistics it can be shown that the corresponding population distribution between the upper energy level and the lower energy level is about 100 000/100 001.[78,102] Setting the energy output of the transmitter too high causes this tiny population difference to be totally erased, since energy is absorbed faster than it can be dissipated into the environment (*signal saturation*). Such a circumstance invalidates one of the fundamental premises underlying quantitative spectrometry (cf. requirement 3 for the Lambert–Beer–Bouguer law as stated in Sec. 3.1). The danger of saturation is much greater in the case of a time-independent transmitter than with an instrument based on the pulse technique (cf. Sec. 18.6.1).

(3) Finally, there is a problem associated with the necessary *initial concentration c_0*. So far, most quantitative studies involving NMR titrations have been conducted at concentrations of 0.1 M to > 1 M in order to obtain reasonable signal-to-noise ratios within an acceptable time period. The improvements in long-term stability accompanying the introduction of sophisticated lock systems, coupled with the sensitivity enhancement offered by PFT techniques, have made significant reductions in sample size a real possibility. For example, reports have recently appeared of ^{31}P-NMR titrations of pyridoxal phosphate-dependent enzymes (e.g., tryptophane synthase,[103] tryptophanase,[104] and D-serum dehydratase[105,106]) in which the initial concentration c_0 of enzyme and pyridoxal phosphate is stated to be $1-5 \cdot 10^{-4}$ M. It should be noted, however, that this work entailed the accumulation and averaging of between 15 000 and 40 000 PFT spectra at each titration step, with a total measurement time per sample of from 8 to 24 hours.

Literature

1. H. Lachmann, *Dissertation*. Tübingen 1973.
2. H. Lachmann, M. Mauser, F. Schneider, H. Wenck, *Z. Naturforsch.* 26 b (1971) 629.
3. H. Lachmann, *Habilitationsschrift*. Tübingen/Würzburg 1982.
4. H. Lachmann, *Z. Anal. Chem.* 301 (1980) 148.
5. G. W. Ewing, A. Maschka, *Physikalische Analysen- und Untersuchungsmethoden der Chemie*. R. Bohrmann Verlag, Wien 1961. Chap. 7.
6. G. Kortüm, *Kolorimetrie, Photometrie und Spektrometrie*. Springer-Verlag, Berlin 1962. Chap. I.
7. G. Gauglitz, *Praktische Spektroskopie*. Attempto Verlag, Tübingen 1983.
8. H. L. Pardue, T. E. Hewitt, J. Milano, *Clin. Chem.* 20 (1974) 1028.
9. H. L. Youmans, H. Brown, *Anal. Chem.* 48 (12976) 1152.
10. B. G. Wybourne, *J. Opt. Soc. Amer.* 50 (1960) 84.
11. S. Ebel, E. Glaser, Saa Abdulla, U. Steffen, V. Walter, *Fresenius Z. Anal. Chem.* 313 (1982) 24.
12. W. Luck, *Ber. Bunsenges. Z. Elektrochem.* 64 (1960) 676.
13. R. B. Cook, A. R. Jankow, *J. Chem. Educ.* 49 (1972) 405.
14. H. H. Perkampus, *UV-VIS-Spektroskopie und ihre Anwendungen*. Springer-Verlag, Berlin 1986.
15. J. Agterdenboos, J. Vlogtman, L. V. Broekhoven, *Talanta* 21 (1974) 225.
16. J. Agterdenboos, J. Vlogtman, *Talanta* 21 (1974) 231.
17. H. Ley, B. Arends, *Z. Phys. Chem.* (B) 6 (1930) 240.
18. S. Udenfriend, *Fluorescence Assay in Biology and Medicine I and II*. Academic Press, New York 1969.
19. C. A. Parker, *Photoluminescence of Solutions*. Elsevier Publishers, Amsterdam 1968.
20. H. V. Drushel, A. L. Sommers, R. C. Cox, *Anal. Chem.* 35 (1963) 2166.
21. G. G. Gilbaut, *Practical Fluorescence: Theory, Methods and Techniques*. Marcel Decker, New York 1973.
22. *Einführung in die Fluoreszenzspektroskopie*. Perkin Elmer Publication. Series Angewandte UV-Spektroskopie, Vol. 4, 1978.
23. G. W. Ewing, *A. Maschka, Physikalische Analysen- und Untersuchungsmethoden der Chemie*. R. Bohrmann Verlag, Wien 1961. Chap. 8.
24. T. Förster, *Fluoreszenz organischer Verbindungen*. Vandenhoek und Ruprecht, Göttingen 1951.
25. G. Gauglitz, *Z. Phys. Chem. N. F.* 88 (1974) 193.
26. J. Eisenbrand, *Fluorimetrie*. Wissenschaftliche Verlagsgesellschaft, Stuttgart 1966.
27. H. H. Perkampus, *Absorptions- und Lumineszenzspektroskopie im ultravioletten und sichtbaren Spektralbereich*. In: *Ullmanns Enzyklopädie der technischen Chemie*. Verlag Chemie, Weinheim 1980, Vol. 5.
28. M. Zander, *Fluorimetrie*. Anleit. Chem. Laboratoriumspraxis. Springer-Verlag, Berlin 1981. Vol. 17.
29. T. Förster, *Z. Elektrochem.* 54 (1950) 531.
30. A. Weller, *Z. Elektrochem.* 56 (1952) 662.
31. T. Shimanouchi, *Tables of Molecular Vibrational Frequencies*. NSRDS-NBS39, Washington 1972. Vol. I, p. 10.
32. M. A. West, D. R. Kemp, *International Laboratory*, May/June (1976) 27.

33. Y. Kanaoka, *Angew. Chem.* 89 (1977) 142.
34. W. Klöppfer, *Adv. Photochem.* 10 (1977) 311.
35. K. Tsutsumi, H. Shizuka, *Z. Phys. Chem. N. F.* 122 (1980) 129.
36. O. S. Wolfbeis, E. Fürlinger, *Z. Phys. Chem.* 129 (1982) 171.
37. D. M. Hercules, *Fluorescence and Phosphorescence Analysis*. Wiley, New York 1966.
38. R. Blume, H. Lachmann, J. Polster, *Z. Naturforsch.* 30 b (1975) 263.
39. H. Lachmann, *Fresenius Z. Anal. Chem.* 317 (1984) 698.
40. J. F. Ireland, P. A. Wyatt, *Adv. Phys. Org. Chem.* 12 (1976) 131.
41. D. M. Hercules, *Anal. Chem.* 30 (1958) 96.
42. W. R. Laws, L. Brand, *J. Phys. Chem.* 83 (1958) 96.
43. L. Brand, J. B. A. Ross, W. R. Laws, *Ann. New York Acad. Sci.* (1981) 197.
44. G. Snatzke, *Optical Rotary Dispersion and Circula Dichroism in Organic Chemistry*. Heyden and Sons Ltd., London 1967. Chapters 3 and 4.
45. G. Snatzke, *Optical Rotary Dispersion and Circular Dichroism in Organic Chemistry*. Heyden and Sons Ltd., London 1967. Chap. 5.
46. J. Flügge, *Grundlagen der Polarimetrie*. Walter De Gruyter & Co., Berlin 1970. Chap. 5.
47. J. Flügge, *Grundlagen der Polarimetrie*. Walter De Gruyter & Co., Berlin 1970. Chap. 4.
48. G. Snatzke, *Optical Rotary Dispersion and Circula Dichroism in Organic Chemistry*. Heyden and Sons Ltd., London 1967. Chap. 1.
49. L. Velluz, M. Legrand, M. Grosjean, *Optical Circular Dichroism*. Verlag Chemie, Weinheim, and Academic Press, New York-London 1965.
50. P. Crabbé, *Optical Rotary Dispersion and Circular Dichroism in Organic Chemistry*. Holden Day Publ., San Fransisco 1965.
51. A. Wollmer, *ORD und CD Spektroskopie*. In: *Biophysik*. Springer-Verlag, Berlin 1982.
52. W. Voelter, E. Bayer, R. Records, E. Bunnenberg, C. Djerassi, *Liebigs Ann. Chem.* 718 (1968) 238.
53. G. Jung, M. Ottnad, P. Hartter, H. Lachmann, *Angew. Chem. Int. Ed. Engl.* 14 (1975) 429.
54. G. Jung, M. Ottnad, W. Voelter, E. Breitmaier, *Z. Anal. Chem.* 259 (1972) 217.
55. H. Eckstein, G. Barth, R. E. Lindner, E. Bunnenberg, C. Djerassi, *Liebigs Ann. Chem.* 6 (1974) 990.
56. G. Blauer, *Optical Activity of Conjugated Proteins*. In: *Structure and Bonding* 18 (1974) 69.
57. *Handbook of Biochemistry*, 2nd. ed. The Rubber Company, Cleveland 1970. Section J 198.
58. H. Günzler, H. Böck, *IR-Spektroskopie*. Verlag Chemie, Weinheim 1975.
59. B. Schrader, *Ullmann Encyklopädie der technischen Chemie*. 4. ed. Verlag Chemie, Weinheim 1980. Vol. 5, p. 303.
60. L. W. Herscher, *Instumentation*. In: D. E. Kendall (Ed.), *Applied Infrared Spectroscopy*. Chapman & Hall, London 1966.
61. P. A. Wilks, *International Laboratory*, July/August (1979) 49.
62. H. Weitkamp, R. Barth, *Einführung in die quatitative Infrarot-Spektroskopie*. Georg Thieme Verlag, Stuttgart 1976.
63. B. Schrader, *Angew. Chem.* 85 (1973) 925.

64. A. E. Martell, *Chemical and Biological Aspects of Pyridoxal Catalysis*. Pergamon Press, London 1963, p. 13.
65. F. J. Anderson, A. E. Martell, *J. Amer. Chem. Soc.* 86 (1964) 715.
66. D. A. Long, *Raman Spectroscopy*. McGraw Hill, New York 1977.
67. W. S. Letchow, *Laserspektroskopie*. Vieweg Verlag, Braunschweig 1977.
68. A. Deffontaine, M. Bridoux, M. Del Haye, *Revue Phys. Appl.* 19 (1984) 415.
69. W. Kiefer, H. J. Bernstein, *Appl. Spectrosc.* 25 (1971) 609.
70. H. J. Sloane, R. B. Cook, *Appl. Spectrosc.* 26 (1972) 589.
71. G. Jung, E. Breitmaier, W. A. Günzler, M. Ottnad, W. Voelter, L. Flohé, in: *Glutathione*. Georg Thieme Verlag, Stuttgart 1974, p. 1.
72. D. Garfinkel, J. T. Edsall, *J. Amer. Chem. Soc.* 80 (1958) 3823.
73. G. Kortüm, *Lehrbuch der Elektrochemie*. Verlag Chemie, Weinheim 1972. Chap. VII.
74. E. J. King, *Acid-Base Equilibria*. Pergamon Press, London 1965.
75. L. Stein, E. Appelman, *Inorg. Chem.* 22 (1983) 3017.
76. A. Deffontaine, M. Bridoux, M. Del Haye, *Revue Phys. Appl.* 19 (1984) 415.
77. H. Günther, *NMR-Spektroskopie*. Georg Thieme Verlag, Stuttgart 1973.
78. H. A. Staab, *Einführung in die theoretische organische Chemie*. 4. ed. Verlag Chemie, Weinheim 1964, Sec. 2.9.
79. H. Suhr, *Anwendungen der kernmagnetischen Resonanz in der organischen Chemie*. Springer-Verlag, Berlin 1965. Chap. I.
80. H. H. Paul, V. Penka, W. Lohmann, in: *Biophysik*. 2. ed. Springer Verlag, Berlin 1982. Sec. 3.4.
81. H. O. Kalimowski, S. Berger, S. Braun, *^{13}C-NMR-Spektroskopie*. Georg Thieme Verlag, Stuttgart 1984.
82. E. Breitmaier, G. Bauer, *^{13}C-NMR-Spektroskopie*. Georg Thieme Verlag, Stuttgart 1977.
83. L. Ernst, *^{13}C-NMR-Spektroskopie, eine Einführung*. Dr. Dietrich Steinkopff Verlag, Darmstadt 1980.
84. *Bruker-Report* 3 (1971) 12.
85. H. Günther, *Chemie in unserer Zeit* 8 (1974) 45.
86. D. Leibfritz, E. Haupt, N. Dubischar, H. Lachmann, R. Oekonomopulos, G. Jung, *Tetrahedron* 38 (1982) 2165.
87. E. Breitmaier, K. H. Spohn, *Tetrahedron* 29 (1973) 1145.
88. M. L. Ahrens, G. Maaß, *Angew. Chem.* 80 (1968) 848.
89. T. L. James, *Nuclear Magnetic Resonance*. In: *Biochemistry*. Academic Press, New York 1975. Chap. 2.
90. F. Kasler, *Quantative Analysis by NMR Spectroscopy*. Academic Press, London 1973, Part B 6.
91. E. Breitmaier, G. Jung, M. Ottnad, W. Voelter, *Fresenius Z. Anal. Chem.* 261 (1972) 328.
92. H. Lachmann, K. D. Schnackerz, *Organic Magnetic Resonnace* 27 (1984) 101.
93. J. P. Gerothanassis, R. Hunston, J. Lauterwein, *Helv. Chim. Acta* 65 (1982) 1764 and 1774.
94. K. D. Schnackerz, R. E. Benesch, S. Kwong, R. Benesch, J. M. Helmreich, *J. Biol. Chem.* 258 (1983) 872.
95. K. D. Schnackerz, E. E. Snell, *J. Biol. Chem.* 258 (1983) 4839.

Literature

96. R. A. Hoffman, S. Forsén, B. Gestblom, in: P. Diehl, E. Fluck, R. Kosfelf (Eds.), *NMR, Basic Principles and Progress*, Vol. 5: *Analysis of NMR Spectra*. Springer Verlag, New York 1971, Chap. IV.
97. M. M. Crutchfield, C. H. Dungan, J. R. van Wazer, in: H. Grayson, E. J. Griffiths (Eds.), *Topics in Phosphorous Chemistry*. Interscience, New York 1967, Vol. 5.
98. H. A. Sober (Ed.), *Handbook of Biochemistry*. 2nd ed. Chemical Rubber Co., Cleveland 1970.
99. R. C. Weast (Ed.), *Handbook of Chemistry and Physics*. 59th ed. CRC Press, Boka Raton 1979.
100. A. M. Chauvet-Monges, M. Hadida, A. Crevat, E. J. Vincent, *Arch. Biochem. Biophys.* 207 (1981) 311.
101. J. P. Taglioni, J. Fournier, *C. R. Acad. Sci. Ser.* C 288 (1979) 141.
102. T. L. James, *Nuclear Magnetic Resonnace in Biochemistry*. Academic Press, New York 1975. Chap. 1.
103. K. D. Schnackerz, P. Bartholmes, *Biochem. Biophys. Res. Commun.* 111 (1983) 817.
104. K. D. Schnackerz, E. E. Snell, *J. Biol. Chem.* 258 (1983) 4839.
105. K. D. Schnackerz, K. Feldmann, W. E. Hull, *Biochemistry* 18 (1979) 1536.
106. K. D. Schnackerz, K. Feldmann, *Biochem. Biophys. Res. Commun.* 95 (1980) 1832.

19 Automation and Data Processing

Completely automated spectrometric titrations were introduced into the literature as long as 30 years ago,[1] although in the early reports all titration curves and their first derivatives were recorded in analog form, and no use was made of data-processing techniques as we understand them today. More recently, a wide variety of microprocessor-controlled titration systems has been described, many of which feature integral data-storage and pK calculation facilities suitable for both potentiometric[2-7] and photometric[8] titrations. Since precision spectrometric titration of a multi-step equilibrium system may involve ≥ 50 titration steps, and information may be required at up to 10 different wavelengths, the amount of experimental data generated is prodigious, and there is ample incentive to strive for a maximum degree of automation coupled with efficient electronic data processing.

As early as the 1960s individual research groups were performing (by hand) point-by-point digitalizations of titration spectra and then transferring the results to punch cards for subsequent processing. However, few of these experiments represented true discontinuous titrations (cf. Sec. 16.2), being based rather on m batches of sample dissolved in appropriate buffers (cf. Sec. 16.1). Not surprisingly, the number of steps investigated per stage of titration was relatively small ($\leq 5-7$). By contrast, the number of wavelengths at which data were acquired was often quite large (ca. 75), since the actual goal of most of the work was computer analysis of individual absorption bands (band-shape analysis, cf. Sec. 12.3.4).[9-13]

The development of a fully-automatic process control system for a *spectrometric titration* presents a greater challenge than does the analogous case of automating a potentiometric titration, since after each titration step it is necessary that a new sample, free of air bubbles, be introduced into the spectrometer. Moreover, a certain amount of control is also necessary during each spectral recording cycle. Even if the latter function is largely assumed by the spectrometer itself, the process must still be initiated by some computer or microprocessor.

Experience to date confirms that the simplest approach to *fully-automatic spectrometric titration* utilizes a system like that illustrated in Fig. 16-2, with additional provision for a suitable flow cell and an inert pumping mechanism. Alternatively, titration may be conducted directly within the spectrometer, as discussed in Sections 16.2.2(a) and 16.2.2(b). The fiber-optics system shown in Fig. 16–3 would appear to constitute a particularly elegant basis for solving the automation problem, but unfortunately this technique is inapplicable in the short-wavelength UV-region.

Virtually all the absorption spectrometric titrations discussed in this book were performed using the recording double-beam spectrometer depicted in Fig. 19–1 (Zeiss model DMR 21 WL), which offers automatic process control and ample data-storage capacity. A cuvette like that described in conjunction with the titration apparatus of Sec. 16.2.1(b) is permanently mounted in the thermostated sample holder of the instrument, where precise temperature control is maintained over both the titration vessel and a reference cuvette. The standard cover for the sample compartment has been replaced by one custom-designed to permit the passage of various tubes and electrical connections. Since the spectrometer comes equipped with a timer and a digital clock (intended to facilitate kinetic studies),[14] accurate records may be maintained of elapsed time during a lengthy precision titration. Spectrometric data and elapsed time may be accumulated in the form of punched tape, but they may also be entered directly into an integral process-control computer.[15] Access is also available

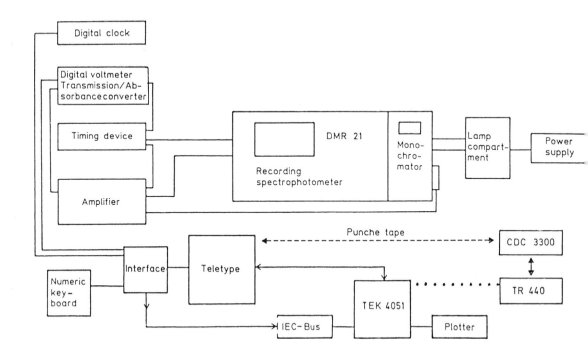

Fig. 19–1. Block diagram of a recording double-beam spectrophotometer with provision for automatic scan control and data acquisition.

to a mainframe computer better suited to lengthy computations. The pH and the volume of titrant added at each step are in some cases entered into the computer during the course of titration with the aid of a keypad, but this information may also be saved for entry at a later time.

The titration itself is conducted in a *semi-automatic mode*. First, an appropriate pH is established in the external titration vessel with the aid of a buret and a pH meter. Sample is then introduced into the cuvette as the culmination of a three-fold back-and-forth pumping operation, after which the spectrometer is set in operation. This instrument is designed in such a way that precision "stops" can be installed anywhere along the recording drum. Each stop is capable of tripping a microswitch that in turn causes the instrument to pause when a particular wavelength has been reached in the course of recording a titration spectrum. This interruption in the scan sequence permits a digital record to be made of the current absorbance A_λ, ascertained as a value acquired over a pre-determined integration time (usually 3–5 s). Scanning resumes automatically as soon such an absorbance value has been established.

As indicated in Sec. 18.1.1, a complete spectrometric data-gathering operation of this type consumes ca. 4–5 minutes at each titration step, a time period comparable to that required for equilibration of a pH electrode. The pH is carefully measured at the conclusion of each cycle, and the corresponding information is then entered into the data record together with the incremental volume of titrant R. Renewed adjustment of the pH takes about one minute, so depending on the number of wavelengths at which digital absorbance measurements are required, a precision titration can be carried out at the rate of about 10–15 titration steps per hour.

Most of the data evaluation procedures recommended in previous chapters are readily adaptable to a personal computer, preferably one equipped with a digital plotter and a graphics screen. The major exception is iterative non-linear regression analysis, for which a larger computer is advantageous. Optimization and rapid interpretation of intermediate results is facilitated by displaying all graphic output directly on the computer screen. Later, such results may be converted to hardcopy form with the help of a plotter. The availability of remote, interactive access to a larger computer is certainly useful, because it provides the operator with the opportunity to choose the hardware best suited to working up a particular set of data. Numerical data and graphics may then be output to a terminal screen, to the plotter, or to one of a variety of printers, depending on the circumstances.

Literature

1. S. Wolf, *Chem. Rundsch.* 18 (1959) 1.
2. L. Pehrsson, F. Ingmann, *Talanta* 24 (1977) 79.
3. S. Ebel, J. Hocke, B. Reyer, *Z. Anal. Chem.* 308 (1980) 437.
4. S. Ebel, A. Seuring, *Angew. Chem.* 89 (1977) 129.
5. H. Gampp, M. Maeder, A. D. Zuberbühler, T. Kaden, *Talanta* 27 (1980) 513.
6. S. Ben-Yaakov, R. Raviv, H. Guterman, A. Dayan, B. Lazar, *Talanta* 29 (1982) 267.
7. G. Velinov, N. Todorov, S. Karamphilov, *Talanta* 30 (1983) 687.
8. L. Pehrsson, F. Ingmann, *Talanta* 24 (1977) 87.
9. K. Nagano, D. E. Metzler, *J. Amer. Chem. Soc.* 89 (1967) 2891.
10. D. E. Metzler, K. Nagano, in: *Pyridoxal Catalysis: Enzymes and Model Systems.* Wiley Interscience, New York 1968, p. 81.
11. D. E. Metzler, C. Harris, R. L. Reeves, W. H. Lawton, M. S. Maggio, *Anal. Chem.* 49 (1977) 864 A.
12. C. M. Harris, R. J. Johnson, D. E. Metzler, *Biochem. Biophys. Acta* 421 (1976) 181.
13. Y. V. Morozov, F. A. Savin, in: L. B. Towsend, R. S. Tipson (Eds.), *Nucleic Acid Chemistry.* Wiley Interscience, New York 1978, Vol. II, p. 1009.
14. G. Gauglitz, A. Reule, *Z. Anal. Chem.* 276 (1975) 97.
15. G. Gauglitz, T. Klink, A. Lorch, *Fresenius Z. Anal. Chem.* 319 (1984) 364.

Appendix

A Source Listing of EDIA

The program for the determination of relative pK-values (ΔpK) and absorbances of dibasic protolytes implements a method which was published in *Anal. Chem.* 48 (1976) 1546. The input data are measurements of absorbances for different pH-values and different wavelenghts. The absorbance values have to be corrected for the effect of dilution in the course of titration. The order and formats of the data are described in the first comment of the main program.

The mathematical procedure is based on the fact that a plot of the absorbances at two different wavelengths and pH-values varying over a sufficient range displays a certain part of a conic section. If the parameters of the conic section are known it is possible to derive the interesting quantities (ΔpK and absorbances). Any pair of wavelengths yields these results and it was found that some of the results are much more reliable than others. In order to estimate the reliability for each pair of wavelengthes the dependence of the results on arificially introduced random errors is detected and only those results are averaged for the final result which showed a low dependence on the aritificial errors.

The program contains some subroutines (LINSYS, RANOR and RANDO) for the solution of systems of linear equations and the generation of normally distributed random numbers respectively. These are given in order to have a complete set of programs. Especially in the case of the random number generator it is recommended to look if these routines can be replaced by standard routines of the respective computing center.

```
10=C**********MAIN PROGRAM ***************************************************
20=C   *    FOR A DESCRIPTION OF THE MATHEMATICAL METHOD USED IN THIS
30=C   *    PROGRAM SEE
40=C   *    GOBBER, F. AND POLSTER J., DETERMINATION OF RELATIVE
50=C   *    PK-VALUES OF DIBASIC PROTOLYTES BY REGRESSION ANALYSIS
60=C   *    OF ABSORBANCE DIAGRAMS, ANALYTICAL CHEMISTRY
70=C   *    Vol. 48(11),P. 1546 (1976)
80=C   *
90=C   *    DATA FORMAT
100=C  *    FIRST THREE LINES (MAX. 80 CHARACTERS) COMMENT
110=C  *    FOURTH LINE (FORMAT (3I5,E10.0)))
120=C  *    LSERIE  NUMBER OF SERIES OF MEASUREMENTS (.LE. 10)
130=C  *            OR NUMBER OF WAVELENGTHS USED
140=C  *    MPOINT  NUMBER OF MEASUREMENTS PER SERIES (.LE. 200)
150=C  *    NTEST   NUMBER OF TESTS WITH PSEUDO-MEASUREMENTS (.LE. 30)
160=C  *    STRAND  FIRST VALUE FOR THE RANDOM-NUMBER GENERATOR
170=C  *            SEE SUBR. PSEUDO
180=C  *
190=C  *    FIFTH AND FOLLOWING LINES (TOTAL OF MPOINT LINES)
200=C  *    LSERIE ABSORBANCES IN EACH LINE, FORMAT(5X,10F7.4)
210=C  *    FIFTH LINE ABSORBANCES OF BH2 (LOW PH)
220=C  *    SIXTH LINE ABSORBANCES OF B-- (HIGH PH)
230=C  *    AFTER LINE NO. 7 ORDER IS ARBITRARY
240=C  *
250=       DIMENSION AK(200),AKL(10,200),AL(200),BK(200),BL(200),CD(5)
260=C  *              =MPOINT  =LSERIE
270=       DIMENSION  CAPPAI(31),DPKI(31),CAPPAN(45),DPKN(45),VARCAN(45)
280=C                 =NTEST+1            =LSERIE*(LSERIE-1)/2
290=C      DIMENSION VARDPN(45),IFALSE(45)
300=       WRITE(9,100)
310=100    FORMAT(1H1)
320=C  *    FIRST THREE LINES OF DATA-SET ARE PRINTED
330=       DO 1 I=1,3
340=       READ(5,110) (AK(J),J=1,40)
350=1      WRITE(9,120) (AK(J),J=1,40)
360=110    FORMAT(40A2)
370=120    FORMAT(1H ,40A2)
380=C  *    READ FOURTH LINE
390=       READ(5,130) LSERIE,MPOINT,NTEST,STRAND
400=130    FORMAT(3I5,E10.0)
410=       WRITE(9,135) LSERIE,MPOINT,NTEST,STRAND
420=135    FORMAT(/I5,23H SERIES OF MEASUREMENTS//
430=      1I5,23H MEASUREMENTS PER SERIE//I5,32H TESTS WITH PSEUDO-RANDOM-ERR
440=      2ORS//45H FIRST NUMBER FOR THE RANDOM-NUMBER-GENERATOR,F8.4//
450=      3I1,4H IF=0 NO FAILURE, =1 SINGULAR NORMAL-EQUATIONS, =2 NEWTON-METH
460=      4OD FOR ERRORLESS DATA FAILED, =3 NEG. KAPPA OCCURRED//)
470=C  *    READ ALL MEASUREMENTS
480=       DO 2 M=1,MPOINT
490=2      READ(5,140) (AKL(L,M),L=1,LSERIE)
500=140    FORMAT(5X,10F7.4)
510=       L1=LSERIE-1
520=       NO=0
530=       NVAL=0
540=       N1=NTEST+1
550=       IF (NTEST .GT. 0)    RN=1./FLOAT(NTEST)
560=       RN1=1./FLOAT(N1)
```

A Source Listing of EDIA

```
 570=            WRITE(9,150)
 580=150   150  FORMAT(//110H NO   K    L    IF    AKBH2    ALBH2    AKB    ALB    AKBH
 590=         1     ALBH    VARAKL    KAPPA    STDKAP    DPK    STDDPK//)
 600=C     *
 610=C     *    THE DO-LOOP TERMINATING AT LABEL 20 CONTAINS
 620=C     *    THE EVALUATION OF RESULTS FOR ALL PAIRS OF
 630=C     *    SERIES OF MEASUREMENTS
 640=C     *
 650=            DO 20 K=1,L1
 660=            K1=K+1
 670=            DO 3 M=1,MPOINT
 680=3          AK(M)=AKL(K,M)
 690=            DO 20 L=K1,LSERIE
 700=            NO=NO+1
 710=            DO 4 M=1,MPOINT
 720=4          AL(M)=AKL(L,M)
 730=            IFALSE(NO)=0
 740=            INDIC=0
 750=            CALL NORMEQ (IERR,INDIC,100,MPOINT,AK,AL,CO)
 760=            IF (IERR .EQ. 0)     GOTO 5
 770=            IFALSE(NO)=1
 780=            GOTO 19
 790=5          CALL NEWCOR (IERR,MPOINT,AK,AL,BK,BL,CO)
 800=            IF (IERR .EQ. 0)     GOTO 6
 810=            IFALSE(NO)=2
 820=            GOTO 19
 830=6          AKBH2=BK(1)
 840=            ALBH2=BL(1)
 850=            AKB=BK(2)
 860=            ALB=BL(2)
 870=            CALL ACAPPA (AKB,AKBH,AKBH2,ALB,ALBH,ALBH2,CAPPA,DPK,CO)
 880=            IF (CAPPA .GT. 0.)    GOTO 7
 890=            IFALSE(NO)=3
 900=            GOTO 19
 910=7          CONTINUE
 920=            CALL STAT (MPOINT,IMAX,VARAKL,DMEAN,DMAX,AK,AL,BK,BL)
 930=            IF (NTEST .EQ. 0)    GOTO 18
 940=C     *    UP TO LABEL 160 THE RESULTS FOR NTEST SERIES OF
 950=C     *    PSEUDO- MEASUREMENTS ARE EVALUATED
 960=            STDDEV=SQRT(VARAKL)
 970=            CAPPAI(1)=CAPPA
 980=            DPKI(1)=DPK
 990=            CAMEAN=CAPPA
1000=            DPMEAN=DPK
1010=            DO 10 I=2,N1
1020=            CALL PSEUDO (MPOINT,STDDEV,STRAND,AK,AL,BK,BL)
1030=            CALL NORMEQ (IERR,INDIC,2,MPOINT,AK,AL,CO)
1040=            IF (IERR .NE. 1)     GOTO 8
1050=            IFALSE(NO)=1
1060=            GOTO 19
1070=8          CONTINUE
1080=            CALL ACAPPA (AKB,DUM1,AKBH2,ALB,DUM2,ALBH2,CAPPA,DPK,CO)
1090=            IF (CAPPA .GT. 0)    GOTO 9
1100=            IFALSE(NO)=3
1110=            GOTO 19
1120=9          CONTINUE
```

```
1130=           CAPPAI(I)=CAPPA
1140=           DPKI(I)=DPK
1150=           CAMEAN=CAMEAN+CAPPA
1160=           DPMEAN=DPMEAN+DPK
1170=10         CONTINUE
1180=           CAMEAN=CAMEAN*RN1
1190=           DPMEAN=DPMEAN*RN1
1200=           VARCAP=0.
1210=           VARDPK=0.
1220=           DO 11 I=1,N1
1230=           VARCAP=VARCAP+(CAPPAI(I)-CAMEAN)**2
1240=           VARDPK=VARDPK+(DPKI(I)-DPMEAN)**2
1250=11         CONTINUE
1260=           VARCAP=VARCAP*RN*RN1
1270=           VARDPK=VARDPK*RN*RN1
1280=           CAPPAN(NO)=CAPPAI(1)
1290=           VARCAN(NO)=VARCAP
1300=           STDCAP=SQRT(VARCAP)
1310=           DPKN(NO)=DPKI(1)
1320=           VARDPN(NO)=VARDPK
1330=           STDDPK=SQRT(VARDPK)
1340=           NVAL=NVAL+1
1350=           WRITE(9,160) NO,K,L,IFALSE(NO),AKBH2,ALBH2,AKB,ALB,AKBH,ALBH,
1360=          1VARAKL,CAPPAI(1),STDCAP,DPKI(1),STDDPK
1370=160        FORMAT(4I3,6F8.4,E10.2,5X,F8.2,E10.2,F8.3,E10.2)
1380=           GOTO 20
1390=18         WRITE(9,165) NO,K,L,IFALSE(NO),AKBH2,ALBH2,AKB,ALB,AKBH,ALBH,
1400=          1VARAKL,CAMEAN,VARCAP,DPMEAN,VARDPK
1410=165        FORMAT(4I3,6F8.4,E10.2,5X,F8.2,10X,F8.3)
1420=           NVAL=NVAL+1
1430=           GOTO 20
1440=19         CONTINUE
1450=           WRITE(9,160) NO,K,L,IFALSE(NO)
1460=20         CONTINUE
1470=           IF (NTEST .EQ. 0)      STOP
1480=C  *       THE STATISTICS OF THE PSEUDO-MEASUREMENTS IS EVALUATED BELOW
1490=           WRITE(9,168) NVAL,NO
1500=168        FORMAT(//10H1THERE ARE,I4,21H VALID RESULTS OUT OF,I4,
1510=          117H POSSIBLE RESULTS//64H BELOW AVERAGING IS APPLIED TO RESULTS WI
1520=          2TH VARIANCE .LT. VARMAX//9X,6HRESULT,6X,6HSTDDEV//)
1530=           VCAMIN=1.E20
1540=           VDPMIN=1.E20
1550=           DO 21 I=1,NO
1560=           IF (IFALSE(I) .NE. 0)    GOTO 21
1570=           VCAMIN=AMIN1(VCAMIN,VARCAN(I))
1580=           VDPMIN=AMIN1(VDPMIN,VARDPN(I))
1590=21         CONTINUE
1600=           VCAMAX=VCAMIN
1610=           VDPMAX=VDPMIN
1620=210        VCAMAX=10.*VCAMAX
1630=           VDPMAX=10.*VDPMAX
1640=           NOCA=0
1650=           NODP=0
1660=           CAMEAN=0.
1670=           DPMEAN=0.
1680=           VARCAP=0.
```

A Source Listing of EDIA

```
1690=          VARDPK=0.
1700=          DO 23 I=1,NO
1710=          IF (IFALSE(I) .NE. 0)        GOTO 23
1720=          IF (VARCAN(I) .GT. VCAMAX)        GOTO 22
1730=          NOCA=NOCA+1
1740=          CAMEAN=CAMEAN+CAPPAN(I)
1750=22        IF (VARDPN(I) .GT. VDPMAX)        GOTO 23
1760=          NODP=NODP+1
1770=          DPMEAN=DPMEAN+DPKN(I)
1780=23        CONTINUE
1790=          CAMEAN=CAMEAN/FLOAT(NOCA)
1800=          DPMEAN=DPMEAN/FLOAT(NODP)
1810=          DO 25 I=1,NO
1820=          IF (IFALSE(I) .NE. 0)        GOTO 25
1830=          IF (VARCAN(I) .GT. VCAMAX)        GOTO 24
1840=          VARCAP=VARCAP+(CAMEAN-CAPPAN(I))**2
1850=24        IF (VARDPN(I) .GT. VDPMAX)        GOTO 25
1860=          VARDPK=VARDPK+(DPMEAN-DPKN(I))**2
1870=25        CONTINUE
1880=          IF (NOCA .GT. 1)     VARCAP=VARCAP/FLOAT(NOCA*(NOCA-1))
1890=          IF (NODP .GT. 1)     VARDPK=VARDPK/FLOAT(NODP*(NODP-1))
1900=          STDCAP=SQRT(VARCAP)
1910=          STDDPK=SQRT(VARDPK)
1920=          WRITE(9,170) CAMEAN,STDCAP,NOCA,VCAMAX
1930=          WRITE(9,180) DPMEAN,STDDPK,NODP,VDPMAX
1940=170       FORMAT(/7H KAPPA=,F8.0,4H  +-,F8.0,I6,22H RESULTS USED    VARMAX=,
1950=          1E10.2)
1960=180       FORMAT(/7H   DPK=,F8.2,4H  +-,F8.2,I6,22H RESULTS USED    VARMAX=,
1970=          1E10.2/)
1980=          IF (NOCA .LT. NVAL .OR. NODP .LT. NVAL)        GOTO 210
1990=          STOP
2000=          END
2010=C  *
2020=C********AIJBI********************************************************
2030=C  *
2040=C  *      MATRIX A, R.H.S.-VECTOR B OF THE NORMAL EQUATIONS,
2050=C  *      AND FUNCTION Q (FORMULA 16) ARE COMPUTED HERE
2060=C  *      CO(I) ARE THE COEFFICIENTS ALPHA,BETA,..,EPSILON
2070=C  *
2080=          SUBROUTINE AIJBI (INDIC,M,Q2,AK,AL,A,B,CO)
2090=          DIMENSION A(5,5),AK(M),AL(M),B(5),CO(5)
2100=          DO 1 I=1,5
2110=          B(I)=0.
2120=          DO 1 J=1,5
2130=1         A(I,J)=0.
2140=          Q2=0.
2150=          DO 2 I=1,M
2160=          X=AK(I)
2170=          Y=AL(I)
2180=          WI=1.
2190=          IF (INDIC .NE. 0)
2200=          1WI=1./((2.*CO(1)*X+CO(2)*Y+CO(4))**2+(2.*Y+CO(2)*X+CO(3))**2)
2210=          A(5,5)=A(5,5)+WI
2220=          XW=X*WI
2230=          A(4,5)=A(4,5)+XW
2240=          A(2,5)=A(2,5)+XW*Y
```

```
2250=        XW=XW*X
2260=        A(1,5)=A(1,5)+XW
2270=        A(1,3)=A(1,3)+XW*Y
2280=        XW=XW*X
2290=        A(1,4)=A(1,4)+XW
2300=        A(1,1)=A(1,1)+XW*X
2310=        A(1,2)=A(1,2)+XW*Y
2320=        YW=Y*WI
2330=        A(3,5)=A(3,5)+YW
2340=        YW=YW*Y
2350=        A(2,3)=A(2,3)+YW*X
2360=        B(5)=B(5)-YW
2370=        A(2,2)=A(2,2)+YW*X*X
2380=        YW=YW*Y
2390=        B(3)=B(3)-YW
2400=        B(2)=B(2)-YW*X
2410=        IF (INDIC .NE. 1)      GOTO 2
2420=        Q2=Q2+WI*(Y*(Y+CO(3))+X*(X*CO(1)+Y*CO(2)+C0(4))+CO(5))**2
2430=2       CONTINUE
2440=        A(2,1)=A(1,2)
2450=        A(3,1)=A(1,3)
2460=        A(4,1)=A(1,4)
2470=        A(5,1)=A(1,5)
2480=        A(3,2)=A(2,3)
2490=        A(4,2)=A(1,3)
2500=        A(5,2)=A(2,5)
2510=        A(3,3)=-B(5)
2520=        A(4,3)=A(2,5)
2530=        A(5,3)=A(3,5)
2540=        A(4,4)=A(1,5)
2550=        A(5,4)=A(4,5)
2560=        A(2,4)=A(1,3)
2570=        A(3,4)=A(2,5)
2580=        B(1)=-A(2,2)
2590=        B(4)=-A(2,3)
2600=        RETURN
2610=        END
2620=C   *
2630=C********NORMEQ***********************************************
2640=C   *
2650=C   *   THE NORMAL EQUATIONS ARE SOLVED REPEATEDLY AS
2660=C   *   LONG AS Q DECREASES OR MAX TIMES
2670=C   *
2680=        SUBROUTINE NORMEQ (IERR,INDIC,MAX,M,AK,AL,CO)
2690=        DIMENSION AK(M),AL(M),CO(5),A(5,5),B(5),C(5)
2700=        I=0
2710=        Q3=1.E20
2720=1       CALL AIJBI (INDIC,M,Q2,AK,AL,A,B,C)
2730=        IF (INDIC .EQ. 0)      GOTO 2
2740=        IF (Q2 .GE. Q3)        RETURN
2750=        Q3=Q2
2760=2       INDIC=1
2770=        CALL LINSYS (5,A,B,C,DET,1.E-8,IERR)
2780=        IF (IERR .NE. 0)       RETURN
2790=        DO 3 J=1,5
2800=     3  CO(J)=C(J)
```

```
2810=      I=I+1
2820=      IF (I .LT. MAX)        GOTO 1
2830=      IERR=2
2840=      RETURN
2850=      END
2860=C  *
2870=C**********NEWCOR************************************************
2880=C  *
2890=C  *      FROM THE REAL DATA (AK,AL)
2900=C  *      A SET OF SO-CALLED ERRORLESS DATA (BK,BL) IS DERIVED
2910=C  *
2920=      SUBROUTINE NEWCOR (IERR,M,AK,AL,BK,BL,CO)
2930=      DIMENSION AK(M),AL(M),BK(M),BL(M),CO(5)
2940=      IERR=0
2950=      DO 4 I=1,M
2960=      X=AK(I)
2970=      Y=AL(I)
2980=      XMAX=10.*X
2990=      YMAX=10.*Y
3000=      DO 1 J=1,5
3010=      F=Y*(Y+CO(3))+X*(CO(1)*X+CO(2)*Y+CO(4))+CO(5)
3020=      FX=2.*CO(1)*X+CO(2)*Y+CO(4)
3030=      FY=2.*Y+CO(2)*X+CO(3)
3040=      WI=1./(FX*FX+FY*FY)
3050=      FWI=F*WI
3060=      DX=FWI*FX
3070=      DY=FWI*FY
3080=      X=X-DX
3090=      Y=Y-DY
3100=      IF (X .GT. XMAX .OR. X .LT. 0.)     GOTO 2
3110=      IF (Y .GT. YMAX .OR. Y .LT. 0.)     GOTO 2
3120=      IF (DX*DX+DY*DY .LT. 1.E-10)        GOTO 3
3130=1     CONTINUE
3140=2     IERR=1
3150=      RETURN
3160=3     BK(I)=X
3170=      BL(I)=Y
3180=4     CONTINUE
3190=      RETURN
3200=      END
3210=C  *
3220=C*********ACAPPA************************************************
3230=C  *
3240=C  *      COMPUTE ABSORBANCES OF INTERMEDIATES AKBH, ALBH,
3250=C  *      AND VALUES OF KAPPA AND DPK
3260=C  *
3270=      SUBROUTINE ACAPPA (AKB,AKBH,AKBH2,ALB,ALBH,ALBH2,CAPPA,DPK,CO)
3280=      DIMENSION CO(5)
3290=      AM1=-(2.*CO(1)*AKB+CO(2)*ALB+CO(4))/
3300=     1    (2.*ALB+CO(2)*AKB+CO(3))
3310=      AM2=-(2.*CO(1)*AKBH2+CO(2)*ALBH2+CO(4))/
3320=     1    (2.*ALBH2+CO(2)*AKBH2+CO(3))
3330=      AKBH=(AM2*AKBH2-AM1*AKB+ALB-ALBH2)/(AM2-AM1)
3340=      ALBH=AM1*(AKBH-AKB)+ALB
3350=      CAPPA=4.-2.*(AKB+AKBH2-2.*AKBH)/
3360=     1    ((CO(2)*CO(3)-2.*CO(4))/(4.*CO(1)-CO(2)**2)-AKBH)
```

```
3370=      DPK=0.
3380=      IF (CAPPA .GT. 0.)      DPK=ALOG10(CAPPA)
3390=      RETURN
3400=      END
3410=C *
3420=C********STAT   *************************************************
3430=C *
3440=C *    COMPUTE MEAN AND VARIANCE OF DEVIATIONS BETWEEN ORIGINAL
3450=C *      AND ERRORLESS DATA
3460=C *
3470=      SUBROUTINE STAT (M,IMAX,VAR,DMEAN,DMAX,AK,AL,BK,BL)
3480=      DIMENSION AK(M),AL(M),BK(M),BL(M)
3490=      VAR=0.
3500=      DMEAN=0.
3510=      DMAX=0.
3520=      IMAX=0
3530=      DO 1 I=1,M
3540=      D2=(AK(I)-BK(I))**2+(AL(I)-BL(I))**2
3550=      VAR=VAR+D2
3560=      D=SQRT(D2)
3570=      DMEAN=DMEAN+D
3580=      IF (DMAX .GT. D)      GOTO 1
3590=      DMAX=D
3600=      IMAX=I
3610=1     CONTINUE
3620=      VAR=VAR/FLOAT(M-1)
3630=      DMEAN=DMEAN/FLOAT(M)
3640=      RETURN
3650=      END
3660=C *
3670=C********PSEUDO***************************************************
3680=C *
3690=C *    PSEUDO-MEASUREMENTS ARE GENERATED BY CHANGING THE ERRORLESS
3700=C *      DATA BY RANDOM-NUMBERS
3710=C *    THE NUMBERS XNOR OF SUBR. RANOR ARE NORMALLY DISTRIBUTED
3720=C *      RANDOM-NUMBERS WITH STANDARD-DEVIATION 1
3730=C *
3740=      SUBROUTINE PSEUDO (M,STDDEV,XUNI,AK,AL,BK,BL)
3750=      DIMENSION AK(M),AL(M),BK(M),BL(M)
3760=      DO 1 I=1,M
3770=      CALL RANOR (XUNI,XNOR)
3780=      AK(I)=BK(I)+STDDEV*XNOR
3790=      CALL RANOR (XUNI,XNOR)
3800=1     AL(I)=BL(I)+STDDEV*XNOR
3810=      RETURN
3820=      END
3830=C *
3840=C********RANDO****************************************************
3850=C *    RANDO GENERATES PSEUDO-RANDOM-NUMBERS ACCORDING TO A
3860=C *      CONGRUENCE-METHOD AS GIVEN IN
3870=C *    M.ABRAMOWITZ, HANDBOOK OF MATHEMATICAL FUNCTIONS,
3880=C *      DOVER, NEW YORK, 1965
3890=C *    P.950, FORMULA 26.8.5 WITH S=4, Q=6
3900=C *
3910=C *    THE FIRST NUMBER MUST BE POSITIVE
3920=C *    THE RESULTING RANDOM-NUMBERS ARE UNIFORMLY DISTRIBUTED IN (0,1)
```

A Source Listing of EDIA

```
3930=C    *
3940=         SUBROUTINE RANDO (X)
3950=         DOUBLE PRECISION AUX
3960=         AUX=DBLE(X)*129140163.D-02
3970=C    *                           =3**17/100
3980=         IAUX=AUX
3990=         X=AUX-DBLE(FLOAT(IAUX))
4000=         RETURN
4010=         END
4020=C    *
4030=C********RANOR***************************************************
4040=C    *
4050=C    *   RANOR TRANSFORMS UNIFORMLY DISTRIBUTED RANDOM-NUMBERS TO
4060=C    *   NORMALLY DISTRIBUTED RANDOM-NUMBERS BY METHOD (3) P. 953
4070=C    *   IN THE HANDBOOK OF MATHEMATICAL FUNCTIONS
4080=C    *
4090=C    *   THE MEAN IS ZERO, THE VARIANCE ONE
4100=C    *
4110=         SUBROUTINE RANOR (X,XNOR)
4120=         CALL RANDO (X)
4130=         U1=X
4140=         CALL RANDO (X)
4150=         XNOR=SQRT(-2.*ALOG(U1))*COS(6.2831853*X)
4160=         RETURN
4170=         END
4180=C    *
4190=C********LINSYS***************************************************
4200=C    *
4210=         SUBROUTINE LINSYS (N,A,B,X,DET,EPS,IERR)
4220=         DIMENSION A(N,N),B(N),X(N)
4230=C    *
4240=C    *   LINSYS SOLVES A SYSTEM OF LINEAR EQUATIONS A*X=B
4250=C    *   GAUSSIAN ELIMINATION WITH PIVOTING IN THE LOWER PART OF A
4260=C    *   COLUMN
4270=C    *   A   N*N-MATRIX, N=1 ALLOWED
4280=C    *   B   R.H.S., N-VECTOR
4290=C    *   X   SOLUTION, N-VECTOR, B AND X MAY BE THE SAME ARRAY
4300=C    *   DET   DETERMINANT OF A
4310=C    *   EPS   IF A PIVOT-ELEMENT IS SMALLER THAN EPS*PIVOTMAX, A IS
4320=C    *   CONSIDERED TO BE SINGULAR
4330=C    *   IERR=0  IF A IS NOT SINGULAR, IERR=1 ELSE
4340=C    *
4350=         IERR=0
4360=         IF (N .GT. 1)      GOTO 1
4370=         IF (A(1,1) .EQ. 0.)      GOTO 9
4380=         X(1)=B(1)/A(1,1)
4390=         DET=A(1,1)
4400=         RETURN
4410=1        NM=N-1
4420=         DET=1.
4430=         PIVMAX=0.
4440=         VZ=1.
4450=         DO 6 K=1,NM
4460=         KP=K+1
4470=         PIV=ABS(A(K,K))
4480=         II=K
```

```
4490=           DO 2 I=KP,N
4500=           S=ABS(A(I,K))
4510=C  *          PIVOT-SEARCH IN THE LOWER PART OF THE K-TH COLUMN
4520=           IF (S .LE. PIV)       GOTO 2
4530=           II=I
4540=           PIV=S
4550=2          CONTINUE
4560=           PIVMAX=AMAX1(PIVMAX,PIV)
4570=           IF (PIV .LE. EPS*PIVMAX)      GOTO 9
4580=           IF (II .EQ. K)       GOTO 4
4590=C  *          EXCHANGE OF ROWS IF NECESSARY
4600=           VZ=-VZ
4610=           DO 3 J=K,N
4620=           S=A(K,J)
4630=           A(K,J)=A(II,J)
4640=3          A(II,J)=S
4650=           S=B(K)
4660=           B(K)=B(II)
4670=           B(II)=S
4680=4          DET=DET*A(K,K)
4690=           DO 5 I=KP,N
4700=           S=A(I,K)/A(K,K)
4710=           B(I)=B(I)-S*B(K)
4720=C  *          ELEMENTS OF THE LOWER PART OF THE K-TH COLUMN ARE ELIMINATED
4730=           DO 5 J=KP,N
4740=5          A(I,J)=A(I,J)-S*A(K,J)
4750=6          CONTINUE
4760=           IF (ABS(A(N,N)) .LE. EPS*PIVMAX)      GOTO 9
4770=C  *          END OF ELIMINATION, START OF BACKWARD-SUBSTITUTION
4780            DET=VZ*DET*A(N,N)
4790=           X(N)=B(N)/A(N,N)
4800=           DO 8 K=1,NM
4810=           II=N-K
4820=           KP=II+1
4830=           S=B(II)
4840=           DO 7 J=KP,N
4850=7          S=S-A(II,J)*X(J)
4860=8          X(II)=S/A(II,II)
4870=           RETURN
4880=9          IERR=1
4890=C  *          EXIT FOR SINGULAR MATRIX A
4900=           DET=0.
4910=           RETURN
4920=           END
```

B Source Listing of TIFIT

```
C PROGRAM TIFIT (F.GOEBBER, 13.3.78)
C ====================================                                 MAIN  10
C - STRUCTURE OF THE WHOLE PACKAGE                                     MAIN  10
C                                                                      MAIN  20
C -   SUBROUTINE                        COMMONS (U=USED, C=CHANGED)    MAIN  30
C -   NAME      NO.    CALLS            LAFIT    LAFIT0    DISPAR      MAIN  40
C-------------------------------------------------------------------   MAIN  50
C                                                                      MAIN  60
C -   TIFIT    (1.1)  INTRNS MINVAR CALFUN   U        -         -      MAIN  70
C -                   PRTRES PLTRES ERLESS                             MAIN  80
C -                   MCARLO PRSTAT                                    MAIN  90
C -   INTRNS   (2.1)   -                    U C       -         -      MAIN 100
C -   PRTRES   (2.2)   -                     U        -         -      MAIN 110
C -   PLTRES   (2.3)  (PLOT-PROGRAMS)        U        -         -      MAIN 120
C -   ERLESS   (2.4)  MINVA1 FUNSET         U C      U C       U C     MAIN 130
C -   MCARLO   (2.5)  MINVAR CALFUN RANOR   U C       U         -      MAIN 140
C -   PRSTAT   (2.6)   -                     U        -         -      MAIN 150
C -   MINVAR   (3.1)  CALFUN                 -        -         -      MAIN 160
C -   MINVA1   (3.2)  DISTAN                 -        -         -      MAIN 170
C -   RANOR    (3.3)  RANDO                  -        -         -      MAIN 180
C -   CALFUN   (4.1)  FUNSET SINV PMAVEC     U        -         -      MAIN 190
C -   DISTAN   (4.2)  FUNSET                 U        -         U      MAIN 200
C -   RANDO    (4.3)   -                     -        -         -      MAIN 210
C -   FUNSET   (5.1)   -                     -        -         -      MAIN 220
C -   SINV     (5.2)  MFSD                   -        -         -      MAIN 230
C -   PMAVEC   (5.3)   -                     -        -         -      MAIN 240
C -   MFSD     (6.1)   -                     -        -         -      MAIN 250
C                                                                      MAIN 260
C-------------------------------------------------------------------   MAIN 270
C                                                                      MAIN 280
C                                                                      MAIN 290
C***** TIFIT (1.1)                                                     MAIN 300
C                                                                      MAIN 310
C -   MAIN-PROGRAM FOR CURVE-FITTING OF ABSORBANCE-MEASUREMENTS         MAIN 320
C -   FROM TITRATION-SYSTEMS                                           MAIN 330
C                                                                      MAIN 340
      COMMON /LAFIT/ IVAL(10),MEASA,MEASZ,NFUNCT,NLAMDA,NOMEAS,NPAR,    MAIN 350
     1AH(7,150),EW(150,10),EPSA0(7,10),H(150),PH(150)
      DIMENSION COVRKL(10),LAMDA(10),RKL(6),STDEPS(7,10)                MAIN 370
C                                                                      MAIN 380
C -   INTRNS: READ AND TRANSFORM (IF NECESSARY) DATA                   MAIN 390
C                                                                      MAIN 400
      OPEN(UNIT=4,FILE='AUS.4',STATUS='NEW')
      OPEN(UNIT=8,FILE='EIN.8',STATUS='OLD')
      OPEN(UNIT=10,FILE='EIN.10',STATUS='OLD')

      CALL INTRNS                                                      MAIN 410
     1(IEREST,IPLOT,IPRINT,LAMDA,MAXFUN,NPSEUD,DX0,DXEND,RKL,START)     MAIN 420
C                                                                      MAIN 430
C -   MINVAR: MINIMIZATION OF THE GOODNESS-OF-FIT                      MAIN 440
C -   FUNCTION Q. Q IS A NONLINEAR FUNCTION OF RKL.                    MAIN 450
C                                                                      MAIN 460
```

```
      NFUN=MAXFUN                                                    MAIN  470
      CALL MINVAR (DX0,DXEND,IERR,IPRINT,NFUN,NPAR,QMIN,RKL)         MAIN  480
C                                                                    MAIN  490
C  -  AFTER THE FOLLOWING CALL OF CALFUN THE VALUES IN EPSA0 ARE     MAIN  500
C  -  REALLY THOSE CORRESPONDING TO THE ACTUAL RKL                   MAIN  510
      CALL CALFUN (NPAR,QMIN,RKL)                                    MAIN  520
C                                                                    MAIN  530
C  -  PRTRES: PRINT RESULTS                                          MAIN  540
C                                                                    MAIN  550
      CALL PRTRES (IEREST,IERR,LAMDA,MAXFUN,NFUN,DXEND,QMIN,RKL)     MAIN  560
C                                                                    MAIN  570
C  -  PLTRES: PLOT RESULTS                                           MAIN  580
C                                                                    MAIN  590
C     CALL PLTRES (IPLOT)                                            MAIN  600
      IF (IEREST .EQ. 0) STOP                                        MAIN  610
C                                                                    MAIN  620
C  -  ERLESS GENERATES A DATA SET, WHICH IS CLOSE TO THE ORIGINAL DATA  MAIN 630
C  -  AND FREE OF ERRORS, I.E. DERIVED FROM THE PARAMETERS COMPUTED ABOVE  MAIN 640
      CALL ERLESS (IERR,NOEW,NOSIGN,RKL,STDPH,STDEW)                 MAIN  650
      IF (IERR .NE. 0) STOP                                          MAIN  660
C                                                                    MAIN  670
C  -  MCARLO: ESTIMATE ERRORS BY A MONTE-CARLO-METHOD                MAIN  680
C                                                                    MAIN  690
      CALL MCARLO (COVRKL,IERR,NPSEUD,RKL,START,STDEPS,STDEW,STDPH)  MAIN  700
      IF (IERR .NE. 0) STOP                                          MAIN  710
C                                                                    MAIN  720
C  -  PRSTAT: PRINT STATISTICAL RESULTS                              MAIN  730
      CALL PRSTAT (COVRKL,NOEW,NOSIGN,NPSEUD,STDEPS,STDEW,STDPH)     MAIN  740
      STOP                                                           MAIN  750
      END                                                            MAIN  760
C                                                                    INTR   10
C*****  INTRNS  (2.1)                                                INTR   20
C                                                                    INTR   30
C  -  READ AND TRANSFORM DATA                                        INTR   40
C                                                                    INTR   50
      SUBROUTINE INTRNS                                              INTR   60
     1(IEREST,IPLOT,IPRINT,LAMDA,MAXFUN,NPSEUD,DX0,DXEND,RKL,START)  INTR   70
      COMMON /LAFIT/ IVAL(10),MEASA,MEASZ,NFUNCT,NLAMDA,NOMEAS,NPAR, INTR   80
     1AH(7,150),EW(150,10),EPSA0(7,10),H(150),PH(150)                INTR   90
      DIMENSION LAMDA(10),RKL(6),FORM1(80),FORM2(80)                 INTR   10
C                                                                    INTR  110
C  -  IEREST =0 NO ERROR-ESTIMATES                                   INTR  120
C  -  IPLOT  =0 NO PLOT, .GT. 0 SEE SUBR. PLTRES                     INTR  130
C  -  IPRINT CONTROLS OUTPUT FROM SUBR. MINVAR (0: NO OUTPUT)        INTR  140
C  -  MAXFUN MAX. NUMBER OF CALLS OF CALFUN FROM SUBR. MINVAR        INTR  150
C  -  NPSEUD NUMBER OF MONTE-CARLO-TESTS                             INTR  160
C  -  DX0 AND DXEND ARE USED BY MINVAR TO CONTROL MINIMIZATION       INTR  170
C  -  RKL(I), I=1,NPAR : INDEPENDENT VARIABLES                       INTR  180
C  -  START FIRST NUMBER FOR THE RANDOM-NUMBER PROGRAM RANOR         INTR  190
C                                                                    INTR  200
C  -  IVAL ALLOWS TO IGNORE SPECIFIED SERIES OF MEASUREMENTS         INTR  210
C  -  MEASA, MEASZ ARE THE FIRST AND THE LAST SUBSCRIPT OF           INTR  220
C  -  THE MEASUREMENTS TO BE USED                                    INTR  230
C  -  NFUNCT NUMBER OF FUNCTIONS IN THE FIT (HERE NFUNCT=NPAR+1)     INTR  240
C  -  NLAMDA NUMBER OF SERIES OF MEASUREMENTS                        INTR  250
C  -  NOMEAS NUMBER OF MEASUREMENTS (DIFFERENT PH-VALUES)            INTR  260
```

B Source Listing of TIFIT

```
C -   NPAR NUMBER OF INDEPENDENT VARIABLES                          INTR 270
C -   AH FUNCTIONS IN THE FIT                                       INTR 280
C -   EW MEASURED ABSORPTIONS                                       INTR 290
C -   EPSA0 ABSORPTION-COEFFICIENTS                                 INTR 300
C -   H H+-CONCENTRATIONS (COMPUTED FROM PH)                        INTR 310
C -   PH MEASURED PH-VALUES                                         INTR 320
C                                                                   INTR 330
C     DATA FORM1/'I3,7X,10F7.0'/
C     DATA FORM2/'3X,2F7.0'/
      READ(10,1000) (H(I),I=1,20)                                   INTR 340
1000  FORMAT(20A4)                                                  INTR 350
      WRITE(4,2000) (H(I),I=1,20)                                   INTR 360
2000  FORMAT('1',20X,20A4)                                          INTR 370
      READ(10,1010) NLAMDA                                          INTR 380
1010  FORMAT(I2)                                                    INTR 390
      READ(10,1020) (LAMDA(L),L=1,NLAMDA)                           INTR 400
1020  FORMAT(10(I3,1X))                                             INTR 410
      READ(10,1030) V0,S0,P0                                        INTR 420
1030  FORMAT(3F6.0)                                                 INTR 430
      WRITE(4,2005) V0,S0,P0                                        INTR 440
2005  FORMAT('0CORRECTIONS WITH V0 =',1P,E10.3,' S0 =',E10.3,       INTR 450
     1' P0 =',E10.3)                                                INTR 460
C     VARIABLE FORMAT STATEMENTS
C     WRITE (9,1243)
C1243 FORMAT('0ENTER FORMAT FOR K,EW / PH,ML: STD=0, ELSE=1')
C     READ (8,1245) FORM0
C1245 FORMAT (B)
C     IF (FORM0.EQ.0.0) GOTO 1040
C     READ (8,1248)FORM1
C1248 FORMAT (80A1)
C     READ (8,1248)FORM2
      M=1
10    READ(10,1040) K,(EW(M,L),L=1,NLAMDA)                          INTR 480
1040  FORMAT(I3,7X,10F7.0)
      IF (K .EQ. 999) GOTO 20                                       INTR 500
      M=M+1                                                         INTR 510
      GOTO 10                                                       INTR 520
20    NOMEAS=M-1                                                    INTR 530
      DO 30 M=1,NOMEAS                                              INTR 540
      READ(10,1050) PH(M),H(M)                                      INTR 550
1050  FORMAT(3X,2F7.0)
30    CONTINUE                                                      INTR 570
C                                                                   INTR 580
C -   UP TO (H(M)=) H(M) IS THE AMOUNT OF ACID (BASE) ADDED IN COURSE OF   INTR 590
C -   THE TITRATION. THESE VALUES ARE USED TO COMPUTE A CORRECTION WHICH   INTR 600
C -   ACCOUNTS FOR DILUTION                                         INTR 610
      DELTA=S0*P0-7.                                                INTR 620
      DO 50 M=1,NOMEAS                                              INTR 630
C                                                                   INTR 640
C -   CORRECTION OF THE PH-VALUES                                   INTR 650
      PH(M)=(PH(M)+DELTA)/S0                                        INTR 660
      FACTOR=(V0+H(M))/V0                                           INTR 670
      H(M)=10.**(-PH(M))                                            INTR 680
      DO 40 L=1,NLAMDA                                              INTR 690
      EW(M,L)=EW(M,L)*FACTOR                                        INTR 700
40    CONTINUE                                                      INTR 710
```

```
50    CONTINUE                                                        INTR 720
      WRITE(4,2010) NLAMDA                                             INTR 730
2010  FORMAT('0',I3,' WAVELENTGHS/SERIES OF MEASUREMENTS')             INTR 740
      WRITE(4,2020) (LAMDA(L),L=1,NLAMDA)                              INTR 750
2020  FORMAT(10I4)                                                     INTR 760
      WRITE(4,2030) NOMEAS                                             INTR 770
2030  FORMAT('0',I4,' VALUES PER SERIES')                              INTR 780
      DO 100 L=1,NLAMDA                                                INTR 790
      IVAL(L)=1                                                        INTR 800
100   CONTINUE                                                         INTR 810
      WRITE(9,2100)                                                    INTR 820
2100  FORMAT('0ENTER IVAL (DEFAULT=1), IVAL(L)=0: LEAVE OUT SERIES L'/)INTR 830
      READ(8,1100,ERR=110,END=110) (IVAL(L),L=1,NLAMDA)                INTR 840
1100  FORMAT(20I1)                                                     INTR 850
110   WRITE(4,2110) (IVAL(L),L=1,NLAMDA)                               INTR 860
2110  FORMAT('0IVAL ',10I2)                                            INTR 870
      WRITE(4,2120)                                                    INTR 880
2120  FORMAT('0ENTER IPLOT (DEFAULT=0), IPRINT(=0), MAXFUN (=200)'/    INTR 890
     1' MEASA(=1), MEASZ(=NOMEAS)'/)                                   INTR 900
      IPLOT=0                                                          INTR 910
      IPRINT=0                                                         INTR 920
      MAXFUN=200                                                       INTR 930
      MEASA=1                                                          INTR 940
      MEASZ=NOMEAS                                                     INTR 950
      READ(8,1100,ERR=120,END=120) IPLOT,IPRINT,MAXFUN,MEASA,MEASZ     INTR 960
120   WRITE(4,2130) IPLOT,IPRINT,MAXFUN,MEASA,MEASZ                    INTR 970
2130  FORMAT('0IPLOT =',I2,' IPRINT =',I2,' MAXFUN =',I4/              INTR 980
     1' MEASUREMENTS',I3,' - ',I4,' USED')                             INTR 990
      WRITE(4,2135)                                                    INTR1000
2135  FORMAT('0ENTER IEREST (=0) NPSEUD (=0) START (=0.5)'/)           INTR1010
      IEREST=0                                                         INTR1020
      NPSEUD=0                                                         INTR1030
      START=0.5                                                        INTR1040
      READ(8,1100,ERR=125,END=125) IEREST,NPSEUD,START                 INTR1050
125   WRITE(4,2136) IEREST,NPSEUD,START                                INTR1060
2136  FORMAT('0IEREST =',I3,' NPSEUD =',I4,' START =',1P,E10.2)        INTR1070
      WRITE(9,2140)                                                    INTR1080
2140  FORMAT('0ENTER DX0 (=0.5), DXEND (=1.E-4)'/)                     INTR1090
      DX0=0.5                                                          INTR1100
      DXEND=1.E-4                                                      INTR1110
      READ(8,1100,ERR=130,END=130) DX0,DXEND                           INTR1120
130   WRITE(4,2150) DX0,DXEND                                          INTR1130
2150  FORMAT('0MINIMIZATION: DX0 =',1P,E11.3,' DXEND =',E11.3)         INTR1140
140   WRITE(9,2160)                                                    INTR1150
2160  FORMAT('0ENTER NO. AND STARTING VALUES OF INDEP. VARIABLES'/)    INTR1160
      READ(8,1100,ERR=140,END=140) NPAR ,(RKL(I),I=1,NPAR )            INTR1170
      WRITE(4,2170) NPAR ,(RKL(I),I=1,NPAR )                           INTR1180
2170  FORMAT('0NO. OF VARIABLES NPAR =',I2/                            INTR1190
     1' STARTING VALUES :',1P,6E11.3)                                  INTR1200
      NFUNCT=NPAR+1                                                    INTR1210
      RETURN                                                           INTR1220
      END                                                              INTR1230
C                                                                      PRTR  10
C***** PRTRES (2.2)                                                    PRTR  20
C                                                                      PRTR  30
C -   PRINT RESULTS                                                    PRTR  40
```

B Source Listing of TIFIT

```
C                                                                       PRTR  50
      SUBROUTINE PRTRES (IEREST,IERR,LAMDA,MAXFUN,NFUN,DXEND,QMIN,RKL)   PRTR  60
      COMMON /LAFIT/ IVAL(10),MEASA,MEASZ,NFUNCT,NLAMDA,NOMEAS,NPAR,     PRTR  70
     1AH(7,150),EW(150,10),EPSA0(7,10),H(150),PH(150)                    PRTR  80
      DIMENSION LAMDA(10),RKL(6)                                         PRTR  90
      WRITE(4,1000) QMIN,(RKL(I),I=1,NPAR )                              PRTR 100
 1000 FORMAT('0QMIN =',1P,E11.3,' PK-VALUES',6E12.5)                     PRTR 110
      WRITE(4,1010) NFUN                                                 PRTR 120
 1010 FORMAT('0',I5,' CALLS OF CALFUN')                                  PRTR 130
      NP1=NPAR-1                                                         PRTR 140
      WRITE(4,1020) NPAR,NP1                                             PRTR 150
 1020 FORMAT(///' ABSORBANCES*(TOTAL CONC. A0)'//                        PRTR 160
     127X,'AH',I1,9X,'AH',I1,'  ...'//)                                  PRTR 170
      DO 10 L=1,NLAMDA                                                   PRTR 180
      IF (IVAL(L) .EQ. 0) GOTO 10                                        PRTR 190
      WRITE(4,1030) LAMDA(L),L,(EPSA0(J,L),J=1,NFUNCT)                   PRTR 200
 1030 FORMAT(I4,' NM = LAMDA(',I2,')',1P,3X,7E11.3)                      PRTR 210
   10 CONTINUE                                                           PRTR 220
      IF (IERR .EQ. 0) GOTO 99                                           PRTR 230
      WRITE(4,2000)                                                      PRTR 240
 2000 FORMAT('0MINIMIZATION INCOMPLETE')                                 PRTR 250
   99 RETURN                                                             PRTR 260
      END                                                                PRTR 270
C                                                                       PLTR  10
C***** PLTRES (2.3)                                                      PLTR  20
C                                                                       PLTR  30
C  -  PLOT OF MESURED DATA AND COMPUTED CURVES                           PLTR  40
C                                                                       PLTR  50
      SUBROUTINE PLTRES (IPLOT)                                          PLTR  60
      COMMON /LAFIT/ IVAL(10),MEASA,MEASZ,NFUNCT,NLAMDA,NOMEAS,NPAR,     PLTR  70
     1AH(7,200),EW(200,10),EPSA0(7,10),H(200),PH(200)                    PLTR  80
C                                                                       PLTR  90
C  -  IPLOT = 0 NO PLOT                                                  PLTR 100
C  -        1 PLOT OF COMPUTED CURVES (MIN/MAX = 0./1.)                  PLTR 110
C  -        2 SAME AS 1 PLUS MEASURED POINTS                             PLTR 120
C  -        3,4 SAME AS 1,2 WITH MIN/MAX FROM THE MEASUREMENTS           PLTR 130
      IF (IPLOT .EQ. 0) RETURN                                           PLTR 140
C     CALL WAIT                                                          PLTR 150
C     CALL NEWPAG                                                        PLTR 160
C     CALL WINDOW (5.,30.,2.,22.)                                        PLTR 170
C     CALL SCALE (0.,14.,0.,1.)                                          PLTR 180
C     CALL AXES (0.,0.,7.,0.2)                                           PLTR 190
C     CALL AXNUM ('X',0,1,'PH',2)                                        PLTR 200
C     CALL AXNUM ('Y',0,1,'A',1)                                         PLTR 210
C     CALL RASTER (1.,.1)                                                PLTR 220
      IPLT1=MOD(IPLOT,2)                                                 PLTR 230
      IPLT2=IPLOT-2                                                      PLTR 240
      DO 60 L=1,NLAMDA                                                   PLTR 250
      IF (IVAL(L) .EQ. 0) GOTO 60                                        PLTR 260
      IF (IPLT2 .LE. 0) GOTO 20                                          PLTR 270
      EMIN=EW(MEASA,L)                                                   PLTR 280
      EMAX=EMIN                                                          PLTR 290
      DO 10 M=MEASA,MEASZ                                                PLTR 300
      EMIN=AMIN1(EMIN,EW(M,L))                                           PLTR 310
      EMAX=AMAX1(EMAX,EW(M,L))                                           PLTR 320
   10 CONTINUE                                                           PLTR 330
```

```
C     CALL SCALE (0.,14.,EMIN,EMAX)                              PLTR 340
20    DO 40 M=MEASA,MEASZ                                        PLTR 350
      AHEPS=0.                                                   PLTR 360
      DO 30 I=1,NFUNCT                                           PLTR 370
      AHEPS=AHEPS+AH(I,M)*EPSA0(I,L)                             PLTR 380
30    CONTINUE                                                   PLTR 390
C     IF (M .EQ. MEASA) CALL MOVE (PH(M),AHEPS)                  PLTR 400
C     CALL DRAW (PH(M),AHEPS)                                    PLTR 410
40    CONTINUE                                                   PLTR 420
      IF (IPLT1 .NE. 0) GOTO 60                                  PLTR 430
      DO 50 M=MEASA,MEASZ                                        PLTR 440
C     ALL MOVE (PH(M),EW(M,L))                                   PLTR 450
C     CALL DRAW (PH(M),EW(M,L))                                  PLTR 460
50    CONTINUE                                                   PLTR 470
60    CONTINUE                                                   PLTR 480
C     CALL SIZE (4)                                              PLTR 490
C     CALL WAIT                                                  PLTR 500
      RETURN                                                     PLTR 510
      END                                                        PLTR 520
C                                                                ERLE  10
C***** ERLESS (2.4)                                              ERLE  20
C                                                                ERLE  30
C  -  COMPUTE FROM THE PREVIOUSLY DETERMINED PARAMETERS AN ERRORLESS  ERLE  40
C  -  SET OF DATA CLOSE TO THE ORIGINAL DATA                     ERLE  50
C                                                                ERLE  60
      SUBROUTINE ERLESS (IERR,NOEW,NOSIGN,RKL,STDPH,STDEW)       ERLE  70
      COMMON /LAFIT/ IVAL(10),MEASA,MEASZ,NFUNCT,NLAMDA,NOMEAS,NPAR,  ERLE 80
     1AH(7,150),EW(150,10),EPSA0(7,10),H(150),PH(150)            ERLE  90
      COMMON /LAFIT0/ EPSA00(7,10),EW0(150,10),PH0(150),RKL0(6)  ERLE 100
      COMMON /DISPAR/ M,RK(6),PHM0                               ERLE 110
C                                                                ERLE 120
C  -  COMMON /LAFIT0/ STORES THE ERRORLESS DATA                  ERLE 130
C  -  COMMON /DISPAR/ SERVES FOR COMMUNICATION BETWEEN SUBR. ERLESS AND  ERLE 140
C  -  SUBR. DISTAN                                               ERLE 150
      EXTERNAL DISTAN                                            ERLE 160
      DIMENSION RKL(6)                                           ERLE 170
      DO 10 I=1,NPAR                                             ERLE 180
      RKL0(I)=RKL(I)                                             ERLE 190
      RK(I)=10.**(-RKL(I))                                       ERLE 200
10    CONTINUE                                                   ERLE 210
      NOVAL=0                                                    ERLE 220
      DO 15 L=1,NLAMDA                                           ERLE 230
      IF (IVAL(L) .NE. 0) NOVAL=NOVAL+1                          ERLE 240
15    CONTINUE                                                   ERLE 250
      DELTA0=0.01                                                ERLE 260
      DELTAZ=0.0001                                              ERLE 270
      MAXFUN=50                                                  ERLE 280
      DO 20 M=MEASA,MEASZ                                        ERLE 290
      PHM=PH(M)                                                  ERLE 300
      PHM0=PH(M)                                                 ERLE 310
      CALL MINVA1 (DELTA0,DELTAZ,FMIN,IERR,MAXFUN,NFUN,PHM,DISTAN)  ERLE 320
C                                                                ERLE 330
C  -  PHM IS THAT PH-VALUE, WHICH YIELDS MINIMUM DISTANCE FROM THE  ERLE 340
C  -  COMPUTED ABSORPTION TO THE MEASURED VALUE WITH SUBSCRIPT M  ERLE 350
C  -  PH0(M) IS USED BY SUBR. MCARLO AS AN ERRORLESS VALUE       ERLE 360
      IF (IERR .NE. 0) GOTO 999                                  ERLE 370
```

B Source Listing of TIFIT

```
            PH0(M)=PHM                                              ERLE 380
      20    CONTINUE                                                ERLE 390
C                                                                   ERLE 400
C     -     IN THE FOLLOWING THE STANDARD-DEVIATIONS OF THE PH- AND ABSORPTION-   ERLE 410
C     -     MEASUREMENTS ARE COMPUTED.                              ERLE 420
C     -     THE NUMBER OF CHANGES OF SIGN OF RESIDUALS (NOSIGN) IS COUNTED   ERLE 430
C     -     AS A CRUDE TEST FOR UNCORRELATED ERRORS                 ERLE 440
            NOPH=MEASZ-MEASA+1                                      ERLE 450
            NOEW=0                                                  ERLE 460
            NOSIGN=0                                                ERLE 470
            ISIGNO=0                                                ERLE 480
            VARPH=0.                                                ERLE 490
            VAREW=0.                                                ERLE 500
            DO 50 M=MEASA,MEASZ                                     ERLE 510
            VARPH=VARPH+(PH(M)-PH0(M))**2                           ERLE 520
            H(M)=10.**(-PH0(M))                                     ERLE 530
            CALL FUNSET (NFUNCT,NPAR,AH(1,M),RK,H(M))               ERLE 540
            DO 40 L=1,NLAMDA                                        ERLE 550
            IF (IVAL(L) .EQ. 0) GOTO 40                             ERLE 560
            NOEW=NOEW+1                                             ERLE 570
            AHEPS=0.                                                ERLE 580
            DO 30 I=1,NFUNCT                                        ERLE 590
            AHEPS=AHEPS+AH(I,M)*EPSA0(I,L)                          ERLE 600
      30    CONTINUE                                                ERLE 610
C                                                                   ERLE 620
C     -     EW0(M,L): ERRORLESS ABSORPTION-VALUE                    ERLE 630
            EW0(M,L)=AHEPS                                          ERLE 640
            RESIDU=EW(M,L)-AHEPS                                    ERLE 650
            VAREW=VAREW+RESIDU**2                                   ERLE 660
            ISIGN=1                                                 ERLE 670
            IF (RESIDU .LT. 0.) ISIGN=-1                            ERLE 680
            IF (ISIGNO .EQ. ISIGN) GOTO 40                          ERLE 690
            NOSIGN=NOSIGN+1                                         ERLE 700
            ISIGNO=ISIGN                                            ERLE 710
      40    CONTINUE                                                ERLE 720
      50    CONTINUE                                                ERLE 730
            VARPH=VARPH/FLOAT(NOPH-1)                               ERLE 740
            VAREW=VAREW/FLOAT(NOEW-NOVAL*NPAR)                      ERLE 750
C                     =NUMBER OF PARAMETERS IN EPSA0                ERLE 760
C     -     FOR THIS FORMULA SEE                                    ERLE 770
C     -     J.V. BECK, K.J. ARNOLD, PARAMETER ESTIMATION IN ENGINEERING   ERLE 780
C     -     AND SCIENCE (1977), P. 241 (6.2.19), WILEY, NEW YORK    ERLE 790
            STDPH=SQRT(VARPH)                                       ERLE 800
            STDEW=SQRT(VAREW)                                       ERLE 810
            DO 70 L=1,NLAMDA                                        ERLE 820
            IF (IVAL(L) .EQ. 0) GOTO 70                             ERLE 830
            DO 60 I=1,NFUNCT                                        ERLE 840
C                                                                   ERLE 850
C     -     EPSA00 REFERENCE-VALUES OF EPSA0 FOR SUBR. MCARLO       ERLE 860
            EPSA00(I,L)=EPSA0(I,L)                                  ERLE 870
      60    CONTINUE                                                ERLE 880
      70    CONTINUE                                                ERLE 890
            RETURN                                                  ERLE 900
     999    WRITE (4,1000) M                                        ERLE 910
    1000    FORMAT('0SUBR. ERLESS: MINIMIZATION INCOMPLETE AT POINT NO.',I4//)   ERLE 920
            RETURN                                                  ERLE 930
```

```
              END                                                       ERLE 940
C                                                                       MCAR  10
C***** MCARLO (2.5)                                                     MCAR  20
C                                                                       MCAR  30
C -   COMPUTE ESTIMATES OF VARIANCES BY A MONTE-CARLO-METHOD            MCAR  40
C                                                                       MCAR  50
      SUBROUTINE MCARLO(COVRKL,IERR,NPSEUD,RKL,START,STDEPS,STDEW,STDPH) MCAR  60
      COMMON /LAFIT/ IVAL(10),MEASA,MEASZ,NFUNCT,NLAMDA,NOMEAS,NPAR,     MCAR  70
     1AH(7,150),EW(150,10),EPSA0(7,10),H(150),PH(150)                   MCAR  80
      COMMON /LAFIT0/ EPSA00(7,10),EW0(150,10),PH0(150),RKL0(6)         MCAR  90
      DIMENSION COVRKL(10),STDEPS(7,10),RKL(6)                          MCAR 100
      NCOV=NPAR*(NPAR+1)/2                                              MCAR 110
      DO 10 I=1,NCOV                                                    MCAR 120
      COVRKL(I)=0.                                                      MCAR 130
   10 CONTINUE                                                          MCAR 140
      DX0=0.05                                                          MCAR 150
      DXEND=1.E-4                                                       MCAR 160
      DO 30 L=1,NLAMDA                                                  MCAR 170
      DO 20 I=1,NFUNCT                                                  MCAR 180
      STDEPS(I,L)=0.                                                    MCAR 190
   20 CONTINUE                                                          MCAR 200
   30 CONTINUE                                                          MCAR 210
      RAND=START                                                        MCAR 220
      DO 100 NP=1,NPSEUD                                                MCAR 230
      DO 50 M=MEASA,MEASZ                                               MCAR 240
C                                                                       MCAR 250
C -   PSEUDO-MEASUREMENTS OF PH ARE COMPUTED FROM PH0                   MCAR 260
      CALL RANOR (RAND,RNOR)                                            MCAR 270
      PH(M)=RNOR*STDPH+PH0(M)                                           MCAR 280
      H(M)=10.**(-PH(M))                                                MCAR 290
      DO 40 L=1,NLAMDA                                                  MCAR 300
      IF (IVAL(L) .EQ. 0) GOTO 40                                       MCAR 310
C                                                                       MCAR 320
C -   PSEUDOMEASUREMENTS OF EW (ABSORPTIONS) ARE COMPUTED FROM EW0      MCAR 330
      CALL RANOR (RAND,RNOR)                                            MCAR 340
      EW(M,L)=RNOR*STDEW+EW0(M,L)                                       MCAR 350
   40 CONTINUE                                                          MCAR 360
   50 CONTINUE                                                          MCAR 370
C                                                                       MCAR 380
C -   FIT RKL FOR THE PSEUDO-MEASUREMENTS                               MCAR 390
      CALL MINVAR (DX0,DXEND,IERR,0,100,NPAR,QMIN,RKL)                  MCAR 400
      IF (IERR .NE. 0) GOTO 999                                         MCAR 410
C                                                                       MCAR 420
C -   CALFUN CACULATES THE EPSA0 TOO                                    MCAR 430
      CALL CALFUN (NPAR,QMIN,RKL)                                       MCAR 440
      K=0                                                               MCAR 450
      DO 70 I=1,NPAR                                                    MCAR 460
      DO 60 J=1,I                                                       MCAR 470
      K=K+1                                                             MCAR 480
C                                                                       MCAR 490
C -   COVRKL: PACKED SYMMETRIC COVARIANCE-MATRIX OF RKL                 MCAR 500
      COVRKL(K)=COVRKL(K)+(RKL(I)-RKL0(I))*(RKL(J)-RKL0(J))             MCAR 510
   60 CONTINUE                                                          MCAR 520
   70 CONTINUE                                                          MCAR 530
      DO 90 L=1,NLAMDA                                                  MCAR 540
      IF (IVAL(L) .EQ. 0) GOTO 90                                       MCAR 550
```

B Source Listing of TIFIT

```
            DO 80 I=1,NFUNCT                                        MCAR 560
            STDEPS(I,L)=STDEPS(I,L)+(EPSA0(I,L)-EPSA00(I,L))**2     MCAR 570
      C                                                             MCAR 580
      C - STDEPS: STANDARD-DEVIATIONS OF THE EPSA0-VALUES           MCAR 590
   80       CONTINUE                                                MCAR 600
   90       CONTINUE                                                MCAR 610
  100       CONTINUE                                                MCAR 620
            K=0                                                     MCAR 630
            DO 120 I=1,NPAR                                         MCAR 640
            DO 110 J=1,I                                            MCAR 650
            K=K+1                                                   MCAR 660
            COVRKL(K)=COVRKL(K)/FLOAT(NPSEUD)                       MCAR 670
  110       CONTINUE                                                MCAR 680
  120       CONTINUE                                                MCAR 690
            DO 140 L=1,NLAMDA                                       MCAR 700
            IF (IVAL(L).EQ.0)   GOTO 140                            MCAR 710
            DO 130 I=1,NFUNCT                                       MCAR 720
            STDEPS(I,L)=SQRT(STDEPS(I,L)/FLOAT(NPSEUD-1))           MCAR 730
  130       CONTINUE                                                MCAR 740
  140       CONTINUE                                                MCAR 750
            RETURN                                                  MCAR 760
  999       WRITE(4,1000)                                           MCAR 770
 1000       FORMAT('0SUBR. MCARLO: MINIMIZATION FAILED'///)         MCAR 780
            RETURN                                                  MCAR 790
            END                                                     MCAR 800
      C                                                             PRST  10
      C***** PRSTAT (2.6)                                           PRST  20
      C                                                             PRST  30
      C - PRINT RESULTS OF STATISTICAL ESTIMATIONS                  PRST  40
      C                                                             PRST  50
            SUBROUTINE PRSTAT (COVRKL,NOEW,NOSIGN,NPSEUD,STDEPS,STDEW,STDPH) PRST 60
            COMMON /LAFIT/ IVAL(10),MEASA,MEASZ,NFUNCT,NLAMDA,NOMEAS,NPAR,  PRST 70
           1AH(7,150),EW(150,10),EPSA0(7,10),H(150),PH(150)         PRST  80
            DIMENSION COVRKL(10),STDEPS(7,10)                       PRST  90
            WRITE(4,1000)                                           PRST 100
 1000       FORMAT('1'//20X,'ESTIMATIONS CONCERNING ERRORS'//)      PRST 110
            EXVAL=0.5*FLOAT(NOEW+1)                                 PRST 120
            WRITE(4,1010) NOSIGN,EXVAL                              PRST 130
 1010       FORMAT('0SIGN OF RESIDUALS CHANGES',I5,' TIMES. EXPECTED VALUE IS' PRST 140
           1,F7.1,/,'0IF THE EXPECTED VALUE IS MUCH GREATER, THE ERRORS' PRST 150
           2,' ARE CORRELATED OR THE MODEL IS BAD',/)               PRST 160
            WRITE(4,1020) NPSEUD,STDPH,STDEW                        PRST 170
 1020       FORMAT(/,I4,' SETS OF PSEUDO-MEASUREMENTS USED'//       PRST 180
           11P,E10.2,' ESTIMATED STANDARD-DEVIATION OF PH-MEASUREMENTS'// PRST 190
           21P,E10.2,' ESTIMATED STANDARD-DEVIATION OF ABSORPTIONS') PRST 200
            WRITE(4,1030)  (I,I=1,NPAR)                             PRST 210
 1030       FORMAT(///'0COVARIANCE-MATRIX OF THE PK-VALUES FROM THE ', PRST 220
           1'MONTE-CARLO-METHOD'/                                   PRST 230
           2'0LARGE OUTER DIAGONAL-ELEMENTS MAY BE CAUSED BY'/      PRST 240
           3'0TOO MANY PARAMETERS OR BAD INITIAL GUESS (LARGE QMIN)'/// PRST 250
           44X,4I10)                                                PRST 260
            K0=1                                                    PRST 270
            DO 10 I=1,NPAR                                          PRST 280
            K1=K0+I-1                                               PRST 290
            WRITE(4,1040) I,(COVRKL(J),J=K0,K1)                     PRST 300
 1040       FORMAT(I4,1P,5E10.2)                                    PRST 310
```

```
      K0=K1+1                                                   PRST 320
10    CONTINUE                                                  PRST 330
      WRITE(4,1050) (I,I=1,NFUNCT)                              PRST 340
1050  FORMAT(//'0STANDARD-DEVIATIONS OF ABSORBANCES FROM THE ', PRST 350
     1'MONTE-CARLO-METHOD'///4X,5I10)                           PRST 360
      DO 20 L=1,NLAMDA                                          PRST 370
      IF (IVAL(L) .NE. 0)   WRITE(4,1040) L,(STDEPS(J,L),J=1,NFUNCT)  PRST 380
20    CONTINUE                                                  PRST 390
      RETURN                                                    PRST 400
      END                                                       PRST 410
C                                                               MVAR 10
C***** MINVAR (3.1)                                             MVAR 20
C                                                               MVAR 30
      SUBROUTINE MINVAR (DX0,DXEND,IERR,IPRINT,MAXFUN,N,QMIN,X) MVAR 40
C                                                               MVAR 50
C -   MINVAR SEARCHES FOR A MINIMUM OF THE FUNCTION CALCULATED  MVAR 60
C -   IN CALFUN BY VARIING ALONG THE AXES WITH DECREASING STEP-LENGTH  MVAR 70
C -                                                             MVAR 80
C -   PARAMETERS                                                MVAR 90
C -                                                             MVAR 100
C -   DX0    INITIAL STEP-LENGTH                                MVAR 110
C -   DXEND  TERMINATION IF STEP-LENGTH IS SMALLER THAN DXEND   MVAR 120
C -   IERR =0 MINIMUM HAS BEEN FOUND AS DESIRED                 MVAR 130
C -        =1 TERMINATION BECAUSE OF TOO MANY CALLS OF CALFUN   MVAR 140
C -   IPRINT =0 NO OUTPUT FROM MINVAR                           MVAR 150
C -   IPRINT =I OUTPUT AFTER EACH I STEPS                       MVAR 160
C -   MAXFUN MAXIMUM NUMBER OF CALLS OF CALFUN                  MVAR 170
C -   AFTER RETURN FROM MINVAR NUMBER OF CALLS ACTUALLY USE     MVAR 180
C -   N  NUMBER OF INDEPENDENT VARIABLES                        MVAR 190
C -   QMIN  MINIMUM AS DETERMINED BY MINVAR                     MVAR 200
C -   X    VECTOR OF VARIABLES(DIM.=N)                          MVAR 210
C -   WHEN MINVAR IS CALLED, X HAS TO BE THE INITIAL            MVAR 220
C -   ESTIMATE OF X.                                            MVAR 230
C -   AFTER RETURN X CORRESPONDS TO QMIN                        MVAR 240
C                                                               MVAR 250
      DIMENSION X(N)                                            MVAR 260
      CALL CALFUN (N,QMIN,X)                                    MVAR 270
      DX=DX0                                                    MVAR 280
      NFUN=1                                                    MVAR 290
      IPRI=IPRINT                                               MVAR 300
10    IMPROV=0                                                  MVAR 310
C                                                               MVAR 320
C -   BEGIN OF A GREAT LOOP (VARIATION OF ALL VARIABLES)        MVAR 330
      DO 100 I=1,N                                              MVAR 340
C                                                               MVAR 350
C -   VARIATION OF X(I)                                         MVAR 360
      IMPR=0                                                    MVAR 370
      D=DX                                                      MVAR 380
      XI=X(I)                                                   MVAR 390
30    IF (NFUN .LT. MAXFUN) GOTO 40                             MVAR 400
C                                                               MVAR 410
C -   TERMINATION BECAUSE OF TOO MANY CALLS OF CALFUN           MVAR 420
      IERR=1                                                    MVAR 430
      GOTO 999                                                  MVAR 440
40    IPRI=IPRI-1                                               MVAR 450
      IF (IPRI .NE. 0) GOTO 50                                  MVAR 460
```

B Source Listing of TIFIT

```
C                                                                       MVAR 470
C -   CONDITIONAL OUTPUT OF INTERMEDIATE RESULTS                        MVAR 480
C -   IPRINT .LE. 0 CAUSES NO OUTPUT                                    MVAR 490
      IPRI=IPRINT                                                       MVAR 500
      WRITE(4,1000) QMIN,(X(J),J=1,N)                                   MVAR 510
1000  FORMAT(1P,E14.6,3X,5E14.6)                                        MVAR 520
50    X(I)=XI+D                                                         MVAR 530
      NFUN=NFUN+1                                                       MVAR 540
      CALL CALFUN (N,Q,X)                                               MVAR 550
      IF (QMIN .LT. Q) GOTO 70                                          MVAR 560
C                                                                       MVAR 570
C -   SUCCESSFUL STEP                                                   MVAR 580
      QMIN=Q                                                            MVAR 590
      XI=X(I)                                                           MVAR 600
      IMPR=1                                                            MVAR 610
      GOTO 30                                                           MVAR 620
C                                                                       MVAR 630
70    X(I)=XI                                                           MVAR 640
C                                                                       MVAR 650
C -   NO SUCCES                                                         MVAR 660
      IF (IMPR .EQ. 0)GOTO 80                                           MVAR 670
C                                                                       MVAR 680
C -   IMPROVEMENTS HAVE BEEN FOUND IN THIS DIRECTION                    MVAR 690
      IMPROV=1                                                          MVAR 700
      GOTO 100                                                          MVAR 710
80    IF (D .LT. 0.) GOTO 100                                           MVAR 720
C                                                                       MVAR 730
C -   CHANGE OF DIRECTION                                               MVAR 740
      D=-D                                                              MVAR 750
      GOTO 30                                                           MVAR 760
100   CONTINUE                                                          MVAR 770
      IF (IMPROV .GT. 0) GOTO 10                                        MVAR 780
C                                                                       MVAR 790
C -   IMPROV .GT. 0 MEANS, THAT THERE WERE IMPROVEMENTS WITH THIS DX    MVAR 800
C -   THEN DX IS TRIED AGAIN, ELSE DX IS DECREASED                      MVAR 810
      DX=0.1*DX                                                         MVAR 820
      IF (DX .GT. DXEND) GOTO 10                                        MVAR 830
      IERR=0                                                            MVAR 840
      MAXFUN=NFUN                                                       MVAR 850
999   RETURN                                                            MVAR 860
      END                                                               MVAR 870
C                                                                       MVA1  10
C***** MINVA1(3.2)                                                      MVA1  20
C                                                                       MVA1  30
C -   MINIMIZATION OF A FUNCTION OF ONE VARIABLE T                      MVA1  40
C -   COMPUTED IN SUBR. FUNCTN (T,F) (RESULT F)                         MVA1  50
C                                                                       MVA1  60
C -   METHOD: TRIAL AND ERROR AND QUADRATIC INTERPOLATION               MVA1  70
C                                                                       MVA1  80
C -   PARAMETERS:                                                       MVA1  90
C                                                                       MVA1 100
C -   DELTA0, DELTAZ: INITIAL AND ULTIMATE STEP-LENGTH                  MVA1 110
C -   FMIN: MINIMUM FOUND                                               MVA1 120
C -   IERR: =1 IF TERMINATION BECAUSE OF TOO MANY CALLS OF FUNCTN, =0 ELSE MVA1 130
C -   MAXFUN, NFUN NUMBER OF CALLS OF FUNCTN ALLOWED AND ACTUALLY NEEDED MVA1 140
C -   T: INDEPENDENT VARIABLE: AFTER LEAVING MINVA1 CORRESPONDING TO FMIN MVA1 150
```

```
C                                                                     MVA1 160
C -    FUNCTN SEE ABOVE. IT MUST BE DECLARED BY EXTERNAL IN THE CALLING   MVA1 170
C -    PROGRAM                                                        MVA1 180
C                                                                     MVA1 190
       SUBROUTINE MINVA1 (DELTA0,DELTAZ,FMIN,IERR,MAXFUN,NFUN,T,FUNCTN) MVA1 200
       DELTA=DELTA0                                                   MVA1 210
       IERR=0                                                         MVA1 220
       NFUN=1                                                         MVA1 230
       CALL FUNCTN (T,FMIN)                                           MVA1 240
       NFUN=NFUN+1                                                    MVA1 250
10     KOUNT=0                                                        MVA1 260
C                                                                     MVA1 270
C -    A NEW TRIAL AND ERROR LOOP                                     MVA1 280
20     T0=T+DELTA                                                     MVA1 290
       CALL FUNCTN (T0,F0)                                            MVA1 300
       NFUN=NFUN+1                                                    MVA1 310
       IF (NFUN .LT. MAXFUN) GOTO 25                                  MVA1 320
C                                                                     MVA1 330
C -    TOO MANY CALLS OF FUNCTN                                       MVA1 340
       IERR=1                                                         MVA1 350
       GOTO 999                                                       MVA1 360
25     IF (F0 .GE. FMIN) GOTO 30                                      MVA1 370
C                                                                     MVA1 380
C -    F DECREASES, THEREFORE SOME STEPS WITH INCREASING STEP-LENGTH  MVA1 390
C -    ARE TRIED IN THIS DIRECTION                                    MVA1 400
       DELTA=2.*DELTA                                                 MVA1 410
       FMIN=F0                                                        MVA1 420
       T=T0                                                           MVA1 430
       KOUNT=0                                                        MVA1 440
       GOTO 20                                                        MVA1 450
30     IF (KOUNT .GT. 0) GOTO 100                                     MVA1 460
C                                                                     MVA1 470
C -    KOUNT .GT. 0 IF + AND - DELTA HAVE BEEN TRIED                  MVA1 480
C -    ELSE THE OTHER SIGN OF DELTA HAS TO BE USED                    MVA1 490
       T2=T0                                                          MVA1 500
       F2=F0                                                          MVA1 510
       DELTA=-DELTA                                                   MVA1 520
       KOUNT=1                                                        MVA1 530
       GOTO 20                                                        MVA1 540
100    S0=0.5*DELTA*(F2-F0)/(F2+F0-FMIN-FMIN)                         MVA1 550
C                                                                     MVA1 560
C -    QUADRATIC INTERPOLATION IN ORDER TO ACHIEVE FAST ULTIMATE      MVA1 570
C -    CONVERGENCE                                                    MVA1 580
       T0=T+S0                                                        MVA1 590
       CALL FUNCTN (T0,F0)                                            MVA1 600
       NFUN=NFUN+1                                                    MVA1 610
       IF (F0 .GE. FMIN) GOTO 110                                     MVA1 620
C                                                                     MVA1 630
C -    THE RESULT OF THE QUADRATIC INTERPOLATION IS USED ONLY IF      MVA1 640
C -    IT IS REALLY BETTER THAN THE FORMER BEST VALUE                 MVA1 650
       T=T0                                                           MVA1 660
       FMIN=F0                                                        MVA1 670
110    DELTA=AMIN1 (ABS(S0),ABS(0.1*DELTA))                           MVA1 680
C                                                                     MVA1 690
C -    DECREASE DELTA AND TERMINATE IF IT IS LESS THAN DELTAZ         MVA1 700
       IF (DELTA .GT. DELTAZ) GOTO 10                                 MVA1 710
```

B   Source Listing of TIFIT

```
999   RETURN                                                          MVA1 720
      END                                                             MVA1 730
C                                                                     RANO  10
C*****  RANOR (3.3)                                                   RANO  20
C                                                                     RANO  30
C -   RANOR TRANSFORMS UNIFORMLY DISTRIBUTED RANDOM-NUMBERS            RANO  40
C -   TO NORMALLY DISTRIBUTED RANDOM-NUMBERS BY METHOD (3)             RANO  50
C -   P. 953 IN THE HANDBOOK OF MATHEMATICAL FUNCTIONS                 RANO  60
C                                                                     RANO  70
C -   THE MEAN IS ZERO, THE VARIANCE UNITY                            RANO  80
C                                                                     RANO  90
      SUBROUTINE RANOR (X,XNOR)                                       RANO 100
      CALL RANDO (X)                                                  RANO 110
      U1=X                                                            RANO 120
      CALL RANDO (X)                                                  RANO 130
      XNOR=SQRT(-2.*ALOG(U1))*COS(6.2831853*X)                        RANO 140
      RETURN                                                          RANO 150
      END                                                             RANO 160
C                                                                     CALF  10
C*****  CALFUN(4.1)                                                   CALF  20
C                                                                     CALF  30
      SUBROUTINE CALFUN (NPARA ,Q,RKL)                                CALF  40
      COMMON /LAFIT/ IVAL(10),MEASA,MEASZ,NFUNCT,NLAMDA,NOMEAS,NPAR,  CALF  50
     1AH(7,150),EW(150,10),EPSA0(7,10),H(150),PH(150)                 CALF  60
      DIMENSION B(42),C(7,10),RK(6),RKL(6)                            CALF  70
C                                                                     CALF  80
C -   COMPUTATION OF THE GOODNESS-OF-FIT FUNCTION Q                   CALF  90
C -                                                                   CALF 100
C -   MEANING OF THE PARAMETERS:                                      CALF 110
C -                                                                   CALF 120
C -   Q VALUE OF Q (RESULT)                                           CALF 130
C -   NPAR STAGES OF THE TITRATION-SYSTEM (.LE. 4)                    CALF 140
C -   NOMEAS NUMBER OF MEASUREMENTS                                   CALF 150
C -   NLAMDA NUMBER OF DIFFERENT WAVELENGTHS                          CALF 160
C -   EW MEASUREMENTS                                                 CALF 170
C -   H CONCENTRATIONS OF H-ION (COMPUTED FROM PH-VALUES)             CALF 180
C -   EPSA0 ABSORPTION-COEFFICIENTS*TOTAL CONCENTRATION OF            CALF 190
C -   ALL FORMS OF A (RESULT)                                         CALF 200
C -   RKL PK-VALUES                                                   CALF 210
C -                                                                   CALF 220
C -   AH CONCENTRATIONS OF CHEMICAL SPECIES A....AH(NPAR)             CALF 230
C -   B PACKED SYMMETRIC MATRIX COMPUTED FROM AH                      CALF 240
C -   C RECTANGULAR MATRIX COMPUTED FROM AH AND EW                    CALF 250
C -   RK EQUILIBRIUM-CONSTANTS                                        CALF 260
C                                                                     CALF 270
C -   COMPUTE THE PARAMETERS DERIVED FROM THE RKLS                    CALF 280
      DO 10 I=1,NPARA                                                 CALF 290
      RK(I)=10.**(-RKL(I))                                            CALF 300
   10 CONTINUE                                                        CALF 310
C                                                                     CALF 320
C -   COMPUTE MATRIX AH IN SUBR. FUNSET                               CALF 330
      DO 40 M=MEASA,MEASZ                                             CALF 340
      CALL FUNSET (NFUNCT,NPARA,AH(1,M),RK,H(M))                      CALF 350
   40 CONTINUE                                                        CALF 360
C                                                                     CALF 370
C -   COMPUTE MATRIX B                                                CALF 380
```

```
      K=0                                                       CALF 390
      DO 60 I=1,NFUNCT                                          CALF 400
      DO 60 J=1,I                                               CALF 410
      K=K+1                                                     CALF 420
      B(K)=0.                                                   CALF 430
      DO 50 M=MEASA,MEASZ                                       CALF 440
   50 B(K)=B(K)+AH(I,M)*AH(J,M)                                 CALF 450
   60 CONTINUE                                                  CALF 460
C                                                               CALF 470
C  -  INVERT B                                                  CALF 480
      CALL SINV(B,NFUNCT,1.E-10,IER)                            CALF 490
      IF (IER .NE. 0) STOP 'INVERSION OF MATRIX B IN CALFUN FAILED'  CALF 500
C                                                               CALF 510
C  -  COMPUTE MATRIX C                                          CALF 520
      DO 85 L=1,NLAMDA                                          CALF 530
      IF (IVAL(L) .EQ. 0) GOTO 85                               CALF 540
      DO 80 I=1,NFUNCT                                          CALF 550
      C(I,L)=0.                                                 CALF 560
      DO 70 M=MEASA,MEASZ                                       CALF 570
   70 C(I,L)=C(I,L)+EW(M,L)*AH(I,M)                             CALF 580
   80 CONTINUE                                                  CALF 590
   85 CONTINUE                                                  CALF 600
C                                                               CALF 610
C  -  THE COLUMNS OF EPSA0 ARE THE PRODUCT OF INVERTED B TIMES A  CALF 620
C  -  COLUMN OF C                                               CALF 630
      DO 90 L=1,NLAMDA                                          CALF 640
      IF (IVAL(L) .EQ. 0) GOTO 90                               CALF 650
      CALL PMAVEC (B,C(1,L),EPSA0(1,L),NFUNCT)                  CALF 660
   90 CONTINUE                                                  CALF 670
C                                                               CALF 680
C  -  COMPUTE THE GOODNESS-OF-FIT FUNCTION Q                    CALF 690
      Q=0.                                                      CALF 700
      DO 115 L=1,NLAMDA                                         CALF 710
      IF (IVAL(L) .EQ. 0) GOTO 115                              CALF 720
      DO 110 M=MEASA,MEASZ                                      CALF 730
      AHEPS=0.                                                  CALF 740
      DO 100 I=1,NFUNCT                                         CALF 750
  100 AHEPS=AHEPS+AH(I,M)*EPSA0(I,L)                            CALF 760
      Q=Q+(AHEPS-EW(M,L))**2                                    CALF 770
  110 CONTINUE                                                  CALF 780
  115 CONTINUE                                                  CALF 790
      RETURN                                                    CALF 800
      END                                                       CALF 810
C                                                               DIST 10
C***** DISTAN(4.2)                                              DIST 20
C                                                               DIST 30
      SUBROUTINE DISTAN (PHM,D)                                 DIST 40
      COMMON /LAFIT/ IVAL(10),MEASA,MEASZ,NFUNCT,NLAMDA,NOMEAS,NPAR,  DIST 50
     1AH(7,150),EW(150,10),EPSA0(7,10),H(150),PH(150)           DIST 60
      COMMON /DISPAR/ M,RK(6),PHM0                              DIST 70
      DIMENSION AHM(5)                                          DIST 80
      HM=10.**(-PHM)                                            DIST 90
      CALL FUNSET (NFUNCT,NPAR,AHM,RK,HM)                       DIST 100
      D=(PHM-PHM0)**2                                           DIST 110
      DO 20 L=1,NLAMDA                                          DIST 120
      IF (IVAL(L) .EQ. 0) GOTO 20                               DIST 130
```

B Source Listing of TIFIT

```
            AHEPS=0.                                          DIST 140
            DO 10 I=1,NFUNCT                                  DIST 150
            AHEPS=AHEPS+AHM(I)*EPSA0(I,L)                     DIST 160
   10       CONTINUE                                          DIST 170
            D=D+(AHEPS-EW(M,L))**2                            DIST 180
   20       CONTINUE                                          DIST 190
            RETURN                                            DIST 200
            END                                               DIST 210
C                                                             RAND 10
C***** RANDO (4.3)                                            RAND 20
C                                                             RAND 30
C -    RANDO GENERATES PSEUDO-RANDOM-NUMBERS ACCORDING TO A   RAND 40
C -    CONGRUENCE-METHOD AS GIVEN IN                          RAND 50
C -    M. ABRAMOWITZ, HANDBOOK OF MATHEMATICAL FUNCTIONS,     RAND 60
C -    DOVER, NEW YORK, 1965                                  RAND 70
C -    P. 950, FORMULA 26.8.5 WITH S=4, Q=6                   RAND 80
C                                                             RAND 90
C -    THE FIRST NUMBER MUST BE POSITIVE                      RAND 100
C -    THE RESULTING RANDOM-NUMBERS ARE UNIFORMLY DISTRIBUTED IN (0,1)  RAND 110
C                                                             RAND 120
            SUBROUTINE RANDO (X)                              RAND 130
            DOUBLE PRECISION AUX                              RAND 140
            AUX=DBLE(X)*129140163.D-02                        RAND 150
C           =3**17/100                                        RAND 160
            IAUX=AUX                                          RAND 170
            X=AUX-DBLE(FLOAT(IAUX))                           RAND 180
            RETURN                                            RAND 190
            END                                               RAND 200
C                                                             FSET 10
C***** FUNSET(5.1)                                            FSET 20
C                                                             FSET 30
C -    COMPUTE THE FUNCTIONS F(1)...(NFUNCT) OF THE INDEPENDENT  FSET 40
C -    VARIABLE X AND THE SET OF PARAMETERS PARAM(1)...(NPAR)    FSET 50
C                                                             FSET 60
C -    THESE FUNCTION ARISE FROM A TITRATION-SYSTEM           FSET 70
C -    X IS THE CONCENTRATION OF H+(HYDROGEN-ION) AND         FSET 80
C -    THE PARAMETERS ARE K-VALUES (10.**(-PK))               FSET 90
C                                                             FSET 100
            SUBROUTINE FUNSET (NFUNCT,NPAR,F,PARAM,X)         FSET 110
            DIMENSION F(1),PARAM(1)                           FSET 120
            PROD=PARAM(1)                                     FSET 130
            DENOM=X+PROD                                      FSET 140
            IF (NPAR .EQ. 1) GOTO 15                          FSET 150
            DO 10 I=2,NPAR                                    FSET 160
            PROD=PROD*PARAM(I)                                FSET 170
            DENOM=DENOM*X+PROD                                FSET 180
   10       CONTINUE                                          FSET 190
   15       CONTINUE                                          FSET 200
            F(NFUNCT)=PROD/DENOM                              FSET 210
            J=NFUNCT                                          FSET 220
            DO 20 I=2,NFUNCT                                  FSET 230
            J=J-1                                             FSET 240
            F(J)=F(J+1)*X/PARAM(J)                            FSET 250
   20       CONTINUE                                          FSET 260
            RETURN                                            FSET 270
            END                                               FSET 280
```

```
C                                                               SINV   10
C***** SINV(5.2)                                                SINV   20
C                                                               SINV   30
C -   FROM IBM/360 SSP,VERSION III, AUG. 1970                   SINV   40
C                                                               SINV   50
C                                                               SINV   60
C -   SUBROUTINE SINV                                           SINV   70
C                                                               SINV   80
C -   PURPOSE                                                   SINV   90
C -      INVERT A GIVEN SYMMETRIC POSITIVE DEFINITE MATRIX      SINV  100
C                                                               SINV  110
C -   USAGE                                                     SINV  120
C -      CALL SINV(A,N,EPS,IER)                                 SINV  130
C                                                               SINV  140
C -   DESCRIPTION OF PARAMETERS                                 SINV  150
C -      A- UPPER TRIANGULAR PART OF THE GIVEN SYMMETRIC        SINV  160
C -         POSITIVE DEFINITE N BY N COEFFICIENT MATRIX.        SINV  170
C -         ON RETURN CONTAINS THE RESULTANT UPPER              SINV  180
C -         TRIANGULAR MATRIX.                                  SINV  190
C -      N- THE NUMBER OF ROWS (COLUMNS) IN GIVEN MATRIX.       SINV  200
C -      EPS- AN INPUT CONSTANT WHICH IS USED AS RELATIVE       SINV  210
C -         TOLERANCE FOR TEST ON LOSS OF SIGNIFICANCE.         SINV  220
C -      IER- RESULTING ERROR PARAMETER CODED AS FOLLOWS        SINV  230
C -         IER=0- NO ERROR                                     SINV  240
C -         IER=-1 - NO RESULT BECAUSE OF WRONG INPUT PARAME-   SINV  250
C -              TER N OR BECAUSE SOME RADICAND IS NON-         SINV  260
C -              POSITIVE (MATRIX A IS NOT POSITIVE             SINV  270
C -              DEFINITE, POSSIBLY DUE TO LOSS OF SIGNI-       SINV  280
C -              FICANCE)                                       SINV  290
C -         IER=K- WARNING WHICH INDICATES LOSS OF SIGNIFI-     SINV  300
C -              CANCE. THE RADICAND FORMED AT FACTORIZA-       SINV  310
C -              TION STEP K+1 WAS STILL POSITIVE BUT NO        SINV  320
C -              LONGER GREATER THAN ABS(EPS*A(K+1,K+1)).       SINV  330
C                                                               SINV  340
C -   REMARKS                                                   SINV  350
C -      THE UPPER TRIANGULAR PART OF GIVEN MATRIX IS ASSUMED TO BE SINV 360
C -      STORED COLUMNWISE IN N*(N+1)1/2 SUCCESSIVE STORAGE LOCATIONS SINV 370
C -      IN THE SAME STORAGE LOCATIONS THE RESULTING UPPER TRIANGU-  SINV 380
C -      LAR MATRIX IS STORED COLUMNWISE TOO.                   SINV  390
C -      THE PROCEDURE GIVES RESULTS IF N IS GREATER THAN 0 AND ALL SINV 400
C -      CALCULATED RADICANDS ARE POSITIVE.                     SINV  410
C                                                               SINV  420
C -   SUBROUTINES AND FUNCTION SUBPROGRAMS REQUIRED             SINV  430
C -      MFSD                                                   SINV  440
C                                                               SINV  450
C -   METHOD                                                    SINV  460
C -      SOLUTION IS DONE USING THE FACTORIZATION BY SUBROUTINE MFSD. SINV 470
C                                                               SINV  480
C                                                               SINV  490
C                                                               SINV  500
      SUBROUTINE SINV(A,N,EPS,IER)                              SINV  510
C                                                               SINV  520
C                                                               SINV  530
      DIMENSION A(21)                                           SINV  540
      DOUBLE PRECISION DIN,WORK                                 SINV  550
C                                                               SINV  560
```

B Source Listing of TIFIT

```
C   -   FACTORIZE GIVEN MATRIX BY MEANS OF SUBROUTINE MFSD             SINV 570
C   -   A = TRANSPOSE(T) * T                                           SINV 580
        CALL MFSD(A,N,EPS,IER)                                         SINV 590
        IF(IER) 9,1,1                                                  SINV 600
C                                                                      SINV 610
C       INVERT UPPER TRIANGULAR MATRIX T                               SINV 620
C       PREPARE INVERSION-LOOP                                         SINV 630
1       IPIV=N*(N+1)/2                                                 SINV 640
        IND=IPIV                                                       SINV 650
C                                                                      SINV 660
C       INITIALIZE INVERSION-LOOP                                      SINV 670
        DO 6 I=1,N                                                     SINV 680
        DIN=1.D0/DBLE(A(IPIV))                                         SINV 690
        A(IPIV)=DIN                                                    SINV 700
        MIN=N                                                          SINV 710
        KEND=I-1                                                       SINV 720
        LANF=N-KEND                                                    SINV 730
        IF(KEND) 5,5,2                                                 SINV 740
2       J=IND                                                          SINV 750
C                                                                      SINV 760
C       INITIALIZE ROW-LOOP                                            SINV 770
        DO 4 K=1,KEND                                                  SINV 780
        WORK=0.D0                                                      SINV 790
        MIN=MIN-1                                                      SINV 800
        LHOR=IPIV                                                      SINV 810
        LVER=J                                                         SINV 820
C                                                                      SINV 830
C       START INNER LOOP                                               SINV 840
        DO 3 L=LANF,MIN                                                SINV 850
        LVER=LVER+1                                                    SINV 860
        LHOR=LHOR+L                                                    SINV 870
3       WORK=WORK+DBLE(A(LVER)*A(LHOR))                                SINV 880
C       END OF INNER LOOP                                              SINV 890
C                                                                      SINV 900
        A(J)=-WORK*DIN                                                 SINV 910
4       J=J-MIN                                                        SINV 920
C       END OF ROW-LOOP                                                SINV 930
C                                                                      SINV 940
5       IPIV=IPIV-MIN                                                  SINV 950
6       IND=IND-1                                                      SINV 960
C       END OF INVERSION-LOOP                                          SINV 970
C                                                                      SINV 980
C       CALCULATE INVERSE(A) BY MEANS OF INVERSE(T)                    SINV 990
C       INVERSE(A) = INVERSE(T) * TRANSPOSE(INVERSE(T))                SINV1000
C       INITIALIZE MULTIPLICATION-LOOP                                 SINV1010
        DO 8 I=1,N                                                     SINV1020
        IPIV=IPIV+I                                                    SINV1030
        J=IPIV                                                         SINV1040
C                                                                      SINV1050
C       INITIALIZE ROW-LOOP                                            SINV1060
        DO 8 K=I,N                                                     SINV1070
        WORK=0.D0                                                      SINV1080
        LHOR=J                                                         SINV1090
C                                                                      SINV1100
C       START INNER LOOP                                               SINV1110
        DO 7 L=K,N                                                     SINV1120
```

```
      LVER=LHOR+K-I                                        SINV1130
      WORK=WORK+DBLE(A(LHOR)*A(LVER))                      SINV1140
7     LHOR=LHOR+L                                          SINV1150
C     END OF INNER LOOP                                    SINV1160
C                                                          SINV1170
      A(J)=WORK                                            SINV1180
8     J=J+K                                                SINV1190
C     END OF ROW- AND MULTIPLICATION-LOOP                  SINV1200
C                                                          SINV1210
9     RETURN                                               SINV1220
      END                                                  SINV1230
C                                                          PMAV  10
C***** PMAVEC(5.3)                                         PMAV  20
C                                                          PMAV  30
C                                                          PMAV  40
C                                                          PMAV  50
C     SUBROUTINE PMAVEC                                    PMAV  60
C                                                          PMAV  70
C     PURPOSE                                              PMAV  80
C     MULTIPLY VECTOR B BY MATRIX A WHICH IS A PACKED SYMMETRIC   PMAV  90
C     MATRIX, STORED LINEARLY A(1,1),A(1,2),A(2,2),A(1,3),...     PMAV 100
C                                                          PMAV 110
C     VECTOR(C) = MATRIX(A)*VECTOR(B)                      PMAV 120
C ......                                                   PMAV 130
C                                                          PMAV 140
      SUBROUTINE PMAVEC (A,B,C,N)                          PMAV 150
      DIMENSION A(1),B(1),C(1)                             PMAV 160
      IND0=0                                               PMAV 170
      DO 40 I=1,N                                          PMAV 180
      C(I)=0.                                              PMAV 190
      JND=IND0                                             PMAV 200
      DO 30 J=1,N                                          PMAV 210
      IF (J .GT. I) GOTO 10                                PMAV 220
      JND=JND+1                                            PMAV 230
      GOTO 20                                              PMAV 240
10    JND=JND+J-1                                          PMAV 250
20    C(I)=C(I)+A(JND)*B(J)                                PMAV 260
30    CONTINUE                                             PMAV 270
      IND0=IND0+I                                          PMAV 280
40    CONTINUE                                             PMAV 290
      RETURN                                               PMAV 300
      END                                                  PMAV 310
C                                                          PMAV  10
C***** MFSD(6.1)                                           PMAV  20
C                                                          PMAV  30
                                                           PMAV  40
C                                                          PMAV  50
C     SUBROUTINE MFSD                                      PMAV  60
C                                                          PMAV  70
C     PURPOSE                                              PMAV  80
C     FACTOR A GIVEN SYMMETRIC POSITIVE DEFINITE MATRIX    PMAV  90
C                                                          PMAV 100
C     USAGE                                                PMAV 110
C     CALL MFSD(A,N,EPS,IER)                               PMAV 120
C                                                          PMAV 130
C     DESCRIPTION OF PARAMETERS                            PMAV 140
```

B Source Listing of TIFIT

```
C         A -UPPER TRIANGULAR PART OF THE GIVEN SYMMETRIC           PMAV 150
C         POSITIVE DEFINITE N BY N COEFFICIENT MATRIX.              PMAV 160
C         ON RETURN A CONTAINS THE RESULTANT UPPER                  PMAV 170
C         TRIANGULAR MATRIX.                                        PMAV 180
C         N -THE NUMBER OF ROWS (COLUMNS) IN GIVEN MATRIX.          PMAV 190
C         EPS -AN INPUT CONSTANT WHICH IS USED AS RELATIVE          PMAV 200
C         TOLERANCE FOR TEST ON LOSS OF SIGNIFICANCE.               PMAV 210
C         IER - RESULTING ERROR PARAMETER CODED AS FOLLOWS          PMAV 220
C         IER=0- NO ERROR                                           PMAV 230
C         IER=-1 - NO RESULT BECAUSE OF WRONG INPUT PARAME-         PMAV 240
C         TER N OR BECAUSE SOME RADICAND IS NON-                    PMAV 250
C         POSITIVE (MATRIX A IS NOT POSITIVE                        PMAV 260
C         DEFINITE, POSSIBLY DUE TO LOSS OF SIGNI-                  PMAV 270
C         FICANCE)                                                  PMAV 280
C         IER=K- WARNING WHICH INDICATES LOSS OF SIGNIFI-           PMAV 290
C         CANCE. THE RADICAND FORMED AT FACTORIZA-                  PMAV 300
C         TION STEP K+1 WAS STILL POSITIVE BUT NO                   PMAV 310
C         LONGER GREATER THAN ABS(EPS*A(K+1,K+1)).                  PMAV 320
C                                                                   PMAV 330
C         REMARKS                                                   PMAV 340
C         THE UPPER TRIANGULAR PART OF GIVEN MATRIX IS ASSUMED TO BE PMAV 350
C         STORED COLUMNWISE IN N*(N+1)/2 SUCCESSIVE STORAGE LOCATIONS. PMAV 360
C         IN THE SAME STORAGE LOCATIONS THE RESULTING UPPER TRIANGU- PMAV 370
C         LAR MATRIX IS STORED COLUMNWISE TOO.                      PMAV 380
C         THE PROCEDURE GIVES RESULTS IF N IS GREATER THAN 0 AND ALL PMAV 390
C         CALCULATED RADICANDS ARE POSITIVE.                        PMAV 400
C         THE PRODUCT OF RETURNED DIAGONAL TERMS IS EQUAL TO THE    PMAV 410
C         SQUARE-ROOT OF THE DETERMINANT OF THE GIVEN MATRIX.       PMAV 420
C                                                                   PMAV 430
C         SUBROUTINES AND FUNCTION SUBROUTINES REQUIRED             PMAV 440
C         NONE                                                      PMAV 450
C                                                                   PMAV 460
C         METHOD                                                    PMAV 470
C         SOLUTION IS DONE USING THE SQUARE-ROOT METHOD OF CHOLESKY. PMAV 480
C         THE GIVEN MATRIX IS REPRESENTED AS PRODUCT OF TWO TRIANGULAR PMAV 490
C         MATRICES. WHERE THE LEFT HAND FACTOR IS THE TRANSPOSE OF  PMAV 500
C         THE RETURNED RIGHT HAND FACTOR.                           PMAV 510
C                                                                   PMAV 520
C                                                                   PMAV 530
C                                                                   PMAV 540
          SUBROUTINE MFSD(A,N,EPS,IER)                              PMAV 550
C                                                                   PMAV 560
C                                                                   PMAV 570
          DIMENSION A(1)                                            PMAV 580
          DOUBLE PRECISION DPIV,DSUM                                PMAV 590
C                                                                   PMAV 600
C         TEST ON WRONG INPUT PARAMETER N                           PMAV 610
          IF(N-1) 12,1,1                                            PMAV 620
1         IER=0                                                     PMAV 630
C                                                                   PMAV 640
C         INITIALIZE DIAGONAL-LOOP                                  PMAV 650
          KPIV=0                                                    PMAV 660
          DO 11 K=1,N                                               PMAV 670
          KPIV=KPIV+K                                               PMAV 680
          IND=KPIV                                                  PMAV 690
          LEND=K-1                                                  PMAV 700
```

```
      C                                                             PMAV 710
      C     CALCULATE TOLERANCE                                     PMAV 720
            TOL=ABS(EPS*A(KPIV))                                    PMAV 730
      C                                                             PMAV 740
      C     START FACTORIZATION-LOOP OVER K-TH ROW                  PMAV 750
            DO 11 I=K,N                                             PMAV 760
            DSUM=0.D0                                               PMAV 770
            IF(LEND) 2,4,2                                          PMAV 780
      C                                                             PMAV 790
      C     START INNER LOOP                                        PMAV 800
      2     DO 3 L=1,LEND                                           PMAV 810
            LANF=KPIV-L                                             PMAV 820
            LIND=IND-L                                              PMAV 830
      3     DSUM=DSUM+DBLE(A(LANF)*A(LIND))                         PMAV 840
      C     END OF INNER LOOP                                       PMAV 850
      C                                                             PMAV 860
      C     TRANSFORM ELEMENT A(IND)                                PMAV 870
      4     DSUM=DBLE(A(IND))-DSUM                                  PMAV 880
            IF(I-K) 10,5,10                                         PMAV 890
      C                                                             PMAV 900
      C     TEST FOR NEGATIVE PIVOT ELEMENT AND FOR LOSS OF SIGNIFICANCE PMAV 910
      5     IF(SNGL(DSUM)-TOL) 6,6,9                                PMAV 920
      6     IF(DSUM) 12,12,7                                        PMAV 930
      7     IF(IER) 8,8,9                                           PMAV 940
      8     IER=K-1                                                 PMAV 950
      C                                                             PMAV 960
      C     COMPUTE PIVOT ELEMENT                                   PMAV 970
      9     DPIV=DSQRT(DSUM)                                        PMAV 980
            A(KPIV)=DPIV                                            PMAV 990
            DPIV=1.D0/DPIV                                          PMAV1000
            GO TO 11                                                PMAV1010
      C                                                             PMAV1020
      C     CALCULATE TERMS IN ROW                                  PMAV1030
      10    A(IND)=DSUM*DPIV                                        PMAV1040
      11    IND=IND+I                                               PMAV1050
      C                                                             PMAV1060
      C     END OF DIAGONAL-LOOP                                    PMAV1070
            RETURN                                                  PMAV1080
      12    IER=-1                                                  PMAV1090
            RETURN                                                  PMAV1100
            END                                                     PMAV1110
```

Index

Absorbance
  association equilibria 279, 282, 292
  Lambert-Beer-Bouguer law 13
  metal-complex equilibria 245, 254, 256, 261
  protolysis equilibria
    multi-step (non-overlapping) 51
    single-step 14, 15, 33, 40
    s-step 184, 201 f.
    three-step 134
    two-step 69
  redox equilibria 302, 305
Absorbance (A) diagrams
  association equilibria 279-287, 290
  metal complex-equilibria 245-254
  protolysis equilibria
    multi-step (non-overlapping) 48 ff., 64
    multi-step (overlapping) 57 ff.
    simultaneous 219
    single-step 17, 18, 35 ff.
    s-step 185 ff.
    three-step 136-155, 173 ff.
    two-step 73 f., 80-92, 115 ff.
  redox equilibria 305 f., 308-312
Absorbance difference
  protolysis equilibria
    simultaneous 215 ff.
    single-step 17 f., 35 ff.
    s-step 184, 188 ff.
    three-step 135 f., 174 ff.
    two-step 70 ff., 115 f., 122 ff.
Absorbance difference (AD) diagrams
  protolysis equilibria
    simultaneous 214 f., 218-221
    single-step 17, 18, 35 ff., 115 f.
Absorbance difference quotient (ADQ) diagrams
  association equilibria 290
  metal-complex equilibria 264
  protolysis equilibria
    principle 18 f.
    s-step (generalized form) 189 ff.
    three-step 148 ff., 173 ff.
    two-step 69 ff., 118, 121-126
Absorbance polyhedron 191-197
Absorbance tetrahedron 141-148, 160 ff., 173 ff., 254
Absorbance triangle
  association equilibria 280
  metal complex-equilibria 244-251
  protolysis equilibria 85-92
  redox equilibria 308-312
Acetoacetate 311 f.
Acidic end of the titration (titration limit)
  metal-complex equilibria 248 ff.
  protolysis equilibria
    simultaneous 215
    single-step 34, 35
    s-step 209
    three-step 135
    two-step 69, 88, 108 f.
Acridin orange 277
Activity 23, 270 f., 300 f., 329
Activity coefficient 23 ff., 231, 248, 329
ADQ3-diagrams 19, 135 f.
ADQs-diagrams 19, 189 f.
ADQz-diagrams 190 ff., 264, 290

Affine transformation (distortion) 68, 88, 105, 116 f., 161 f., 173, 196, 222, 245, 260, 262, 279, 282, 284, 305, 309
Alkaline end of the titration (titration limit)
  protolysis equilibria
    simultaneous 215, 219
    single-step 34
    s-step 209
    three-step 135
    two-step 69, 88, 108 f.
p-Aminobenzoic acid 223, 226
Ampholytes 48, 55, 65, 69, 80, 86, 89, 133, 135, 148 ff., 193 f., 224
Analyte 3
Angle of rotation 357, 359 ff.
Association 277 ff.
Association constant 278, 281-290, 292 ff.

Band shape analysis 232 ff.
Barycentric coordinates 105, 197
Basis vectors 115, 162
Benesi-Hildebrand equation 293
1,2,4-Benzenetricarboxylic acid
  A-diagrams 137 ff.
    determiation of $A_{\lambda BH_2}$, $A_{\lambda BH}$ 138 ff., 147 ff.
    tangent method 148-155
    uniform subregions 138 f., 147 f.
  ADQ-diagrams 148 ff.
  branched system 136 f.
  dissociation diagram 85
  pK-values 141, 147, 155, 157 ff.
  titration curves 139, 158
  titration spectra 138
1,3,5-Benzenetricarboxylic acid 224
Biot's law 359
Bisecting-line method, see Bisection of side
Bisecting points
  protolysis equilibria
    three step
      A-diagrams 145 f., 160 f.
      concentration diagrams 170 f.
    two step
      A-diagrams 69
      concentration diagrams 110 f.
  redox equilibria 308
Bisection of side (method)
  protolysis equilibria
    A-diagrams 88 ff., 96 ff.
    concentration diagrams 109 ff.
    simultaneous 218 ff.
  redox equilibria 306-309, 312
Bloch theory 379
Branched system 223 ff., 371
Burets 324-327

CD spectrometer 357 ff.
Cells with (without) transference 334 f.
Center 99 f., 106 f.
Central projection 123
Characteristic concentration diagram, see Concentration diagram
Charge-transfer complexes, see Electron donor-acceptor complexes
Chemical shift 230 f., 374 ff., 379 ff.
Chemical shift diagrams 381 ff.
Chromaticity diagram 89
Circular dichroism (CD) 356 ff.
Classical dissociation constants 24, 47 f., 248 f.
Complementary tristimulus colorimetry 89
Complex formation function 241
^{13}C-NMR spectroscopy 376 f., 379 ff.
Concentration diagram
  association equilibria 278 ff.
  metal complex equilibria 243 ff., 252 ff., 259 ff.
  protolysis equilibria
    simultaneaous 212 f.
    s-step 185 ff., 194 ff.
    three-step 143, 162-173
    two-step 68, 105-117
  redox equilibria 303 f., 306 ff.
Concentration difference quotient (CDQ) diagrams 117-125
Concentration polyhedron 194 f.
Concentration tetrahedron 161 f., 171 ff., 253 f.
Concentration triangle
  metal complexes 245
  protolysis equilibria 88, 105, 108-117
  redox equilibria 308

Conditioning problems 45 f.
Confidence interval 104 f.
Conic section 86, 99, 105-109, 116
Control simulation
   protolysis equilibria
      single-step 46
      s-step 206
      three-step 157 ff.
      two-step 92 ff.
Conventional pH-scale 47, 211, 329-335
Cramer's rule
   protolysis equilibria
      single-step 45
      s-step 185, 189
      three-step 170
Curvature 165 f.
Curvature vector 165 f.
Cysteine
   A-diagrams 227 f.
   branched system 227 ff.
   Laser-Raman spectrum 229
   pK values 227
   titration spectra 227 f.
Cystine 361 ff.

Daphnetin 258 f., 265 ff.
Debye-Hückel equation 47
Degree of dissociation
   protolysis equilibria
      single-step 24, 33, 44 f.
4-Deoxypyridoxine
   A-diagrams 51
   formula 49
   pK-values 51, 206
   titration curves 49, 52
   titration spectra 50
Difference spectroscopy 366
Differential method 41
Diffusion potentials 330, 335
7,8-Dihydroxycoumarin, see Daphnetin
Dilution effect, correction function 14
Dimer 278 ff.
Dinitrophenolate 293 f.
Directional vector
   protolysis equilibria
      s-step 187, 193
      three-step 166-169, 174
      two-step 124 f.

Dissociation constants, see Mixed dissociation constants
Dissociation diagram
   protolysis equilibria
      single-step 44 f.
      three-step 141 ff.
      two-step 84 ff.
Dissociation-degree diagram 213, 215-221
Distance relationships
   metal complex-equilibria 246 f., 266, 268
   protolysis equilibria
      A-diagrams 96 ff., 160 f.
      concentration diagrams 112 ff., 171 ff.
   redox equilibria 306, 309 f.
Dopamine 229
Double-beam spectrophotometer
   UV-VIS 337 ff., 392
   IR 365
   Raman 368
Double nonochromators 345, 357, 360, 369

EDIA (program)
   application 86
   description 104 f.
   source listing 395 ff.
Effect of solvent on spectra 233 ff.
Electrical neutrality 47
Electrical potential, see Electromotive force
Electromotive force (emf) 300 ff.
Electromotive force series 301
Electron donor-acceptor complexes 277, 292 ff.
Electronic excited states 350 f.
Electronic ground state 350 f.
Ellipticity 356, 360 ff.
Ellipticity (θ)-diagrams 363 f.
Emission spectra 344 f.
Enzyme electrode 271
Eosine 285-290
Equivalence point 40
Error functions
   protolysis equilibria
      single-step 44, 48
      s-step 203 ff.
      two-step 79 f.

Ethidium bromide 277
Excess absorbance
  association equilibria 287, 291-295
  metal- complex equilibria 256 ff., 261
Excess absorbance (EA) diagram
  association equilibria 287 ff., 294
  metal complex-equilibria 242, 260 ff., 265 ff.
Excess absorbance difference 264
Excess absorbance difference (EAD) diagram 264
Excess absorbance difference quotient (EADQ- and EADQz-)diagram 264
Excess absorbance quotient diagrams 291
Excess absorbance triangle 262 ff.
Excitation spectra 344, 346
External standard 211 ff.

Factor analysis 16
Faraday effect 358
Fiber-optics 322 ff., 337, 392
Filter 344
Flow system 322
Fluorescence 344 ff.
Fluorescence spectrum 344, 352
Fluorescence standard 350
Förster cycle 350 f.
Factor analysis 16
Galvanic cells 299, ff., 306
Gaussian error summation 103
Gauss-Newton algorithm 93, 205
Gibbs triangle, see Absorbance triangle or
  Concentration triangle
Glass electrode 334 f.
Glycine 226
Glycine methyl ester, pK value 226
Goodness 203 ff.

Halving method
  protolysis equilibria
    single-step 42, 61, 64
Henderson-Hasselbalch equation 24, 34
Histidine 381 f.
Hydronium ion activity
  abbreviation 14
  measuring 329-335

  protolysis equilibria 23, 68,95 f., 134, 329 ff.
  redox equilibria 301
D-3-Hydroxybutyrate 311
D-3-Hydroxybutyrate: NAD$^+$ oxidoreductase 311 f.
2-Hydroxy-5-nitro-α-toluenesulfonic acid 339 f.
3-Hydroxypyridine 233 ff., 237

Ill-conditioning 92, 205
Individual stability constant, see Stability constant
Inflection point analysis
  protolysis equilibria
    single-step 39 ff., 60 f.
    three-step 136
    two-step 75 ff.
Integral absorption 234 ff., 366
Intensity (I) diagrams 352 ff.
Integral molar absorptivity 235
Ionic strength 47, 95, 159, 231, 322
Ion-selective electrodes 270 f.
IR-spectroscopy
  branched system 229 f., 236
  measurement 365 ff.
Isoelectric point 233
Isosbestic points
  CD 363
  general example 10
  matrix-rank analysis 16
  protolysis equilibria
    branched 234
    multi-step (non-overlapping) 49 ff.
    multi-step (overlapping) 57 ff.
    simultaneous 215 ff.
    single-step 18, 35 ff., 322, 340, 351, 353
    three-step 138
    two-step 70
Isostilb 354

Job method, see Method of continuous variation

Knoblauch's law 346

Lambert-Beer-Bouguer law

association equilibria 282, 292
equation 13
error function 340 ff.
fluorescence 346 f.
generalized form 13
integrated form 234 f.
IR 366
metal-complex equilibria 245, 256, 261
prerequisites 13
protolysis equilibria
 simultaneous 213
 single-step 14, 33
 s-step 184
 three-step 134
 two-step 69
redox equilibria 302, 305
Law of mass action 23
Least square analysis
 protolysis equilibria
  s-step 203
  two-step 82
Linear conversions 116
Linearization of titration curves
 protolysis equilibria
  one-step 42 ff., 61, 64
  three-step 138 ff., 147 f., 155
  two-step 92 f.
 redox equilibria 302
Linearly (in)dependent absorbances (differences)
 protolysis equilibria
  single-step 17, 35
  s-step 185
  three-step 135 f.
  two-step 18 f., 69 ff., 122 ff.
Linearly (in)dependent concentration variables
 association equilibria 282, 288
 metal complex-equilibria 245, 260 f.
 protolysis equilibria
  single-step 15
  s-step 184, 202
  three-step 135
  two-step 70, 80, 105
 redox equilibria 306
 simultaneous 213 f.
Linear regression (analysis)
 protolysis equilibria

  single-step 45
  three-step 155 ff.
  two-step 92 f.
Linear segments, see Uniform subregions
Lines bisecting the sides, see Bisection of side

Macrotricyclic ammonium salts 293 f.
Macroscopic dissociation constants 137, 224 f.
Macroscopic pK values, see Macroscopic dissociation constants
Marquardt method 93, 205
Mass analytical titration 11
Matrix-rank analysis
 association equilibria 290
 definition 15 f.
 graphic methods 16 ff.
 protolysis equilibria
  simultaneous 214
  single-step 34 ff., 60
  s-step 190, 205, 208 f.
  three-step 135 f.
  two-step 69 ff.
Medium effects 34, 42, 48, 52, 318
Metal-complex equilibria
 A-diagrams 245 ff., 254
 concentration diagrams 241 ff.
 excess-absorbance diagrams 261 ff., 266 ff.
 method of continuous variation 256 ff.
Metalloporphyrins 311
Method of bisecting lines, see Bisection of side
Method of continuous variation 243, 256 ff., 306-312
S-Methylcysteine 229
Microscopic dissociation constants 137, 224 ff., 233 ff.
Microscopic pK values, see Microscopic dissociation constants
Microtitration 324-327
Midpoints, see Bisecting points
Mixed dissociation constants
 metal- complex equilibria 242, 253, 259
 protolysis equilibria
  single-step 24 f., 33 f.
  three-step 134

two-step 68
simultaneous 212
Molar absorptivity
CD 356
definition 13
Molar angle of rotation 356, 360 ff.
Molar ellipticity 359 f.
Monochromator
fluorescence 344 f.
Raman 368 f.
UV-VIS 338, 342
Monte Carlo approach
A-diagrams 103 ff.
titration curves 46, 207 ff.
Multiple shooting method 205
Multiple wavelength analysis 31, 205, 208
Multiplex spectroscopy 377 f.

NADH (NAD+) 311 f.
Naphthalene 277
ß-Naphthol 351 ff.
Newton's method 104, 205
NIR 367
o-Nitrophenol
pK value 214
simultaneous titration 214-220
titration spectra 216
p-Nitrophenol
A-diagrams 37
AD-diagrams 37
iteration process 203 f.
linearization of titration curve 43 ff.
(mixed) dissociation constant 204, 206, 214
simultaneous titration 214-220
thermodynamic dissociation constant 48
titration curves 41
titration spectra 36 f.
Non-linear curve-fitting
association equilibria 283-287
protolysis equilibria
multi-step (non-overlapping) 51 ff., 64
multi-step (overlapping) 60 f.
single-step 46
s-step 202 ff.

three-step 157 ff.
two-step 87, 93 ff., 98-105
Non-linear regression analysis, see Non-linear curve-fitting
Non-overlapping titration systems 48 ff., 64
Normal hydrogen electrode (NHE) 301
Nuclear magnetic resonance (NMR)
branched system 230 ff.
chemical shift diagrams 381 ff.
Fourier-transform 377 f.
spectrometer 372 ff.
titrations 379 ff.

Optical rotatory dispersion 356 ff.
ORD spectrometer 357 ff.
Osculating planes
A-diagrams 142, 173 ff.
concentration diagrams 165 f., 170
Overall stability constant 242 ff., 257, 265 ff.
Overlapping protolysis system
four-step 57 ff.
simultaneous 211 ff.
s-step 183 ff.
two-step 67, 98-105

Parallel projection 116
Parametric form 167, 169 f., 173 f., 187 f., 253
Pathlength 13
Pencil of planes (method)
A-diagrams 145 ff.
concentration diagrams 168 ff.
Pencil of rays (method)
association equilibria 278 ff.
metal-complex equilibria 244-255
protolysis equilibria
s-step 191, 194
A-diagrams 188 f.
concentration diagrams 187 f.
three-step
A-diagrams 144 f., 152-156
ADQ-diagrams 148 ff., 176 f.
concentration diagrams 166 ff.
two step
A-diagrams 90 ff.
concentration diagrams 108-111

Index   431

Photometric determination of pK 47 f., 61
Photometric error function   341 f.
Photometric reproducibility   339
Photometry   3
Photometric titration, definition   3, 11
o-Phthalic acid
   A-diagrams   74, 81
     determination of $A_{\lambda BH}$ 87
     tangent method   86 f.
     uniform subregions   82
   ADQ-diagrams   73, 75, 126
   dissociation-diagram   85
   pK-difference   98, 105
   pK values   83, 85, 89, 93 ff.
   testing system   68
   titration curves   73, 84, 94
   titration spectra   72
   unbranched system   224
pK difference
   sequential   48 ff., 64, 68, 77 ff., 95-105, 134, 160 f.
   simultaneous   211 ff., 219 f., 335
pK values
   absolute   39
   determination
     single-step equilibria   39 ff., 47 f.
     three-step   136-159
     two-step equilibria   75-95
     excited states   350 ff.
   relative   39, 96 ff., 113 ff., 171 ff.
Pockels cell   358
Poggendorf compensatory circuit   300
Point of contact   18, 48
Polar
   metal-complex equilibria   260
   protolysis equilibria
     A-diagram   86 f., 100
     concentration diagram   105, 108 f.
   redox equilibria   308
Polarimeter   357
Polarizer   357
Pole, see Polar
Polynuclear complexes   242
Position vector
   A-diagrams   124 f.
   concentration diagrams   115, 162, 166-170, 186 ff., 194 f.
   protolysis equilibria

Powell method   205
Projective transformation   122 f.
Protolyte   68 f., 74, 92, 96, 99, 176, 211, 241, 252, 259, 317 ff., 343, 350
Proton-magnetic resonance (PMR)   373 ff.
Pulse Fourier-transform (PFT) spectrometer   377 f.
Pulse interferogram   377
Pyridine   379
Pyridoxal
   A-diagrams   52, 55
   curve fitting   206
   cyclization product with histamine   49, 57 ff., 206
   formula   49
   hydrate and hemiacetal   232 f.
   pK values   52, 56
   tautomeric equilibrium   55 f.
   titration curves   52, 55
   titration spectra   52 ff.
Pyrophosphoric acid   384

Quenching   347 ff.
Quotient K_1/K_2
   metal-complex equilibria   260 ff.
   protolysis equilibria
     three-step   171 ff.
     two-step   67 f., 95-105

Raman-Spectroscopy
   branched system   229 f., 236
   quantitative   370 ff.
   spectrometer   368 f.
Raman scattering   348 f.
Ratio of concentrations   88, 91 f., 96 ff.
Ratio spectrometer   345, 350, 368
Rayleigh scattering   348, 369
Reducing of the titration system, see Pencil of rays and Pencil of planes
Reduced absorbance   289
Reduced absorbance diagrams   287, 289 f., 294
Reduced absorbance difference quotient diagram   291
Reduced excess absorbance diagram   294 f.
Regression (pH) calibration   331 ff.

Relative (reduced) concentration 84 f., 88, 212 f., 288 f.
Relative pK values, see pK difference
Resonance frequency 373 ff.
Rotation-angle (α)-diagram 363 f.

Scatchard plots 277
Schiff base 57
Schottky effect 341
Scott equation 293
Selectivity coefficient 270
Side-bisecting line, see Bisection of side
Simulation, see Control simulation
Simultaneous titration 39, 211 ff.
Single-beam spectrophotometer
   fluorescence 344
   NMR 375
   Raman 368
   UV-VIS 337 f.
Single point (pH) calibration 331
Single-step subsystems 48 ff.
Specific rotation 356, 359
Spectral bandwith 342
Spectrofluorometric titration 352 f.
Spectrometric titration
   definition 3 ff.
   procedure 317-327
   registration 338 ff., 391 ff.
Spectrometric titration curve, example (schematic) 11
Spectroscopically uniform 80, 214
Spin-spin coupling 232
Spline functions 234
SPSS NONLINEAR (program)
   protolysis equilibria
      three-step 157 ff.
      two-step 93 ff.
s-system 187 ff., 194
s´-system 187 ff., 194
Stability constant 242 ff., 252 ff., 259 ff.
Standard buffers 330-335
Standard deviation
   protolysis equilibria
      single-step 45
      s-step 207
      three-step 157 ff.
      two-step 87, 100, 104 f.

Standard electrode potential 300 ff., 306, 312, 330
Standard error 104 f.
Stereospectrogram 9 ff.
Stoichiometric boundary conditions
   association equilibria 278, 281, 290
   metal complexes 243, 253, 257
   protolysis equilibria
      simultaneous 212
      single-step 14, 23
      s-step 184
      three-step 134
      two-step 68
   redox equilibria 302 f., 306
Stokes Raman spectrum 369 f.
Stray light 338, 342 f., 345 f., 357, 360, 369
Sultone 340

Tangents
   association equilibria 278
   metal- complex equilibria 260 ff.
   protolysis equilibria
      A-diagrams 188 f., 193 f.
      concentration diagrams 186 f.
      simultaneous 219 ff.
      single- step
         A-diagrams (non-overlapping s-step) 51-59
         titration curve 40
      s-step
         concentration diagrams 186 f.
      three-step
         A-diagrams 142, 150 ff.
         ADQ-diagrams 149 ff.
         concentration diagrams 162 ff.
      two-step
         A-diagrams 84 ff.
         concentration diagrams 108 f.
   redox equilibria 311
Tangent vector 162 ff., 186 f.
Tautomeric constant 224, 235 ff.
Tautomeric equilibrium 224, 234, 281
Ternary systems 88
Tetrahydropyridine, derivative 57 ff.
Thermodynamic dissociation constant, single-step protolysis equilibria 23, 47

Thermodynamic equilibrium constant
   definition 23
   metal-complex equilibria 248 f.
   redox equilibria 300, 302
Thiamine pyrophosphate 382 ff.
TIFIT (program)
   protolysis equilibria
      description 202-208
      multi-step 51,60
      source listing 405
      three-step 159
      two-step 95, 384
Titrant 3
Titration
   b_0-titration 260 ff., 265 ff., 281-295
   continuous (analog) 7
   discontinuous (digital) 7, 317, 391
   m_0-titration 260, 265, 267 ff.
   microprocessor-controlled 391
Titration apparatus 319-327
Titration curves
   protolysis equilibria
      multi-step (non-overlapping) 52, 55
      multi-step (overlapping) 59 f.
      simultaneous 211 ff.
      single-step 39 ff., 365
      s-step 202 ff.
      three-step 134 f., 381 ff.
      two-step 69, 73, 75 ff., 93 ff., 355
Titration spectra
   association equilibria 285
   definition 4
   general example 10
   metal-complex equilibria 250

   protolysis equilibria
      multi-step (non-overlapping) 49 ff.
      multi-step (overlapping) 57 ff.
      simultaneous 214 ff.
      single-step 35 ff., 339, 351, 363, 381
      two-step 72, 352 ff.
   registration 337 ff.
Tiration variable
   controlled 4
   manipulated 4, 11
1,3,5-Trinitrobenzene 277
Tubbs ring method 40 f.
Two-point (pH) calibration 331
Tyrosine 229

Unbranched protolysis equilibria
   two-step 67, 224
   three-step 137
Unic resonance 231 f.
Uniform subregions (linear segments)
   protolysis equilibria
      multi-step (non-overlapping) 48 ff.
      multi-step (overlapping) 57 ff.
      three-step 136 ff.
      two-step 80 ff.
Universal buffer 318

Vinylogous compounds 310 f.

Wavelength reproducibility 339, 358
Wavenumber 235
Well-conditioning 205
Wheatstone bridge, see Poggendorf compensatory circuit